Structure and functioning of plant populations

Koninklijke Nederlandse Akademie van Wetenschappen
Verhandelingen Afdeling Natuurkunde, Tweede Reeks, deel 70

structure and functioning of plant populations

Editors: A.H.J. Freysen and J.W. Woldendorp
Institute for Ecological Research, The Netherlands

North-Holland Publishing Company,
Amsterdam, Oxford, New York, 1978

Editors' address: Institute for Ecological Research,
Kemperbergerweg 67, 6816 RM Arnhem, The Netherlands

Accepted for publication April 1977
ISBN 0 7204 8444 8

Preface

The present volume contains the papers presented at the congress held on 9-12 May, 1977, in Wageningen, The Netherlands. This congress was organized by a group of Dutch workers in the field of experimental plant ecology and the population biology of plants, on the occasion of the twenty-fifth anniversary of the Weevers' Duin station which, as the Department of Dune Research of the Institute for Ecological Research of the Royal Netherlands Academy of Arts and Sciences, is one of the main centres of plant ecology in The Netherlands. Since the demography of plants was one of the major topics, the congress was organized in co-operation with the *Societas Internationalis de Plantarum Demographia*, which was formally founded during the congress.

The contents of this volume are intended to give a generalized picture of current knowledge on the demography, population distribution ecology, and population genetics of plants on the one hand and on eco-physiology and the interactions between plants, soil, and other components of the ecosystem, such as micro-organisms and insects, on the other. In the opinion of the organizers of the congress, such a joint contribution of the fields of genetics, soil microbiology, phytopathology, and plant physiology, as well as population biology, is indispensable for the understanding of the functioning of the various plant species forming the vegetation. In this multidisciplinary approach the accent lies on the comparison and systematization of those properties of plant species that enable them to function in a particular plant community. Such knowledge about the functioning of the individual species can lead to a better understanding of fundamental ecological questions, for instance problems associated with the diversity and stability of plant communities and the distribution of species in space and time. More insight into this field is not only important from the purely scientific point of view but is also required for the protection of our natural resources.

March, 1978.

A.H.J. Freysen

J.W. Woldendorp

Acknowledgements

The preparation of this book would have been impossible without the financial support provided by the Ministry of Education and Sciences (The Netherlands) and the assistance of the Royal Netherlands Academy of Arts and Sciences.
Mrs. A. Bastian typed the final copy and contributed greatly to the solution of lay-out problems. The English text of a number of papers was read by Mrs. I. Seeger. Some of the Figures were redrawn by W. Verholt. Mr. F. Bos gave editorial advice. Dr. P.J.M. van der Aart, Dr. P.J. van den Bergh, Dr. W. van Delden, Professor W. Ernst, Professor P.J.C. Kuiper, and Dr. J.W. Woldendorp summarized the discussions. Mr. J. van Heeswijk assisted in the correction of the typed manuscript.

List of contributors

R.W. ALLARD, — Department of Genetics, University of California, Davis, California 95616, U.S.A.

J.P. VAN DEN BERGH, — Centre for Agrobiological Research, Wageningen, The Netherlands

W.G. BRAAKHEKKE, — Centre for Agrobiological Research, Wageningen, The Netherlands

R. BROUWER, — Department of Botany, University of Utrecht, The Netherlands

W.H.O. ERNST, — Biological Laboratory, The Free University, Amsterdam, The Netherlands

J.P. GRIME, — Unit of Comparative Plant Ecology (NERC), Department of Botany, University of Sheffield, Sheffield S10 2TN, U.K.

J.L. HARPER, — School of Plant Biology, University College of North Wales, Bangor, Gwynedd LL57 2UW, U.K.

A.L. KAHLER, — Department of Genetics, University of California, Davis, California 95616, U.S.A.

P.J.C. KUIPER, — Department of Plant Physiology, Biological Centre, University of Groningen, The Netherlands

D.A. LEVIN, — Department of Botany, The University of Texas, Austin, Texas, U.S.A.

R.D. MILLER, — Department of Genetics, University of California, Davis, California 95616, U.S.A.

BARBARA MOSSE, — Rothamsted Experimental Station, Harpenden, Herts. AL5 2JQ, U.K.

T.A. RABOTNOV, — Department of Geobotany, 117234 Moscow University, Moscow, U.S.S.R.

F.W. WENT, — Desert Research Institute, University of Nevada System, Reno, Nevada 89507, U.S.A.

J.B. WILSON, — Botany Department, University of Otago, Dunedin, New Zealand

J.W. WOLDENDORP, — Institute for Ecological Research, Arnhem, The Netherlands

J.C. ZADOKS, — Laboratory of Phytopathology, Agricultural University, Wageningen, The Netherlands

Contents

D.A. LEVIN & J.B. WILSON

THE GENETIC IMPLICATIONS OF ECOLOGICAL ADAPTATIONS IN PLANTS

J.P. GRIME

INTERPRETATION OF SMALL-SCALE PATTERNS IN THE DISTRIBUTION OF
PLANT SPECIES IN SPACE AND TIME

J.P. VAN DEN BERGH & W.G. BRAAKHEKKE

COEXISTENCE OF PLANT SPECIES BY NICHE DIFFERENTIATION

F.W. WENT

THE EXPERIMENTAL APPROACH TO ECOLOGICAL PROBLEMS

W. ERNST

CHEMICAL SOIL FACTORS DETERMINING PLANT GROWTH

R. BROUWER

SOIL PHYSICAL CONDITIONS AND PLANT GROWTH

P.J.C. KUIPER

*MECHANISMS OF ADAPTATION TO PHYSICAL AND CHEMICAL FACTORS
IN PLANTS*

J.W. WOLDENDORP

THE RHIZOSPHERE AS PART OF THE PLANT-SOIL SYSTEM

BARBARA MOSSE

MYCORRHIZA AND PLANT GROWTH

J.C. ZADOKS

THE BIOTIC ENVIRONMENT OF PLANTS

On coenopopulations of plants reproducing by seeds

1. THE CONCEPT OF THE COENOTIC POPULATION

The plants reproducing exclusively or predominantly by means of seeds largely determine the composition, structure, productivity, and dynamics of many types of phytocoenosis. This form of reproduction has of course received much attention in many countries. The results of the studies on the reproduction by seeds of herbaceous plants -mainly meadow plants- carried out in the USSR and Finland (SHENNIKOV & BARATYNSKAYA 1924; SHENNIKOV 1930, 1941; BOGDANOVSKAYA-GIENEF 1926, 1941; LINKOLA 1930, 1936; PERTTULA 1941; SÖYRINKI 1938, 1939; see also the review given by RABOTNOV 1969d), as well as the data accumulated during five years of observation and experimentation in subalpine meadows in the northern part of the Caucasus, formed the basis for the development of the concept of the coenotic population as a sum of individuals of any species or of an intraspecific taxon within a particular phytocoenosis* (RABOTNOV 1945, 1950a, 1950b). It was postulated that each species in different phytocoenoses is represented by a particular coenopopulation (the term coenopopulation was proposed by PETROVSKII in 1961).

The scope of the studies on coenopopulations increased significantly after A.A. Uranov organized a laboratory in the Lenin Pedagogical Institute in Moscow in 1964. Under his guidance, a large number of species belonging to different life forms were investigated (URANOV 1967, 1968, 1973, 1974, 1975; URANOV et al. 1970; URANOV & SMIRNOVA 1969; URANOV & SEREBRYAKOVA 1976). The results of the studies on coenopopulations were published mainly in Russian, and are therefore scarcely known to ecologists in other countries. Because it seems to me that these results form an important addition to the valuable studies on the demography of plants carried out by Professor J. Harper and his collaborators, the data collected in the USSR on coenopopulations of plants reproduced by seeds are reported in this paper.

The study of coenopopulations includes the determination of the number of individuals of a particular species per unit area belonging to different age groups, which makes it necessary to follow the change in the morphological

---

* A phytocoenosis is the assemblage of plants living in a particular locality. Both the phytocoenosis and the coenopopulation are concrete entities.

character of individuals with age. Coenopopulations can be studied by two methods: (1) by observations on all individuals of a species on permanent plots over a period of many years, and (2) by single estimations of all individuals of a species within plots situated in different environments and their classification into definite age groups, which is possible only after a preliminary study of the ontogenesis of that species under natural conditions. The first of these methods provides valuable data on the numerical dynamics of a population (appearance of new individuals, transition of individuals from one age-state to another, dying off), the second method yields data on differences between the numbers of specimens and the composition of populations in various types of phytocoenosis.

A coenopopulation is characterized by the number of individuals belonging to different age groups per unit area. Classification into age groups is based on the concept of the four main periods of life of plants reproducing by seeds: 1) the period of primary dormancy - existence of individuals for a variable period as viable seeds under natural conditions in the soil or on its surface; 2) the virginal period - from germination to flowering and fructification, subdivided into four age groups: seedling, juvenile, immature, mature virginal (mature virginal plants are individuals that have not yet flowered, but have the appearance of mature plants. Immature plants are intermediate between juveniles and mature virginal plants);
3) the generative period - covering reproduction by seeds, subdivided into three states: initial phase of increasing vegetative and generative vigour, maximal vigour representing the period of life culmination, finally the onset of senescence with decreasing vegetative and generative vigour; and
4) the senile period, when due to senescence plants lose the ability to reproduce by seeds (RABOTNOV 1945, 1950a, 1950b, 1969a).

The peculiarities in the composition of coenopopulations proved to be the result of the coevolution of the various species of autotrophic plants with other organisms in a certain environment. Each coenopopulation is characterized by a specific combination of relationships with organisms trophically or topically associated with the individuals forming it. This system of organisms is called a consortium (a term introduced by BEKLEMISHEV 1951 and RAMENSKII 1952). The coenopopulations of independently existing plants (non-epiphytes) are considered to form the centre (determinant) of consortiums (LAVRENKO 1959; RABOTNOV 1969b). The organisms directly associated either trophically or topically with a determinant of a consortium (consorts) form the first circle of the consortium (phytophages, phytoparasites, symbionts, etc.). The organisms associated with the consorts of the first circle (zoophages, zooparasites, etc.) form the second circle, and so on (MASING 1966; RABOTNOV 1969b, 1972). Individual autotrophs or animals and the heterotrophs associated with them are sometimes referred to as consortiums.

The organisms consortively associated with autotrophs (phytophages, phyto-parasites, symbionts, organisms of rhizospheres and phyllospheres) exert a strong influence on individuals of all age groups, i.e., the components of coenopopulations, and often determine whether they can exist or continue to exist. Some consorts (pollinators, distributors of fruits and seeds) that do not affect individuals are nevertheless of great importance for the fate of the coenopopulation. There can be no doubt that the composition of consorts varies with the age-state of individuals, although data on this problem are very scarce.

The vigour of the organs of individuals and their distribution in the environment above and under the ground varies with age. Since each age group of individuals occupies the ecological niche peculiar to it, the transition of young individuals to more mature states (for polycarpic plants up to the life culmination phase) is only possible if suitable free niches are available. The amount of organic matter produced by individuals (including the share reaching the soil with the dying-off of organs), the effect of individuals on the environment, their ability to reproduce generatively and vegetatively as well as to compete with other plants, their response to ecological factors, the duration of the vegetative period, the position of overwintering buds, etc., as well as the response to different forms of human influence, vary with the age-state of the individual plants. Juvenile plants are known to be more shade-resistant than mature ones. Juvenile individuals of *Bromus inermis* and *Elytrigia repens* are hemicryptophytes, whereas mature ones are geophytes. In experiments in the subalpine meadows in the northern part of the Caucasus the application of NPK affected mature individuals of *Anemone fasciculata* favourably, but its juvenile individuals negatively. Early cutting, on the contrary, favoured juvenile individuals and suppressed mature ones. Therefore, the coenopopulational approach to the autoecological study of plants is of great importance.

Thus, under all conditions a coenopopulation, which represents a form of existence of a species in the phytocoenosis and a form of its adaptation to cohabitation with other organisms, constitutes a highly heterogeneous system of differently aged individuals of the same autotrophic species, differing in vigour and in ecological and biological properties and consortively associated with a large number of species of other, mainly heterotrophic organisms.

2. THE VIABLE SEEDS AS A COMPONENT OF COENOPOPULATIONS

In the concept of the coenopopulation, viable seeds are considered as individuals in the state of primary dormancy. The most important characteristics of a coenopopulation are its ability to produce viable seeds and their presence in it. Also of importance are the number of seeds, the dynamics of their quantity, the distribution in the soil, and the ratio between them and the number of individuals in the active state. Furthermore, there are coenopopulations in which viable seeds are only temporary components, characterizing the seasonal state. This category of coenopopulations includes populations of many species whose seeds are unable to remain viable for a long time because they either germinate very soon or lose their ability to germinate. There are also coenopopulations in which seeds are permanent components. Sometimes seeds are distributed only in the topmost (0-1 or 0-2 cm) layer of the soil; in other cases, a high proportion are distributed at a greater depth (5-10, 10-20 cm and deeper) where they cannot germinate. Some coenopopulations consist solely of viable seeds and others are characterized by alternation between the dormant (consisting only of viable seeds) and the active states either from year to year or over periods of years. Finally, there are many variants with different ratios of the numbers of viable seeds and individuals in the active state.

In most coenopopulations the number of viable seeds varies considerably from year to year or between periods of years, due to differences in the supply and consumption of seeds. The supply of seeds is usually the result of the seeding of generative individuals which are components of the population. Relatively few coenopopulations, i.e., those consisting mainly of individuals in the vegetative state, are unable to produce seeds. The amount of seeds produced is determined by: (1) the number of generative individuals per unit area; (2) the mean number of flowers per individual; (3) the percentage of fruit-bearing flowers (% fruit-setting); and (4) the mean number of seeds per fruit. All these parameters vary from population to population and within the same coenopopulation from year to year and from individual to individual. In demographic studies the assessment of seed production by species has considerable importance.

Of particular interest is the percentage of fruit-setting, which is determined by the vitality of generative individuals, growth conditions during the flowering period, and relationships with consorts (pollinators, parasites, phytophages). The absence of fruit formation from flowers can be due to: (1) death of flowers under the influence of environmental conditions (frost, hail,etc.); (2) absence of pollination; (3) impossibility for the plant to form fruit from all pollinated flowers; and (4) loss of flowers or fruits in the process of formation due to infestation by phytophages or phytoparasites.

The study of anthecology is undoubtly important for the elucidation of the demography of plants reproducing by seeds. Judged from the observations made in the USSR -which apparently can be extrapolated to the whole of extratropical Eurasia- the main dominants of the phytocoenoses of the climax forest (all coniferous species, *Quercus, Fagus, Carpinus,* etc.), steppe (bunch grasses), semi-desert and desert (*Artemisia, Chenopodiaceae*) are pollinated by wind and therefore do not depend on the consort-pollinators. The phytocoenoses with predominantly entomopollinated species (*Tilia, Salix,* etc.) occupy smaller areas and have often arisen as the result of human activity (burned areas with *Chamaenerion angustifolium,* heathlands, etc.) (LAVRENKO 1947). In some phytocoenoses, e.g. in spruce or fir forests, conditions are unfavourable for wind and insect pollination of plants in the lower layers, and these become autogamic (PONOMARYEV & VERETSHAGINA 1973). Nevertheless, the number of entomophilous species in extratropical regions and especially in tropical rain forests is very large, and the seed production of such species depends on the activity of pollinators. For species pollinated by one or only a few species of pollinators, the variation of their numbers in different years -due to the influence of organisms consortively associated with them- is very important, as pointed out by Darwin when he suggested that the distribution of clover in England depended on the number of cats in villages. In phytocoenoses used by man as pastures or hayfields, seed production is limited because of the loss of generative shoots by grazing and mowing.

The seeds produced in coenopopulations can: (1) be removed from a phytocoenosis under the action of different agents; (2) be destroyed by animals or lose their germinating ability due to infestation by parasitic fungi or bacteria; (3) become seedlings; or (4) retain their viability for a variably long period of time, forming a pool of viable seeds in the soil. The proportion of seeds undergoing these fates depends on the life strategy of the species. The seeds of anemochores are transported in the highest numbers. The ability to colonize new substrates (e.g. new deposits in flood valleys) and territories with disturbed vegetation (due e.g. to burning and felling) is part of the life strategy of many of anemochores. Some of them can colonize such areas because they produce large amounts of seeds which are easily carried by the wind over great distances, whereas as a rule they cannot reproduce under a mother cover. Their seeds germinate quickly or soon lose their ability to germinate. For such species the ability of buried seeds to remain viable for a long time is useless, since their seedlings can only become established in areas where competition with mature plants is absent or weakened, for instance on newly formed substrates and in disturbed communities. The seeds can also be transported by animals to places outside the area of the parent coenopopulation.

Of particular interest is endozoochory, i.e., transmission via the eating of fruits by animals in one place followed by the deposition in another place of excrement containing the seeds, which retain their viability after passing through the animal's digestive tract. This mechanism resulting from the co-evolution of some species of plants and animals, ensures not only the distribution of seeds, but also favourable conditions for the establishment of the seedlings arising from them (a certain degree of suppression of mature plants by excrements, better supply of nutrients to seedlings). The possibility for and efficacy of the endozoochorial distribution of seeds of valuable forage plants has long been known to the people of some countries and used by them for the distribution by sheep of seeds of *Trifolium repens* (Western Europe, New Zealand) and of *Mesembryanthemum* (Africa, RIDLEY 1930). The passage of some seeds through the digestive tract of animals is known to stimulate their germination.

A very important adaptive property of many species of plants is the ability of their seeds to retain their germinability for long periods after burial. At present, a substantial amount of data is available on the content of viable seeds in soils of different phytocoenoses (from tropical rain to boreal forests as well as steppes, prairies, meadows, semi-deserts, and deserts). In the USSR this problem has been studied for phytocoenoses of all natural zones except the tundra. Table 1 illustrates the variation in the viable-seed content of the soils of various phytocoenoses. Meadow soils usually contain several thousands and often more than 10,000 viable seeds per m^2. According to CHIPPINDALE & MILTON (1934) and MILTON (1936, 1939, 1943), meadows in Great-Britain contained 4,200-69,500 seeds per square metre.

TABLE 1. *Data on the number of seeds in Russian phytocoenoses*

Phytocoenosis	Region	Number of seeds/m^2	Author(s)
Desert	Uzbekistan	9-438	STRELKOVSKAYA 1941; AMELIN 1947
Desert	Turkmenia	150-600	NECHAEVA 1954
Takyr	Central Asia	irca 1,000	RODIN & SUKHOVERKO 1956
Desert steppe	Kazakhstan	860-4,700	LAVRENKO & BORISOVA 1976
Bunchgrass steppe	Ukraine	14,000-20,000	OSICHNYUK 1973
Meadow steppe	Near Kursk	18,800	GOLUBEVA 1962
Meadow	Oka valley	2,500-17,000	RABOTNOV 1956
Upland meadow	Moscow region	1,300-36,200	PYATIN 1970
Upland meadow	Leningradsky region	13,000-27,600	BOGDANOVSKAYA-GIENEF 1954
Mountain *Nardetum*	Carpathians	700-2,000	MALINOVSKII 1959

In the soils of forests the pool of viable seeds of the true forest plants is usually not large, but after afforestation there may be large amounts of seeds belonging to plants of the formerly open areas (burning and felling, old fields, etc.). The highest amount was found in secondary forests, e.g. in the soil of a birch forest on the site of a felled *Piceetum myrtilloso-oxalidosum* where there were 14,000/m^2 viable seeds, almost exclusively from plants specific to felling areas (KARPOV 1969). The seeds of many plants of old fields and burned-over and felled areas can retain their germinability in the soil for a long time after the open areas become overgrown by forest (OOSTING & HUMPHREYS 1940; LEVINGSTON & ALLESSIO 1968; KARPOV 1969). Usually, however, forest soil contains a relatively small amount of seeds, e.g. 450-1,900/m^2 in nemoral-pine and deciduous forests around Moscow (RYSIN & RYSINA 1965), 660-3,200/m^2 in spruce forests (KARPOV 1969); and 65-1,200/m^2 in pine forests of southern taiga (KAMENETSKAYA 1969). The largest amount of seeds is found in forests with abundant litter.

There are sufficient grounds to assume that the ability of seeds to retain viability for a long time was acquired only by those species for which seed reproduction (seed production, germination, rooting of seedlings) was only possible in certain years under an appropriate combination of conditions. However, for the species reproducing at regular (e.g. annual) intervals by seeds the selection of plants with seeds having prolonged viability could not be effective. Naturally, such selection would be particularly effective for annual plants, since in their case the absence of the conditions for reproduction by seeds in any year would mean that the coenopopulation would disappear if no viable seeds remained in the soil.

Annual plants seem to have originated in arid regions with an open vegetation and a poor water supply. Under such conditions, these plants, which have a small coefficient of transpiration and consume little water for completion of their short life cycle, have some advantages over perennial plants. Since arid regions are characterized by irregular precipitation differing from year to year, annual plants growing in such regions developed adaptations ensuring germination of their seeds only in the years with sufficient water for completion of their life cycle. Judging by the observations made in the USA (WENT, see this volume), one adaptation of this kind consists of the presence of germination inhibitors in the seeds. The seeds of some desert annuals are known to germinate only after the inhibitors have been washed out, which requires an amount of precipitation that is sufficient to bring the water content of the soil to a level adequate for the completion of the life cycle of these plants. Probably, annual plants of arid regions of other continents are also characterized by such adaptations. It should be noted here that observations made in the deserts of Central Asia suggest that the fruits of annual plants have better-developed adaptations against being eaten by animals (awns, burs, etc.) than do the fruits of perennial plants (MOROZOVA 1952).

Pyrophytic annuals are peculiar in that their seeds can only germinate after exposure to fire. Such an adaptation could arise only in areas where the evolution of species was influenced by their periodic exposure to fire, e.g. in the chaparral in California. The fire that damages and destroys mature plants and fertilizes the soil with ash, creates favourable conditions for the establishment of the seedlings. The transition of the seeds of pyrophytes from the dormant to the active state can be triggered by a disturbance of the impermeability of the seed coat to air and water, destruction of the germination inhibitors or, probably, a change in the environment.

In closed communities annual plants are semi-parasites or parasites, and there are also some species which are symbiotically associated with nitrogen-fixing micro-organisms or mycorrhizal fungi. Annual plants in such communities can only compete successfully with perennials if they obtain supplementary nutrition from some other organism. The seeds of parasitic plants germinate only when they are in contact with the roots of their host plants. For instance, the seeds of *Orobanche cumana*, which are extremely small and lack a thick protective coat, retain their viability in the soil for 6 yo 8 years and can start to germinate as soon as they come in contact with sunflower roots. The role of mycorrhizal fungi in the germination of the seeds of obligate annual mycotrophs, e.g. *Gentiana* spp., is not clear. It is possible that their relationships with fungi are the same as in the *Orchidaceae*. Large-scale germination of the seeds of some leguminous annuals in meadows and steppes can only occur in years that are favourable for the establishment of the seedlings and is usually observed in years with especially adequate moisture levels in the soil (inundation by flood water, copious rainfall).

Among annuals there are also species whose seeds lose their viability quickly when buried in soil. These species include cultivated annuals and also some weed annuals. Thus we see, as exemplified by annual plants, that seeds of species with the same life form can differ significantly as to their ability to remain viable when buried in the soil as well as to the conditions necessary for germination.

In the soils of closed phytocoenoses it is mainly or exclusively the seeds of perennial plants which accumulate. The degree of preservation of the viability of these seeds in the soil varies widely. Viable seeds of many species do not occur in the soil or are only briefly preserved there. This is characteristic of the species of trees which are dominant in climax coenoses (conifers, *Quercus, Fagus,* etc.), and also seems to be typical of the dominants of other climax communities (steppe, semi-desert, desert), i.e., long-lived species represented in phytocoenoses by a large number of individuals which on the one hand have sufficient stability and on the other are periodically self-seeding. For such species long-term preservation of viable seeds in soil is of course of no importance.

As a rule, viable seeds of plants vigorously reproducing vegetatively are absent in the soil, and this also holds for some relatively short-lived species which are not dominants and multiply only by seeds. To the latter belong, according to the observations made in the meadows in the flood valley of the Oka (USSR), rather regularly self-seeding monocarpic and oligocarpic *Umbelliferae (Sesili libanotis, Angelica archangelica, Heracleum sibiricum),* whose seedlings almost annually become established in rather large quantities. This ensures sufficient stability in the communities, even in the absence of viable seeds of these species in the soil.

The species with large quantities of viable seeds in their coenopopulations can be assumed on sufficiently good grounds to have been formed under the alternation (over single years or periods of several years) of conditions favourable for germination of their seeds, as well as establishment of their seedlings, with long periods without such conditions. This pattern is characteristic of meadow explerents*, such as *Ranunculus repens* and *Agrostis stolonifera,* and also of some plants occurring in burned and felled areas. It is found also in the species sometimes represented by relict coenopopulations consisting of viable buried seeds, e.g. the coenopopulations of *Calluna vulgaris, Carex* spp., and *Juncus* spp. It is conceivable that the property of prolonged preservation of seed viability was acquired by *Calluna vulgaris* as an adaptation to the periodic emergence of heathlands on former forest sites and their subsequent afforestation, and by *Carex* and *Juncus* under the conditions of alternating periods of drying and waterlogging of swampy areas.

Thus, the difference in the proportion of viable seeds in the composition of coenopopulations characterizes the life strategy of particular species. The existence of viable seeds in any coenopopulation, even in the absence of a distinct periodicity in the reproduction by seed, ensures its survival.

3. THE VIRGINAL INDIVIDUALS AS A COMPONENT OF COENOPOPULATIONS

The total seed pool of phytocoenoses and coenopopulations can fluctuate to a variable degree in different years or periods of several years. These fluctuations are determined by differences in supply and consumption in individual years. Only some of the seeds annually reached the soil surface develop into seedlings. For instance, according to observations made in mountain *Nardetum* in the Carpathians, the ratio between the number of fallen seeds and the number of resulting seedling in 1951 varied from 6:1 for *Potentilla aurea* up to 89:1 for *Nardus stricta,* and in 1953 from 5:1 and up to 144:1, respectively (MALINOVSKII 1959). Investigation of a maritime meadow on the Baltic coast showed an even lower ratio, i.e., 27:1 for *Deschampsia caespitosa,* 40:1 for *Calamagrostis neglecta,* 284:1 for *Carex nigra,* and 8,458:1 for *Juncus gerardii* (SUOMALAINEN 1930). This ratio can vary within one species from one type of phytocoenosis to another. For instance, on the subalpine meadows of South Osetia (USSR) the relevant ratio ranged for *Festuca varia* from 14:1 up to 2,845:1 and for *Geranium ibericum* from 18:1 to 52:1 (GOGINA 1960). Since the viability of the seeds of most wild-growing plants is sufficiently high except in some single years, the small number of seedlings developed from the seeds must be explained by a rapid loss of germinability, death during germination, consumption by animals, and in some cases by incorporation into a potential soil seed pool.

The conditions for germination of the seeds on and in the soil can vary considerably from year to year or between periods of years. Different species and phytocoenoses behave differently. Besides the species giving a relatively stable number of seedlings in different years, there are species and phytocoenoses in which the development of seedlings from seeds has a highly

---

* Derived from the latin verb *expleo* = to fill. This term was introduced by RAMENSKII.

dynamic character. The conditions necessary for the establishment of seedlings also vary from year to year and between periods of years. As a rule, the majority of the seedlings arising in undisturbed phytocoenoses die off in the year of emergence. In certain years all seedlings die off. An extremely high death rate is characteristic of a very large number of species belonging to different life forms and growing under different conditions.

The causes of seedling death include physical factors (drought, extreme temperatures, erosion, etc.), destruction by phytophages, infestation with fungi or bacterial parasites, and competition with mature plants. The last of these factors is universal in nature. The seedlings weakened by competition perish more easily under the action of physical factors, phytophages, and phytoparasites. Therefore, any disturbances -caused mainly by soil animals- that affect a close stand of vegetation and decrease the competitive pressure exerted by mature plants on seedlings, thus vacating appropriate ecological niches for seedlings, create favourable conditions for the establishment of the seedlings. Borrowing an apt term from HARPER (1961), we can say that the more "safe sites" arise in a phytocoenosis, the more successful plant reproduction will be. Because young individuals of perennial plants exist in phytocoenoses under conditions of intense competition for light, water, and nutrients, they develop slowly and their virginal period is prolonged and can be often divided into four age-states: seedling, juvenile, immature and mature virginal.

The dying off of young individuals continues when seedlings change to the juvenile and then to the immature and mature virginal states, but diminishes as the vigour of the plants increases. Nevertheless, only a small percentage of the plants developing from seeds attain the mature state. This is quite natural, because if the numbers of mature individuals of perennial plants are to be kept at a particular level, only a small number of young individuals may reach the mature state yearly or periodically, this number depending on the longevity of the species. In this connection the concept of "the renewal fraction" was introduced (KHITROVO 1915), which was characterized as the ratio of one to the number of years of the mean lifespan of mature individuals, i.e., 1:10 for the plants living 10 years and 1:50 for those living 50 years, etc.

A very important property of perennial plants reproducing by seeds is the ability of the individuals to persist for a long time in the juvenile, immature, and suppressed-mature virginal states i.e., surviving under relatively poor subsistence conditions (little light, water, nutrients). Owing to this property an appreciable pool of individuals is formed, in spite of a considerable annual mortality of young plants, in coenopopulations of many species able to attain the mature state as soon as the necessary conditions arise, namely, as soon as appropriate ecological niches become vacant as the result of the dying-off of neighbouring mature plants, disturbance of the close stand of vegetation by animals, etc. Thus, the actual duration of the virginal period

of the species in question and the total number of their virginal individuals
in coenopopulations are determined not only by the biological properties of
the species but also, and to a higher degree, by the character of the phyto-
environment and the activity of animals. Under conditions less favourable for
the transformation of virginal into mature individuals, their number in
coenopopulations is higher than it is when this transition occurs more quickly.
The other conditions being equal, the rate of transformation of young into
mature plants can be assessed from the number of immature individuals, large
numbers being an indication of a slower rate.

Due to the heterogeneity of the environment in phytocoenoses and the
different effects on the plants of their consorts (useful, harmful), the fate
of the individuals developing simultaneously from a year-class of seeds differs:
some of them die off quickly, other develop very slowly and, even if they
reach considerable age, do not acquire the ability to flower and bear fruit;
they die off in the virginal state. There are also plants that transform
relatively quickly into mature, flowering, and fruit-bearing plants. Thus,
"age-groups" can be formed in coenopopulations by individuals of different age
and vitality, which enhances the heterogeneity of coenopopulations.

4. THE MATURE INDIVIDUALS AS A COMPONENT OF COENOPOPULATIONS

The heterogeneity of the composition of age groups of mature individuals was
discovered by foresters a long time ago. Kraft's scale distinguishes five
classes of vitality among mature trees: I = exceptionally well-developed,
II = dominant, III = codominant, IV = suppressed, V = strongly suppressed
(MOROZOV 1925). In different phytocoenoses, i.e., in different coenopopulations,
the ratio between the trees belonging to different Kraft classes can vary. The
plants belonging to different vitality classes differ not only in vigour and
their effect on the environment and on other plants, but also in their seed
production, which is important for demographic studies. For example, according
to the observations of A.N. Sobolev and A.V. Fomichev (MOROZOV 1925), in a
mature spruce forest in a year of a good spruce-seed yield, the seed production
of the trees of the various Kraft classes was 100% in class I, 88% in class II,
37% in class III, 0.5% in class IV, and 0 in class V. These data show that
individuals of the same species in the same phytocoenosis reach the life
culmination state at different levels of vigour and giving different seed
yields. Moreover, on the basis of these data Morozov arrived at a very
important conclusion, namely, that the struggle for existence leads to natural
selection. Only conquerors in the struggle for existence can promote the
success of the group.

The heterogeneity of the composition of a group of mature individuals is
even greater if the age of the individuals differs. One very important factor
in demographic studies is the heterogeneity of a group of mature individuals

with respect to fruit-bearing periodicity and the volume of seed production. Plants that have become capable of flowering and fruiting often do not flower every year. These interruptions, which determine the periodicity of fruit-bearing, have long been known for trees of the boreal and temperate zones. It is also known that the fruit-bearing periodicity increases with deteriorating growth conditions, in particular from south and north. Later it was found that interruptions in flowering occur in many herbaceous species. Their character (frequency, duration) varies from species to species and from phytocoenosis to phytocoenosis; within coenopopulations it depends on the vitality of individuals, which is associated with their age-state and the degree of competition with surrounding plants. The interruptions in flowering are less pronounced during the life culmination period of the plants and more frequent and longer for senescent individuals.

The observations made in subalpine meadows in the northern part of the Caucasus indicate that interruptions in flowering can be caused by a temporary change in the general growth conditions making them unfavourable for the formation of generative shoots. These changes are mainly associated with the weather conditions prevailing in some years. In such cases the interruptions are not long (one year) and only occur in certain years. Longer interruptions are characteristic of individuals with a low vitality. Furthermore, the flowering ability of mature individuals of meadow plants depends on the kind and intensity of human activity in the locality.

The fruit-bearing periodicity can be considered, at least for some species, as an adaptation of plant serving to suppress the reproduction of phytophages, i.e., the consumers of seeds and fruits (e.g. SALISBURY 1942).

The heterogeneity of a group of adult individuals manifests itself in behaviour differences over a number of years, as shown by the present author's observations, of many years' duration, in permanent plots. Such differences are clearly shown by the result of observations on *Ranunculus acris* in a meadow in the Oka flood valley, where all individuals of this species were marked off in 1950 on a 10 m^2 plot, including 122 adult vegetative and 25 generative specimens. The observations made during the next ten years (1951–1960) showed that on the basis of their behaviour during this period, the vegetative individuals could be classified into 57 categories (Table 2) and the generative ones into 17 categories (Table 3). There is no doubt that the age groups distinguished in 1950 were not homogeneous. This holds especially for the group of vegetative individuals, since that group also contained virginal mature or sex-ripe individuals in the state of interrupted flowering. However, this fact alone cannot account for the varied behaviour of the individuals. Since the effect of phytophages and phytoparasites on *Ranunculus acris* was insignificant, the decisive factor was evidently the competitive influence of the surrounding plants on adult individuals.

TABLE 2.   *Changes in the state of "vegetative" individuals of* Ranunculus acris *on a field plot during the 1951-1960 period*

1950	1951	1952	1953	1954	1955	1956	1957	1958	1959	1960	Number
v	0										10
v	im	0									1
v	v	0									9
v	g	0									1
v	v	im	0								1
v	v	v	0								17
v	v	g	0								3
v	g	v	0								7
v	g	g	0								7
v	v	d	v	0							1
v	v	v	v	0							5
v	v	g	v	0							1
v	g	g	v	0							2
v	v	v	v	im	0						1
v	v	v	v	v	0						6
v	g	v	v	v	0						4
v	g	g	v	v	0						1
v	g	v	g	v	0						1
v	g	v	v	g	0						2
v	v	g	v	v	0						2
v	v	g	g	v	0						1
v	g	v	v	g	0						1
v	v	v	v	v	v	0					2
v	g	v	v	g	g	0					1
v	g	v	g	v	v	0					1
v	g	v	v	v	v	0					2
v	v	v	g	v	v	0					1
v	g	g	g	v	v	0					1
v	v	v	v	g	v	0					1
v	v	v	v	v	v	v	0				1
v	v	v	v	g	v	g	0				1
v	g	g	g	g	v	v	0				1
v	g	g	v	v	v	v	0				1
v	g	g	g	v	v	v	0				1
v	v	v	v	v	v	v	im	0			1
v	g	v	v	v	v	v	v	0			1
v	v	g	g	v	v	v	v	0			1
v	g	g	g	g	g	g	g	0			1
v	v	v	v	v	v	v	v	im	0		1
v	v	v	v	g	v	g	v	im	0		1
v	v	v	v	v	v	g	v	v	0		1
v	v	v	g	g	v	v	v	v	0		1
v	v	v	v	v	v	v	v	v	v	0	1
v	g	g	v	v	v	v	v	v	v	0	1
v	g	g	g	g	v	v	v	v	v	0	1
v	v	g	v	v	v	v	v	v	v	0	2
v	g	g	v	v	v	v	v	g	v	0	1
v	v	v	v	v	v	v	v	v	v	v	2
v	g	g	v	g	v	v	v	v	v	v	1
v	g	g	v	v	v	v	v	v	g	g	1
v	g	g	g	g	v	g	v	v	v	v	1
v	v	v	v	g	v	g	g	v	v	v	1
v	g	v	v	v	v	v	v	v	v	v	1
v	v	g	v	v	v	v	v	v	v	v	1
v	v	v	v	g	v	g	v	v	g	g	1
v	v	g	v	v	v	v	g	v	g	g	1
v	v	v	v	v	v	v	v	v	v	g	1

(v: vegetative; g: generative; im: immature; d: dormant; 0: dead)

13

TABLE 3. *Changes in the state of "generative" individuals of* Ranunculus acris *on a field plot during the 1951-1960 period*

1950	1951	1952	1953	1954	1955	1956	1957	1958	1959	1960	Number
g	0										2
g	v	0									1
g	v	v	0								1
g	g	v	0								5
g	g	v	v	v	0						2
g	g	v	v	v	v	v	0				1
g	g	v	v	v	v	g	v	v	v	v	1
g	g	v	v	g	g	v	v	0			1
g	g	g	0								1
g	g	g	v	v	0						1
g	g	g	v	v	v	v	0				1
g	g	g	v	v	v	v	v	v	0		1
g	g	g	v	v	v	v	v	v	v	v	2
g	g	g	g	0							1
g	g	g	g	g	v	v	v	v	v	v	1
g	g	g	g	g	v	g	v	g	g	g	1
g	g	g	g	g	g	g	g	g	0		1

(g: generative state; v: vegetative state; 0: dead)

The heterogeneity of a group of mature individuals can also be assessed from the differences in their seed production. In a study on moist subalpine meadows with *Bromus variegatus* and various herbs, I classified 197 individuals of *Anemone fasciculata* growing on one plot according to their seed production. The result is shown in Table 4.

TABLE 4. *Differences in seed production of* Anemone fasciculata *on a field plot*

Category	Number of seeds	Number of individuals
I	≤ 10	6
II	11- 20	25
III	21- 30	31
IV	31- 40	41
V	41- 50	35
VI	51- 60	16
VII	61- 70	7
VIII	71- 80	5
IX	81- 90	4
X	91-100	2
XI	101-110	4
XII	111-120	0
XIII	121-130	1
XIV	131-140	1
XV	141-150	0
XVI	151-160	1
XVII	161-170	4
XVIII	171-180	2
XIX	181-190	0
XX	191-200	1

On reaching a certain age, individuals of many species of grasses and semi-shrubs can divide into several parts. This phenomenon, called particulation by VYSOTSKII (1915), is characteristic of plants of different life forms (taprooted, bunched, short-rhyzomatous, etc.) among both mesophytes and -even more so- xerophytes. In most cases particulation is specific for senescent individuals, and since it is accompanied by an increase in the number of individuals, it should be considered as a special form of vegetative reproduction, not involving rejuvenation from newly formed individuals as is typical for vegetative reproduction. Wherever possible, such individuals should be considered as a separate age group.

5. THE INDIVIDUALS IN A DORMANT STATE

In the process of evolution and of coevolution with other organisms, the individuals of some species acquired an ability to pass into the dormant state and to persist in this state for a variably long time in the form of dormant underground organs. This phenomenon is especially common in plants growing under unstable water-regime conditions associated with climatic and hydrological factors, and has therefore been studied the most thoroughly for mesophytes and hygrophytes of depressions (including those around lakes) under the conditions prevailing in arid and semi-arid climates. The dormant states can be both seasonal and long-term (many years). The seasonal state of dormancy and semi-dormancy is common for plants of steppes, prairies, and savannas. The plants capable of seasonal semi-dormancy can pass into the state of long-term dormancy. In the U.S.A. during the drought of the 1930s, both grasses (*Andropogon furcatus*) and herbs passed into the state of long-term dormancy (5-9 years) (ALBERTSON & WEAVER 1944; WEAVER 1968).

Among mesophytes and hygrophytes the transition to the dormant state under an inadequate water supply has been observed for *Eleocharis* spp., *Alopecurus arundinaceus, Hordeum brevisubulatum, Juncus gerardii, Calamagrostis neglecta, Butomus umbellatus, Puccinellia dolicholepis, Bolboschoenus maritimus,* and other species (VORONOV 1943; KURKIN 1971, 1976; RABOTNOV 1974). In some cases this transition is due not so much to insufficient water as to an increased concentration of salts. Thus, at Baraba (Western Siberia) under the conditions of slight salinization, *Alopecurus arundinaceus* passed into the dormant state in the second or third year of the drought, and on highly saline soils most of its individuals did so as early as the first year. *Calamagrostis neglecta* and *Bolboschoenus maritimus* showed similar behaviour (KURKIN 1976). There is also some evidence, albeit insufficiently checked, that on felled areas forest plants pass into the dormant state.

The transition of plants into the dormant state can also be caused by animals. This seems to be quite common, and results from the injury to the underground organs caused by burrowing animals, as described for *Panax ginseng*

(KURENTSOVA 1944). The same phenomenon observed for *Festuca rubra* was due to the repeated eating of shoots by grasshoppers (KURKIN 1976). There are reasons to suppose that the transition of virginal individuals of *Chaerophyllum prescotti* into the dormant state was originally an adaptation to the disturbance of the vegetation by burrowing animals during years of massive reproduction of the latter. *Chaerophyllum prescotti* is a monocarpic whose seeds develop first into taprooted individuals; after some years, however, the taproot changes into a round dormant organ and the plant remains in the dormant state for many years. Then, mainly as the result of disturbance of the close stand of the vegetation by human activity (e.g. by mechanical treatment of the soil), the dormant individuals enter into the active state, a generative shoot is formed, and the plant flowers, bears fruit, and dies. It is highly probable that this peculiar life cycle arose in steppe phytocoenoses, i.e., the natural habitat of this species, as an adaptation to periodic massive reproduction of rodents, which involves extensive burrowing. Spherical dormant organs are less easily injured than long taproots by burrowing animals, and the burrowing produces favourable conditions for the development of dormant individuals into luxuriant, abundantly fruiting, generative plants and for the establishment of seedlings (RABOTNOV 1964).

The transition to dormancy can also be caused by competitive relationships, as exemplified by the transition to dormancy of *Taraxacum koksagyz* seedlings sown too densely (ZAVADSKII 1954). A particular form of transition from the active state to dormancy is an interruption, lasting several years, in the formation of above-ground shoots of obligate mycotrophs (*Orchidaceae, Ophioglossum vulgatum, Monotropa hypopithys*). This phenomenon seems to be associated with the special relationships between a vascular plant and a mycosymbiont. It is possible that during their underground existence these plants are not dormant but lead a marginally active life. The registration of dormant individuals is of great importance in demographic studies, because at certain times in coenopopulations of some species, large numbers of mature individuals pass into dormancy. If only the individuals in the active state are recorded an erroneous impression of the numbers of some species may be obtained, and they may be thought to be scarce or even to have disappeared from a phytocoenosis. The transition to dormancy is observed not only for herbaceous species, but also for arborescent plants (semi-shrubs, dwarf shrubs, shrubs) (BORISOVA *et al.* 1976).

Senile individuals which have lost the ability to reproduce by seeds or vegetatively due to senescence, are absent in coenopopulations of monocarpic plants and generally also in those of di-oligocarpic and many short-lived species of plants (lifespan 10-20 years) and also of many trees. These individuals are characteristic of the coenopopulations of long-lived herbaceous species, but generally their numbers are not large and they have little importance with respect to production and influence on the environment.

Coenopopulations are a dynamic phenomenon. New individuals arise, some individuals die, some pass from one age-state into another. The changes within populations can be seasonal, fluctuating, or successional. The seasonal changeability of populations consists in alterations in the number of individuals during the course of a year and in their age and seasonal states. These changes are particularly well-defined in coenopopulations of annual plants, which are characterized by a sharp reduction in numbers and a rapid change of the age-state of individuals from seedlings to generative plants, followed by self-seeding, dying off, and transition of the population into dormant states (viable seeds). In populations of perennial plants the seasonal changes in seed number are especially large (germination, self-seeding). The changes in the number of seedlings are also substantial (emergence usually followed by a high mortality). Significant changes can result, too, from the transition of juvenile into immature individuals and of immature into mature virginal individuals, as well as from the death of young individuals. Plants with a short vegetative period, e.g. ephemers (*Corydalis halleri, Anemone ranunculoides,* etc.), pass from the active into the dormant state during the vegetative season. The mortality of mature individuals during the vegetative season is usually not very high except in years when phytophages reproduce in large numbers or weather conditions deviate sharply from the average.

From year to year, due to divergent meteorological and other conditions in single years, fluctuations are observed in the supply of seeds. The emergence and death of seedlings, the transition of individuals from one age group to another, and the dying off of mature individuals, all involve corresponding changes in the number of individuals and the age composition of a coenopopulation.

Successions involve orientated changes in coenopopulations, progressive for some species and regressive for others. In addition, new populations are formed due to invasion by new species.

For virginal individuals, the degrees of dying off and transition to the next age-state vary from species to species, from year to year, and from phytocoenosis to phytocoenosis. An idea of these variations can be gained from the results of SAURINA (1973), who observed marked-off individuals of *Ranunculus acris* in three types of upland meadow in the Moscow region over a period of three years (1970-1972). Of the juvenile individuals marked off in 10 m^2 plots in different phytocoenoses in 1970:

> 53.5-85.2% were dead by 1971,
>
> 10.3-32.5% died between 1971 and 1972,
>
> 2.0- 7.0% became immatures by 1971 but died by 1972,
>
> 1.0- 4.5% remained in the juvenile state up to 1972 and
>
> 1.0-11.0% were in the immature state in 1971-1972.

Of the immature individuals marked off in 1970:

> 33.5-50.0% died by 1971,
>
> 9.0-41.5% died in the period from 1971 to 1972,
>
> 10.0-14.5% remained in the immature state during all three years,
>
> 3.0-29.0% passed into the mature virginal state in 1971 but
> died by 1972,
>
> 12.0-14.0% passed into the mature virginal state in 1971 and
> remained in it in 1972, and
>
> 0 - 2.0% passed into the mature virginal state in 1972.

Thus, in this observation period most of the juvenile and immature individuals died off, only a small percentage of the juvenile individuals passed into the immature state, and a small percentage of the immature individuals passed into the mature virginal state.

Of 178 juvenile individuals of *Ranunculus acris* marked off by the present author in 1950 on the 10 m^2 plot in the meadow in the Oka flood valley mentioned above, 80 (45%) had died by 1951 (11 (6%) remained in the juvenile state, 76 (43%) passed into the immature state, and 11 (6%) passed into the mature virginal state). By 1952, 40 (22%) had died and by 1953, 38 (21%); only 4 individuals reached generative state. After 10 years (1960), only 6 individuals (3.4%) were still alive.

Generally, plant mortality decreases with increasing maturity. Thus, on average, during the 9-year period in this meadow the annual mortality of the individuals of *Ranunculus acris* was 40.5% for juveniles, 36.0% for immatures, 20.8% for mature vegetative individuals (mature virginal + generative, temporarily vegetative), and 11.6% for generative individuals (RABOTNOV 1958). Data collected by foresters provide the mortality rates for forest trees according to increasing age. For example, judged from the number of trees per hectare in spruce forests, the mortality according to age groups varied as follows (MOROZOV 1925):

> 20- 40 years: 4.3 %,
>
> 40- 60 years: 2.5 %,
>
> 60- 80 years: 1.8 %,
>
> 80-100 years: 1.4 %, and
>
> 100-120 years: 0.75%.

Thus, the mortality rate decreased with increasing age of the trees from 20 to 120 years. It is reasonable to suppose that with further ageing, when senescence sets in, the number of individuals that die will rise.

The observations on perennial herbaceous plants reproducing only by seeds (e.g. on meadows and steppes) suggest that, depending on the lifespan of the individuals of particular species, from 1-2 to 5-10% mature individuals die off in "normal" years but that in some years the mortality rate can be as high as 20-50% and even more. Mass death occurs in years that are especially unfavourable for particular species and also when rodents and certain insects reproduce excessively. For example, the formation of an ice cover in the late autumn of 1952 in the Oka flood valley led by the spring of 1953 to the large-scale death of mature individuals of *Heracleum sibiricum, Sesili libanotis, Medicago falcata, Campanula glomerata,* etc. (RABOTNOV 1971). Between 1945 and 1947 in subalpine meadows of the northern part of the Caucasus, individuals of many species of plants perished as the result of mass reproduction by *Arvicola terrestris* (RABOTNOV 1950a). In contrast, mass dying off of mono- and oligocarpic species occurs after years with especially favourable conditions for growth. For instance, in 1948, which was particularly favourable for *Trifolium pratense,* 92.5% of the individuals of this species flowered and were luxuriantly developed, but 75% of them had died off by 1949 and the rest by 1950 (RABOTNOV 1961). Mature individuals of species reproducing vegetatively sometimes die off in large numbers, as shown by SARUKHAN (1970) for *Ranunculus repens.*

No species can exist in any place forever. They become established on a new substrate or invade already existing phytocoenoses, sometimes after the latter have been disturbed, and they attain maximum participation in phytocoenosis, after wich participation can decrease to the point of complete disappearance. In a particular place, succession of phytocoenoses can occur and, correspondingly, coenopopulations can change. Two terms, the "great cycle" of coenopopulations (RABOTNOV 1969c) and the "coenopopulational flow" (URANOV & SMIRNOVA 1969), have been suggested for the sum of successive coenopopulations in one place. Under great cycle is understood the subcycles of successive coenopopulations. The subcycles comprise the period of formation of a coenopopulation, the period, usually long, of the persistence of a coenopopulation after it is formed, and the period of gradual or rapid decline of a population to the point of disappearance of the species from the phytocoenosis. The first stage corresponds to invading populations, the second to normal, and the third to regressive coenopopulations (RABOTNOV 1945, 1950b). Normal coenopopulations could preferably be called homeostatic. Within the three main types of coenopopulations many variants can be distinguished (RABOTNOV 1950b; RYSIN & RYSINA 1965; RYSIN & KAZANTZEVA 1975; URANOV & SMIRNOVA 1969).

The period of formation of a coenopopulation can be very long and can be divided into several stages: (1) the first stage, with plants of the same age; (2) the second stage, characterized by a group of mature individuals of the same age in the sex-ripe state and usually also a group of young individuals of different ages, produced by the self-seeding of mature individuals; (3) the third stage, formed by both young and mature individuals differing in age. Homeostatic coenopopulations can be stable, vary from year to year or between periods of years (fluctuating), successional as a whole, and mosaically successional.

Invasion populations can consist solely of viable seeds supplied from outside or such seeds in combination with some groups of virginal individuals. Such populations have various fates. They can: (1) decline to the point of complete disappearance from the phytocoenosis or be preserved in the soil as viable seeds; (2) persist in the invasion stage without becoming homeostatic, owing to the absence of the conditions necessary for transition of virginal individuals to generative, and surviving due to the supply of seeds or other diaspores from outside; and (3) change into homeostatic populations.

When coenopopulations undergo degradation the numbers of their individuals decrease, since no new ones arise. This regression can proceed to the point of complete disappearance from the phytocoenosis or, sometimes, to the point of transition to a "relict" population consisting solely of viable seeds or, sometimes, of mature dormant individuals. Such populations can persist in the relict state for a long time.

7. THE MAIN TYPES OF LIFE STRATEGY OF PLANT SPECIES

The composition and dynamics of a coenopopulation depend on the life strategy of the species and its position in the phytocoenosis. The main types of life strategy can be classified on the basis of the classification of "phytocoenotypes" proposed by RAMENSKII (1938), which includes violents, patients, and explerents. The group of pioneer plants should be added to this classification (RABOTNOV 1975). Ramenskii characterized the suggested phytocoenotypes as follows. Violents: "developing vigorously, they occupy territory and retain it, suppressing their rivals by the energy of their vital activity" and by thoroughness in the utilization of the environmental resources (high absorption capacity of the roots, heavily shading foliage, etc.). Patients: "succeed in the struggle for existence not by virtue of vigorous vital activity and growth but by their tolerance for extreme environmental conditions". Explerents: "have a very low competitive capacity but are able to invade vacant territories quickly, filling the gaps between strong plants, although being as easily displaced by the latter". Pioneer plants include the species forming phytocoenoses of the first stages of primary succession.

Within all these groups of plants there are species that are dominant in phytocoenoses. Whereas violents and patients dominate in climax coenoses and also in some persisting secondary phytocoenoses (e.g. meadows), explerents dominate in phytocoenoses of the first stages of secondary succession.

It is characteristic for the dominants of phytocoenoses of the first stages of primary and secondary succession that their populations are formed mainly by the supply of seeds from outside. The phytocoenoses of old fallows are an exception to this: in their case the coenopopulations of plants of the first stage of succession are formed mainly from the seed pool in the soil. Typically in such coenopopulations, new individuals do not become established under the mother cover as soon as it is formed. These coenopopulations arise, reach the mature state quickly and, since no new individuals develop, decline rather rapidly in the course of the succession. The species will not occur in succeeding phytocoenoses or remain as relict coenopopulations.

Both violents and patients can be dominants in climax communities. The dominating violents include trees of climax forests of temperate and boreal zones (deciduous and coniferous trees). It is characteristic of the forest climax coenoses that the conditions created under the tree layer are unfavourable for young tree plants arising from seeds. For this reason, the tree dominants have developed two modes of life strategy. The first of these is the production of large seeds (e.g. acorns, beechnuts) containing appreciable reserves of the energy and nutrients required for the formation of adequately developed seedlings. Most of such seeds fall on the soil under the mother cover, many are consumed by animals, and only an insignificant percentage are carried by birds and rodents beyond the boundaries of the phytocoenoses which produce them. The many seedlings of dominant trees emerging under the mother cover in seed-rich years can exist for many years, and then perish unless a suitable ecological niche for their further development becomes vacant. Vacancies occur chiefly as the result of the death of mature trees (storm damage, etc.), which can only occur on a large scale in older stand where the trees are approaching the limiting age for the prevailing conditions. This mortality leads to the formation of coenopopulations of dominants of different ages.

Picea abies is a representative of a different life strategy. Some of its seeds fall under the mother cover and others are distributed over considerable distances by the wind, also over a snow cover, and reach sites suitable for the establishment of seedlings that give rise to new coenopopulations. Young individuals of *Picea abies* arising from seeds lying under the mother cover have even less chance of developing into mature individuals than the seedlings of trees with large seeds. *Picea abies* germinates only in places where mature plants have died off, i.e., in gaps in woods. Naturally, the dying off of large numbers of mature trees occurs in old stands because senescent trees are very susceptible to winddamage, diseases, and pests. However, it seems that in the past this state was rarely reached in spruce and other coniferous forests in the boreal zone because of the frequent destruction by fires, whereas now they are often cut down by man. It was due to the destruction of spruce forests by fire in the past that this species acquired the ability to colonize easily those burned and felled areas in which the climatic conditions are favourable for its rejuvenation. In regions with a more severe climate the species rejuvenates itself easily only under the cover of secondary stands of *Betula* spp., *Populus tremula, Alnus incana,* and *Pinus sylvestris*.

The dominants of climax phytocoenoses of steppe, semi-desert, and desert zones are represented by patients, whose effect on seedlings and juveniles is less adverse than that of violents. Germination can occur under the mother cover, although it is limited to occasional years because of the aridity of the climate. This leads to homeostatic (normal) populations of individuals of different ages. The activity of burrowing animals is very important for the seed reproduction of the dominants of steppe and semi-desert phytocoenoses.

Some of these animals periodically reproduce abundantly and burrow through large areas of soil, creating conditions similar to those of old fallows, which are favourable for the establishment of seedlings. For example, on Mongolian steppes in years of mass reproduction of the Brandt vole, burrowing leads to the death of bunch grasses (*Stipa, Festuca*) whereas the rhizomatous plant species *Leumus chinensis* spreads. But later, owing to favourable conditions for the germination and quick development of the individuals of bunch grasses, these populations are rejuvenated and again become dominant (LAVRENKO & YUNATOV 1952). Colonial burrowing animals periodically move their colonies from place to place, leaving behind them disturbed spots in the vegetation offering favourable conditions for the establishment of seedlings and the quick transformation of juvenile into mature plants. Spots with rejuvenated coenopopulations also arise. All this produces a mosaic composed of fragments of coenopopulations with assorted age groups within the phytocoenoses. Thus, the dominants in the phytocoenoses of the corresponding stages of succession series form coenopopulations differing as to mode of formation and existence.

Three main developmental stages of coenopopulations -invastion, maturity, and decline- can easily be distinguished among dominant species present in the initial stages of a primary or secondary succession process. In the first stage the coenopopulation is built up predominantly by an external supply of seeds. There is a rapid transition from the invasive to the mature stage of the coenopopulations. In the latter stage there is no reproduction by seed. The seeds are transported and germinate elsewhere, thus establishing new coenopopulations. Finally, there is a gradual or rapid (depending on the lifespan of individuals of the species) decline leading to the disappearance of the species from the changing phytocoenosis or to survival as a relict coenopopulation.

The coenopopulations of dominants of climax phytocoenoses are characterized by a prolonged mature stage with several variants of the establishment of new individuals, these variants being associated with different effects of the dominants on the environment, the lifespan of their individuals, the effect on the environment of other plant species, and the activity of the zoocomponents of biocoenoses. These coenopopulations are self-seeding, but generally play little or no part in the supply of seeds to new coenopopulations. An exception to this rule is formed by some dominants of climax communities (for example *Picea abies*) whose seed production serves as the basis for the formation of coenopopulations in phytocoenoses of the initial stages of a secondary succession series.

The composition of coenopopulations is strongly affected by human activity (e.g. over sowing, planting, selective felling). The composition of coeno-populations is, however, greatly complicated where species reproduce not only by seed but also vegetatively.

All coenopopulations have a spatial distribution, either bordering on other coenopopulations or separated from them. Depending on the type of boundary of phytocoenoses, the coenopopulations characteristic for them are interconnected by gradual transitions or separated from one another rather sharply. The continuity principle is just as applicable to coenopopulations as to phytocoenoses. Because environmental gradients are gradual, spatial continuums of coenopopulations are formed. In phytocoenoses with a mosaic structure all of the species cannot participate in all of the microphytocoenoses forming the mosaic, and, moreover, fragments of their coenopopulations typical for different microphytocoenoses may differ as to the number of individuals and their age spectrum. If the composition of phytocoenoses is faily homogeneous, the distribution of individuals of different ages can vary depending on the number of mature individuals and on the spatial distribution of favourable conditions for the establishment of young plants. For most species except anemochores, a high percentage of the seeds fall near the mother plant, the number decreasing gradually with the distance from it. If seed-producing individuals are close together, the distribution of young individuals arising from their seeds is fairly uniform; but where there is some distance between seeding plants, the distribution depends on the conditions for the establishment of new individuals. Where mother plants create a more favourable microclimate for the establishment of seedlings, as in phytocoenoses of arid regions (steppe, semi-desert, desert), juveniles and immatures are found near mature individuals. The opposite occurs where not only fruits and seeds lying near the seeding individual but also the seedlings are completely destroyed by phytophages (as known for annually bearing tropical trees): here, young individuals can only become established far from the mother plant, where such phytophages are absent (JANZEN 1970). An irregular distribution of juveniles and immatures is often associated with the activity of animals (which disturb the close stand of vegetation and thus create favourable conditions for the establishment of seedlings) as well as with the irregular distribution of ecological niches vacated by the death of mature individuals.

8. REFERENCES

ALBERTSON, F.W. & J.E. WEAVER, 1944 - Nature and degree of recovery of grassland from the great drought of 1933-1940. *Ecol. Monogr.*, 14, 393-479.

AMELIN, J.S.*, 1947 - (Seeds supply for revegetation of desert in Middle Asia). *Dokl. Vses. Akad. sel'.-Khoz. Nauk,* N 3, 38-41.

BEKLEMISHEV, V.N., 1951 - (On the classification of the biocoenotic (synphysiological) connections). *Byull. mosk. Obshch. Ispyt. Prir. (Otd. biol.),* 61, 3-30.

BOGDANOVSKAYA-GIENEF, J.D., 1926 - (On seed regeneration in meadow communities). *Zap. leningr. Sel' -khoz. Inst.,* 3, 216-253.

BOGDANOVSKAYA-GIENEF, J.D., 1941 - (Seed regeneration in meadow coenoses of the forest zone). *Uchen. Zap. leningr. gos. Univ. (Ser. biol. nauk),* N 20, 93-133.

*Titles of Russian papers and books are given in English between parentheses.

BOGDANOVSKAYA-GIENEF, J.D., 1954 - (Seed regeneration in meadow coenoses of the forest zone). *Uchen. Zap. leningr. gos. Univ. (Ser. biol. nauk),* N 34, 3-47.

BORISOVA, J.V., BESPALOVA, Z.G. & T.A. POPOVA, 1976 - (Peculiarities of the phenological development of steppe and desert plants of Northern Gobi (M.N.R.)), *Trudy mosk. Obshch. ispyt. Prir.,* 42, 239-255.

CHIPPINDALE, H.G. & W.E.I. MILTON, 1934 - On the viable seeds present in the soil beneath pastures. *J. Ecol.,* 22, 508-522.

GOGINA, E.E., 1960 - (Seed productivity of some plant species of the high-montain meadows of South Osetia). *Bot. Zh.,* 45, 131-139.

GOLUBEVA, J.N., 1962 - (Some data on pools of viable seeds in soil under meadow-steppe vegetation). *Byull. mosk. Obshch. Ispyt. Prir.,* 67, 76-89.

HARPER, J.L., 1961 - Approaches to the study of plant competition. *In:* F.L. MILTHORPE (Editor), *Mechanisms in biological competition.* University Press, Cambridge, p. 1-39.

JANZEN, D.H., 1970 - Herbivores and the number of tree species in tropical forests. *Am. Nat.,* 104, 501-528.

KAMENETSKAYA, J.V., 1969 - (Soil seed pool of pine and the main species of herbaceous dwarfshrub layer in *Pineta myrtillosum*). *In: Sosn. lesa podzoni yushn. taigi i puti veden. v nihk lesn. khoz.,* Izd. Nauka, Moscow, p. 252-263.

KARPOV, V.G., 1969 - (*Experimental phytocoenology of dark coniferous taiga*). Izd. Nauka Leningrad, 335 p.

KHITROVO, V.N., 1915 - (On the principles of classification for natural meadows). *Mater. organiz. kultur. kormov. plotshadi,* 12, 48-81.

KURENTSOVA, G.E., 1944 - (Some data on the biology and cultivation of jen-shen (*Panax ginseng* C.A. Mey)). *Sov. Got.,* 1.

KURKIN, K.A., 1971 - (Summer and perennial dormancy of herbaceous perennials of Barabian forest-steppe). *Bot. zh.,* 56, 1564-1581.

KURKIN, K.A., 1976 - (*Investigations on the dynamics of meadow systems*). Izd. Nauka Moscow, 284 p.

LAVRENKO, E.M., 1947 - (On the study of the contribution of some dominant plant species to vegetation cover). *Sov. bot.,* 1, 5-16.

LAVRENKO, E.M., 1959 - (Main principles governing plant communities and the methods of their investigation). *Polev. Geobot.,* 1, 13-75.

LAVRENKO, E.M. & J.V. BORISOVA (Editors), 1976 - (*Biocomplex investigations in Kazakhstan*). 3. Izd. Nauka, Leningrad, 292 p.

LAVRENKO, E.M. & A.A. YUNATOV, 1952 - (Fallow regime in steppes as a result of the influence of the Brandt Vole on steppe sward and soil). *Bot. Zh.,* 37, 128-138.

LEVINGSTON, R.B. & M.L. ALLESSIO, 1968 - Buried viable seed in successional field and forest stands. Harvard Forest. Massachusets. *Bull. Torrey bot. Club,* 95, 58-69.

LINKOLA, K., 1930 - Über das Vorkommen von Samenkeimlingen bei Pollokanthen in den natürlichen Pflanzengesellschaften. *Ann. Soc. zool-botan. Fenn. Vanamo,* 11, 150-172.

LINKOLA, K., 1936 - Die Dauer und Jahresklassen-Verhältnisse des Jugend-stadiums bei Wiesenstauden. *Acta for. fennic.,* 42, 1-56.

MALINOVSKII, K.A., 1959 - (*Pastures with Nardus stricta in the subalpine belt of Ukrainian Carpatiens*). Izd. Akad. Nauk., SSSR Kiev, 205 p.

MASING, V.V., 1966 - (Consortiums as elements of the functional structure of biocoenoses). *Trudy mosk. Obshch. ispyt. Prir.,* 27, 117-127.

MILTON, W.E.J., 1936 - The buried viable seeds of enclosed and unenclosed hill land. *Welsh Plant Breed. St., Ser. H.,* 14, 58-73.

MILTON, W.E.J., 1939 - The occurrence of buried seeds in soils at different elevations and on a salt marsh. *J. Ecol.,* 27, 149-159.

MILTON, W.E.J., 1943 - The buried viable seed content of a midland calcareous clay soil. I. *Emp. J. exp. Agric.,* 11, 43-44.

MILTON, W.E.J., 1943 - The buried viable seed content of a midland calcareous clay soil. II. *Emp. J. exp. Agric.,* 11, 155-167.

MOROZOV, G.F., 1925 - (*Forest science*). Gosizdat Moscow, 2nd ed., 367 p.

MOROZOVA, O.I., 1952 - (Evolutionary role of animal grazing in the origin of species of Angiospermae plants). *Bot. Zh.,* 37, 158-172.

NECHAEVA, N.T., 1954 - (Soil seed pool in pastures of the southeastern part of Kara-Kums and the influence of grazing on the burial of seeds). *"Pustyni SSSR i ikh osvoenie"*, 2, Izd. Akad. Nauk SSSR, Leningrad, 370-391.

OOSTING, H. & M. HUMPHREYS, 1940 - Buried viable seeds in a successional series of old field and forest soils. *Bull. Torrey bot. Club*, 67, 253-273.

OSICHNYUK, V.V., 1973 - (Successions of steppe vegetations). *Rastitel "nost" SSSR Kiev*, 249-333.

PERTTULA, U., 1941 - Untersuchungen über die generative und vegetative Vermehrung der Blütenpflanzen in der Wald-Hain-Wiesen und Hainfelsen-vegetation. *Ann. Acad. Sci. Genn.*, Ser. A, 2, 8, N1, 1-388.

PETROVSKII, V.V., 1961 - (Sinusies as forms of joint existence of plants). *Bot. Zh.*, 46, 615-626.

PONOMARYEV, A.N. & V.A. VERETSHAGINA, 1973 - (Anthecological essay on dark coniferous taiga). *Probl. biogeon., geobot. i botan. geogr.*, Izd. Nauka, Leningrad, 196-207.

PYATIN, A.M., 1970 - (On the viable seed content of soils of meadow and forest pastures). *Byull. mosk. Obshch. Ispyt. Prir., Otd. biolog.*, 75, 85-95.

RABOTNOV, T.A., 1945 - (Biological observations on the subalpine meadows of the northern Caucasus). *Bot. Zh.*, 30, 167-177.

RABOTNOV, T.A., 1950a . (The life cycle of perennial herbaceous plants in meadow coenoses). *Trudy bot. Inst. Akad. Nauk SSSR*, Ser. III, 6, 7-204.

RABOTNOV, T.A., 1950b - (Problems associated with the study of population composition for phytocoenological purposes). *Problemy Bot.*, 1, 466-483.

RABOTNOV, T.A., 1956 - (Some data on the germinable seed content of soils of meadow communities). *Akademiku V.N. Sukachevu k 75-letiju so dnya rozhdeniya*, 481-499.

RABOTNOV, T.A., 1958 - (The life cycle of *Ranunculus acer* L. and *auricomus* L.). *Byull. mosk. Obshch. Ispyt. Prir., Otd. biolog.*, 62, 77-86.

RABOTNOV, T.A., 1961 - Some problems in increasing the proportion of leguminous species in permanent meadows. *Proc. 8th. Int. Grassld. Congr.*, 261-264.

RABOTNOV, T.A., 1964 - (On the biology of monocarpic perennial meadow plants). *Byull. mosk. Obshch. Ispyt. Prir., Otd. biol.*, 69, 47-55.

RABOTNOV, T.A., 1969a - (*Some problems encountered in the study of coenopopulations*). *Byull. mosk. Obshch. Ispyt. Prir., Otd. geol.*, 74, 141-149.

RABOTNOV, T.A., 1969b - (On consortia). *Byull. mosk. Obshch. Ispyt. Prir., Otd. biol.*, 74, 109-116.

RABOTNOV, T.A., 1969c - On coenopopulations of perennial herbaceous plants in natural coenoses. *Vegetatio*, 19, 87-95.

RABOTNOV, T.A., 1969d - Plant regeneration from seed in meadows of the USSR. *Herb. Abstr.*, 39, 269-277.

RABOTNOV, T.A., 1971 - (The effect on meadow plants in ice formation in and on the soil). *Byull. mosk. Obshch. Ispyt. Prir., Otd. biol.*, 76, 120-134.

RABOTNOV, T.A., 1972 - Consortia, the importance of their study for phytocoenology. *Folia geobot. phytotax*, Praha, 7, 1-8.

RABOTNOV, T.A., 1974 - (*Meadow science*). Izd. Moscow University, Moscow, 384 p.

RABOTNOV, T.A., 1975 - On phytocoenotypes. *Phytocoenologia*, 2, 66-72.

RAMENSKII, L.G., 1938 - (*Introduction to the geobotanical study of complex vegetations*). Selkhozgiz, Moscow, 620 p.

RAMENSKII, L.G., 1952 - (On some basic concepts in contemporary geobotany). *Bot. Zh.*, 37, 181-201.

RIDLEY, H.N., 1930 - *The dispersal of plants throughout the world*. Ashford Reeve, 744 p.

RODIN, L.E., & R.V. SUKHOVERKO, 1956 - (Higher plants and soil seed pool in takyrs). *Takyrs zap. Turkmenistan i ikh sel'khoz. osvoeniye*. Izd. Akad. Nauk SSSR, Moscow, 30-37.

RYSIN, L.P. & G.P. RYSINA, 1965 - (An attempt at populational analysis in forest communities). *Byull. mosk. Obshch. Ispyt. Prir., Otd. biol.*, 71, 84-94.

RYSIN, L.P. & T.N. KAZANTZEVA, 1975 - (A method for coenopopulational analysis in geobotanical studies). *Bot. Zh.*, 60, 199-208.

SALISBURY, E.J., 1942 - *The reproductive capacity of plants*. London, 244 p.

SARUKHAN, J.A., 1970 - Study of the population dynamics of three *Ranunculus* species. *Proceed. 10th. British Weed Control Conference*.

SAURINA, N.I., 1973 - Productivity of some coenopopulations of *Ranunculus acris* L. and *R. auricomus* L. *Byull. mosk. Obshch. Ispyt. Prir.*, 78, 1.

SHENNIKOV, A.P., 1930 - *(Volga river meadows in the middle-Volga area)*. Izd. Ulyanov. Okr. zem, uprave. Leningrad, 386 p.

SHENNIKOV, A.P., 1941 - *(Meadow science)*. Izd. Leningrad University, Leningrad, 541 p.

SHENNIKOV, A.P. & E.P. BARATYNSKAYA, 1924 - (Some results of studies on the structure and variability of meadow communities). *Zh. russk. bot. Obshch.*, 9, 75-82.

SÖYRINKI, A., 1938 - Studien über die generative und vegetative Vermehrung der alpinen Vegetation Petsamo-Lapplands. I. Allgemeinen Teil. *Ann. Bot. Soc. Zool.-Bot. Fenn. Vanamo*, 11, 1-323.

SÖYRINKI, A., 1939 - Studien über die generative und vegetative Vermehrung der alpinen Vegetation Petsamo-Lapplands. II. Specieller Teil. *Ebenda*, 14, 1-405.

STRELKOVSKAYA, A.L., 1941 - (Seed content of the soils of Ostrakhan sheep-pasture deserts in Uzbekistan). *Byull. vses. nauchno-issled. Inst. Karakulevodstva*, 4, 65-68.

SUOMALAINEN, P., 1930 - Über die Samenkeimlinge auf einer Meeresstrandwiese in Südfinnland. *Ann. Soc. Zool. Botan., Fenn. Vanamo*, 11, 173-187.

URANOV, A.A., 1960 - (The vitality of species in a plant community). *Byull. mosk. Obshch. Ispyt, Prir.*, 65, 77-92.

URANOV, A.A. (Editor), 1967 - *(Ontogenesis and age composition of populations of flowering plants)*. Izd. Nauka, Moscow, 155 p.

URANOV, A.A. (Editor), 1968 - *(The problems of morphogenesis and structure of flowering plant populations)*. Izd. Nauka, Moscow, 231 p.

URANOV, A.A., 1973 - The great life cycle and age spectrum of coenopopulations of flowering plants. *Tezisy dokladov V delegatskogo s'ezda Vses. bot. obtshestva. Nauka*, Kiev.

URANOV, A.A. (Editor), 1974 - *(Age composition of flowering-plant populations in relation to their ontogenesis)*. Izd. Lenin Mosk. Pedag. Inst., Moscow, 260 p.

URANOV, A.A., 1975 - (Age spectrum of phytocoenopopulations as a function of time and energetic wave processes). *Nauch. Dokl. vyssh. Shk., biol. nauki*, 2, 7-34.

URANOV, A.A., GRIGORYEVA, N.M., EGOROVA, V.N., ERMAKOVA, I.M. & A.R. MATVEEV, 1970 - (Variability and dynamics of the age spectrums of some herbaceous plants). *Trudy mosk. Obshch. ispyt. Prir.*, 38, 194-214.

URANOV, A.A. & O.V. SMIRNOVA, 1969 - (Classification and main developmental features of perennial-plant populations). *Byull. mosk. Obshch. Ispyt. Prir., Otd. biol.*, 74, 119-134.

URANOV, A.A. & T.I. SEREBRYAKOVA (Editors), 1976 - *(Coenopopulations of plants)*. Izd. Nauka, Moscow, 216 p.

VORONOV, A.G., 1943 - (On some adaptations of plants to changes in the water-level of lakes). *Bot. Zh.*, 28, 181-186.

VYSOTSKII, G.N., 1915 - (Ergenya-cultural phytological essay). *Trudy byuro prikl. Bot.*, VIII, N 10-11, 1113-1443.

WEAVER, J.E., 1968 - *Prairie plants and their environment. A fifty-year study in the Midwest*. Lincoln, 276 p.

ZAVADSKII, K.M., 1954 - (On plant death after cluster sowing: density of clusters and supply of nutrients). *Bot. Zh.*, 39, 515-544.

The demography of plants with clonal growth

1. INTRODUCTION

I am a firm believer in Dobzhansky's dictum, "Nothing in biology has meaning
except in the light of evolution" (DOBZHANSKY 1973). Evolution happens because
some organisms leave more descendants than others (note that the emphasis is
on descendants, not progeny) and because some variation between ancestors is
heritable. When we come to explain all and every attribute of living organisms
we are driven to look at the heritable attributes of fitness. THODAY (1953)
has pointed out that, partly because environments are variable in time, one
generation will usually be insufficient to measure fitness properly. For the
same reason we may not see, within a single generation, the real expression of
the qualities that conferred fitness in the longer term. For a farmer, every
year is unusual (and usually bad!) and it is the same for the ecologist, and
for plant ancestors and their descendants.

Most of the heritable variation in fitness is exposed through mutation and
recombination and the zygote is therefore the logical starting point for
studying the biology of any living organism. The life cycle that develops from
the zygote is the expression of the interaction of a particular genotype with
its environment. At some point in the life cycle a reproductive phase
contributes new zygotes that start new life cycles. The life cycle that is
relevant to understanding evolution is the cycle from zygote to zygote. Of
course, there is a wide variety of ways in which a plant or animal may express
its genotype through its life cycle. In its simplest form, the genotypic
instruction specifies a determinate growth sequence, as in man, the rabbit or
Drosophila. Higher plants however, express the genotype in repeated
"reiteration" of modular units, namely leaves with axillary buds, and in
various aggregations of these in repeated units of shoot structure (e.g. the
short and the long shoots of an apple tree are reiterations of a basic
structural unit).

Botanists frequently emphasise the great plasticity of plant growth.
Compared with an animal the size of individual plants is immensely sensitive
to the availability of resources; but the individual module of plant growth is
probably no more variable or plastic than the length of a rabbit's leg or a
Drosophila wing. The higher plant expresses its reaction to environmental

stress mainly by varying the number of its modular units of construction rather than their size or form. It is for this reason that we have argued (HARPER & WHITE 1974) that an adequate description of a population of plants has to take account of two parameters: N, the number of genets being the products of individual zygotes and η, the number of modular units (leaf with axillary bud, branch system or ramet) that reiterate the characteristic expression of that genet.

In annual and other monocarpic plants, the process of reiteration has a defined end, defined either by a seasonally triggered lethal act of seed production or by a seasonally lethal event such as the first killing frost of autumn. In the perennial, the product of a zygote continues, year by year, to reiterate new modules of construction and the act of seed formation is sublethal occurring as a more or less extended series of intermittent episodes set within the perennial process of growth. The genetic instructions coded within the zygote are expressed in this repeating process. The results of reiteration produce the characteristic form of a plant which depends on the arrangement of the repeated modular units, their distance apart, the angles of branching of the connecting structures, which modules develop, remain dormant or die. The resulting growth patterns are often highly species specific and only recently have attempts been made to describe these elements of plant form (e.g. HALLE & OLDEMAN 1970 for tropical trees, BELL 1976 for rhizomatous plants).

The results of reiteration in a perennial depend also on whether the modules of growth remain part of a physically and physiologically integrated whole (as in most trees) or form more or less interdependent units (interconnected ramets or shoots) or whether the plant breaks up into disconnected physiologically independent parts (e.g. stoloniferous herbs). Many aquatic angiosperms are splendid examples of plants in which the product of a single zygote falls to pieces as it grows. A clone of *Lemna* or *Pistia* expresses itself as a genetic individual by continually falling to pieces. A zygote of *Hydra* does just the same and, at the end of a season of growth, many daughter zygotes may be the descendants of a single parental zygote but formed from a fragmented phenotype of independently living polyps. One of the successful strategies of plant life is for the genotype to be expressed as a fragmented phenotype with independent, wandering parts.

It will be apparent from this treatment that I distinguish sharply between growth and reproduction. In higher plants the process of growth, whether it leads to a single connected branched clone of shoots such as a tree or to a fragmented population of physiologically independent modules (as in a clone of *Lemna*), is the result of meristematic activity. It is always the result of development from an organised body of cells, interconnected by plasmodesmata and, for a time, integrated by hormonal control.

Reproduction, by contrast, involves the "re-production" (the production again) of a wholly new organisation from a single cell, formed with renewed and cleaned cytoplasm, lacking protoplasmic continuity with other cells and usually following some process of genetic recombination. This occurs even in inbreeders and some apomicts. The isolation of the new individual from the mother is remarkably complete and the new zygote develops, isolated from maternal tissue by intervening triploid endosperm in the angiosperms or haploid tissue in the gymnosperms. I lay much stress on this very fundamental difference between reproduction and growth because I believe that the concept of "vegetative reproduction" (which is an aspect of growth) has done great harm and continues to hinder any approach to the population biology of perennials. It is a similar hindrance to anyone who tries to study the population dynamics of animals such as *Hydra, Obelia* or corals in which the zygote develops (plant-like) by the repetitive reiteration of modules.

Although it is a great conceptual help for population biologists and evolutionists to distinguish sharply between growth and reproduction it does little to solve problems in the field. It is often impossible to decide whether modules of shoot growth in clonal plants in nature are the repeating modules of a single or of many genets. The problem that bedevils the population biologist is of course that it has been and still is largely impossible to count the number of genetic individuals in a population unless (a) all the modules of a genet remain connected as in a tree or (b) there are abundant genetic markers as e.g. in *Trifolium repens* (HARBERD 1963b). In a forest it is possible (though tedious!) to count (a) the number of zygotes N that contributed to the population as the number of trunks and (b) the number of modular shoot units η that have developed from such zygotes. Problems arise, however, even in trees, when there is suckering as in *Populus tremuloides* (or in many *Ulmus* spp). The difficulties are compounded in rhizomatous and stoloniferous herbs where connections are below ground and decay. Usually the most that the population biologist can attempt is a count of the product N.η, i.e. the total module number. Most often the countable module is the non-persistent aerial shoot (the greater part of Tamm's classic studies of the population biology of grassland and woodland herbs (TAMM 1956, 1972a, 1972b) does not distinguish genets and this could have been done only by destructive sampling and excavation which would have made continuous demographic study impossible).

There are occasional exceptions to the rule that genetic individuals cannot be distinguished in clonal population. *Trifolium repens* is such an exception, bearing within its natural population a wide variety of genetic polymorphisms including a series of multiple alleles responsible for leaf marks and segregation at other loci of genes responsible for different visible leaf characters. This makes it possible to identify quite precisely many genetic individuals and to follow the extent of clones. HARBERD (1963b) has used such markers to demonstrate the very wide extent of single clones in *Trifolium repens*. Harberd has been able to employ similar techniques for a number of grass species (HARBERD 1963, 1967; HARBERD & OWEN 1969) in which the markers are much less readily discovered. OINONEN (1967) has also used fine morphological detail to distinguish clones of *Pteridium aquilinum*. In the case of *Trifolium repens* the precision of the leaf mark polymorphism has made it possible to demonstrate the degree of intermingling of clones even in a very old permanent grassland (more than 60 years since the last ploughing). Clones intermingle on a fine scale (3-6 clones present per dm^2) as has been shown by CAHN & HARPER (1976a). The leaf marks are of selective value and sheep select between the marks (CAHN & HARPER 1976b). The extent of intermingling of clones is indicated in Fig. 1 which shows the frequency distribution of clones per dm^2 in a variety of grasslands. Although the grazing behaviour of sheep may be one force that maintains a polymorphism within a pasture, other factors must clearly operate to maintain the fine scale of leaf mark polymorphism. Cattle and horses are unlikely to be able accurately to select individual leaves of

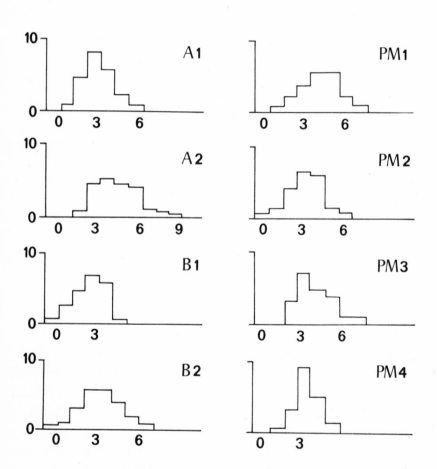

FIG. 1.   *The frequency distribution of white leaf mark morphs of*
           Trifolium repens L. *within 10 cm square quadrats in*
           permanent grasslands.
           A1, A2, B1, B2 *Henfaes Farm, Aber, Gwynedd, North Wales*
           PM1-PM4   *Port Meadow, Oxford*
           (from CAHN & HARPER 1976a*)*

clover. Yet in Port Meadow (Oxford) which has been grazed continuously since at least 1086 by cattle and horses but no sheep, there is still a fine intermingling of leaf mark morphs though perhaps not as strongly expressed as in sheep-grazed pastures. It may be that other predators of clover, e.g. the wood pigeon are selective between morphs.

A single genet of *Trifolium repens* lives long and during its life the parts may wander through a sward as disconnected stolon fragments (like a terrestrial *Lemna*!). The ramets of a single genet intermingle with a variety of potentially competitive neighbours of their own and other species and the fitness of a single genet is tested in its reiterations in a variety of places with a diversity of neighbours. TURKINGTON & HARPER (in press) have examined the ways in which clone-forming plants experience the diversity of the community in which they live. This has been done by determining the frequency of inter- and intra-specific contacts made within the canopy of permanent grassland. It appears that most grasses make predominantly intra-specific contacts but that the wandering growth form of *Trifolium repens* ensures that it meets neighbouring species nearly in the proportion in which they occur in the community. CLEGG (personal communication) describes the morphology of clonal plants as of "phalanx" or "guerilla" type according to whether the clone forms a tight monotonous mass of invading shoots (phalanx) or an intermingling exploring (guerilla) growth form. Clearly, if rhizome or stolon internodes are short or the process of tillering is intravaginal, individual shoots will experience a variety of other species as neighbours. *Trifolium repens* exemplifies par excellence the wandering "guerilla" strategy that maximises inter-specific experience.

The dynamics of white clover populations have been studied within an old permanent pasture in N. Wales where there is a persistent population of buried viable seeds. New seedling recruits, i.e. new genets, appear only rarely and most die. Successful establishment occurs in local bare patches e.g. mole hills, but over the years may represent significant new genetic input to the population. *Trifolium repens* is an obligate outbreeder and is genetically poly-morphic for many features in addition to leaf polymorphism e.g. cyanogenesis, *Rhizobium* strain sensitivity, incompatibility alleles and competitive aggressiveness (TURKINGTON, CAHN & HARPER in preparation). Because of its great economic importance genetic variation within populations of *T. repens* has been deeply explored. We know nothing like so much about other perennial plants at the level of the number of genets yet there is no *a priori* reason to suppose that white clover is in any way exceptional among clonal herbs in the wide diversity of genotypes present in a small area or to suppose that clones of other species do not intermingle in the way found in *T. repens*. Obligate outbreeding is normal in clone forming species and it may well be that a high density of genetic individuals is often maintained. There is no discernible tendency in *T. repens* for single clones to dominate and exclude others even on

a very local scale. SOANE & WATKINSON (personal communication) have used
detailed demographic data obtained by Sarukhán (SARUKHáN & HARPER 1973) and by
Soane to model the rate of loss of genotypes from long lived populations of
Ranunculus repens. They find, even with exceedingly low rates of recruitment
of new genets, that a high diversity of genotypes can be maintained within a
population.

2. THE DEMOGRAPHY OF PLANT PARTS

For most clonal herbs in which there are no studied genetic markers, population
dynamics can be followed only for shoot modules. The modules of growth of higher
plants have many of the qualities that characterise the more "normal" populations
studied by zoologists. A growing plant gains new modules (birth) and loses old
ones (death). The size of the plant is determined by the balance between the
births and deaths of its parts i.e. $N_{t+1} = N_t + B$ (Births) - D (Deaths).
Similarly the size of a population of shoots (irrespective of the number of
genets that contribute to it) is determined by a flux of Births minus Deaths,
plus Immigrants minus Emigrants i.e. $N_{t+1} = N_t + B - D + I - E$. Only
specialised census methods can reveal the magnitude of such a flux. A population
of plants that maintains a constant density of shoots, may do this by a very
rapid turnover or by no turnover at all. Such a flux can only be measured by
mapping or marking modules, preferably at "birth" and repeatedly recording the
fates of the marked modules. In this way it is possible to extract fundamental
parameters of population performance such as the life expectancy of the
modules, e.g. leaves, tillers, shoots. For this sort of demographic analysis
it is essential that the individual shoots be tagged or mapped and followed as
units, not just censused, because if a shoot is lost and another shoot is
gained in a time interval, the two cancel out and a simple census will show no
change.

The dynamics of shoots has been studied in populations of *Carex arenaria*
by NOBLE (1976). Seedlings of this species are rare and most colonisation in the
natural sand dune habitats is by clonal spread. Tillers are produced along
rhizomes which grow rapidly through the sand. During such an invasive phase,
interconnected tillers can often be recognised because they form distinct lines
of shoots extending into uncolonised areas. In later phases of colonisation,
mature and hinterland phases occur with more or less dense stands of tillers
and it is impossible in these, even after excavation, to identify whole genets.

The flux of shoot in populations of *Carex arenaria* is illustrated both for
a juvenile invading population and for a senescent population in dune slacks
(Fig. 2). These figures illustrate how misleading an impression of the activity
of the birth-death flux is given simply by repeated census of the number of
shoots present. It was only because the populations were repeatedly mapped
that the rapid pace of births and deaths could be detected. It is especially

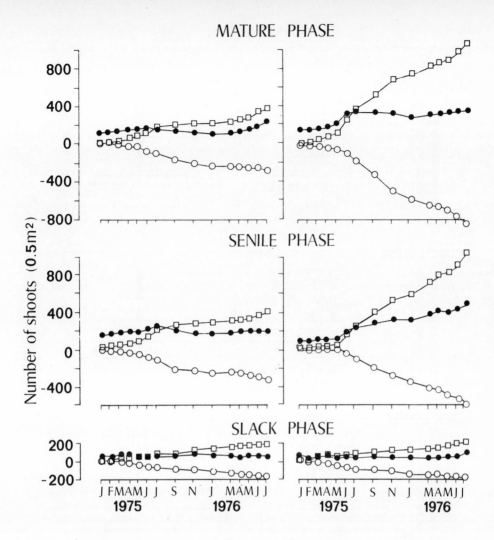

FIG. 2.   *The flux of shoots of* Carex arenaria *in mature, senile and slack*
phases of sand dunes at Aberffraw, Anglesey, N. Wales with (right)
and without (left) applied nitrogen fertiliser (400 kg N/ha)
● *shoot population present;* ○ *cumulative deaths;* □ *cumulative*
births (from NOBLE *1976)*

interesting that in the invasive phase of this species the death rate (as well
as the birth rate) is so high and that in the slack phase, when the population
is not expanding, both the birth and death rates are much lower.

   After a period in which the life of each shoot in an area has been recorded
it is possible to produce age structure diagrams that show the age composition
of the tiller population (Fig. 3). Most new tillers are produced in early summer
and at this stage a series of cohorts of young tillers is injected into an old
population. This period of production of new tillers coincides rather closely
with the period of greatest death risk of the older tillers, (just as in
Sarukhán's studies of *Ranunculus* populations in permanent grassland the greatest

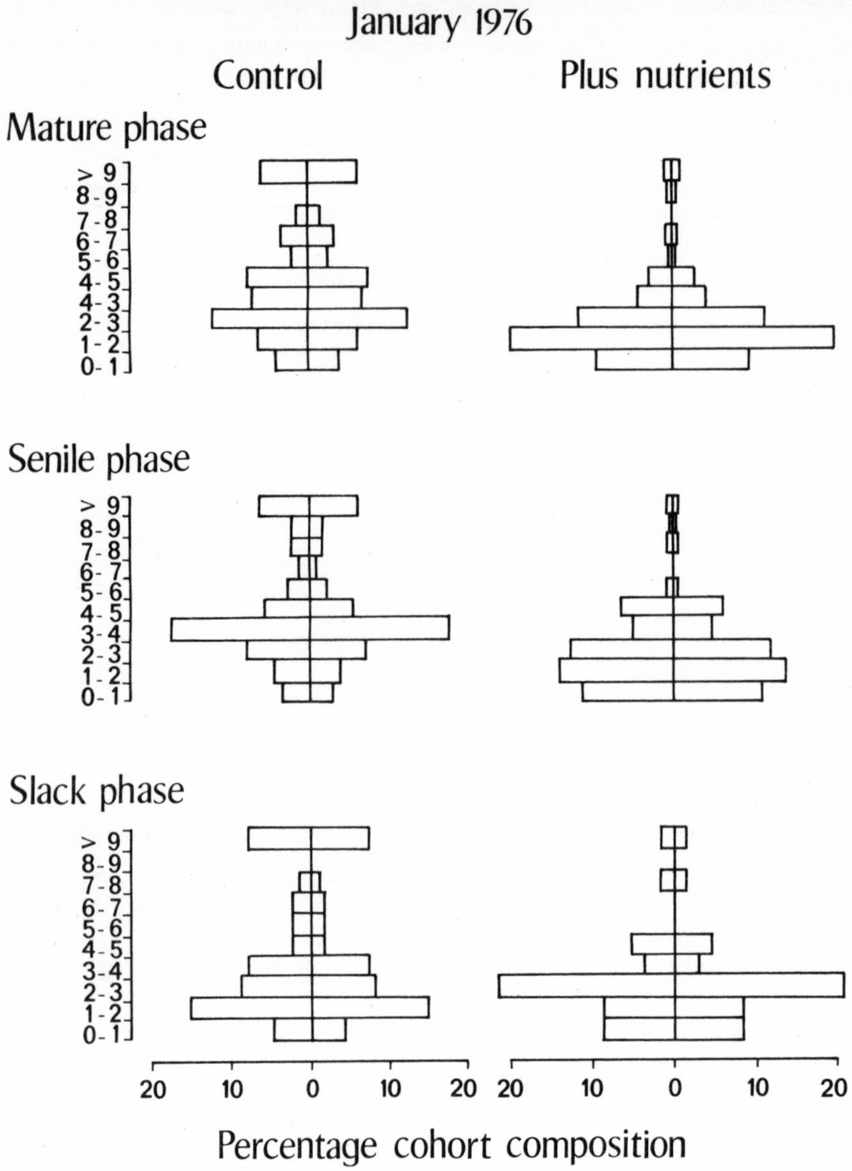

January 1976

FIG. 3.   *The age structure of shoot populations of* Carex arenaria *in*
mature, senile and slack phases of sand dunes at Aberffraw,
Anglesey, N.Wales with and without applied nitrogen fertiliser
(400 kg N/ha) (from NOBLE 1976)

death risk to ramets and to genets occurred at the stage of most growth of the
survivors). Noble applied nitrogenous fertilisers to his studied populations
and the effect of this was to increase the density of tillers present per unit
area. More dramatically however, the age structure of the population was
transformed: the fertilised plants developed many more new tillers and there
was a marked increase in the death rate of older tillers. Only demographic
treatment involving marking each tiller and following its fate could reveal

this sort of change occurring within the communities (Fig. 3).

In clonal systems, some elements of the demography of the population can be obtained by studying leaves. The leaf, with its axillary bud, is the smallest module of organised structure in higher plants and leaves have many of the properties associated with members of populations, e.g. they may increase in number exponentially in a rapidly growing plant, they have juvenile, mature and senescent phases, birth rates, death rates and survivorship curves. BAZZAZ & HARPER (1977) have shown that it is relatively easy and rewarding to describe the growth and reaction of plants to different environments by applying demographic procedures to the analysis of populations of leaves. In natural communities leaf dynamics have been studied by WILLIAMSON (1976) who showed that the leaves of different species of grass within the same chalk grassland had characteristic demographic patterns; leaf survivorship curves were drawn and half lives could be calculated (Fig. 4). The survivorship curves have interesting sigmoid

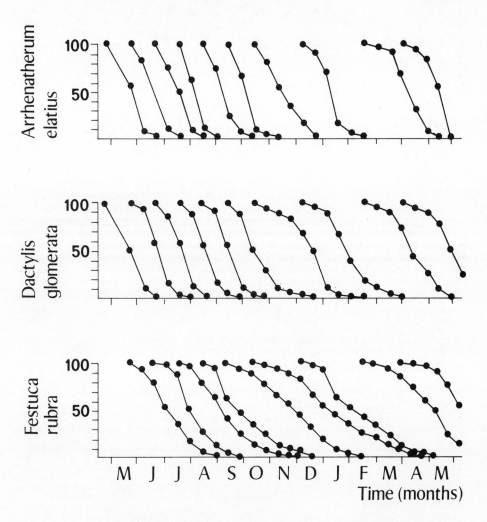

FIG. 4.   *Survivorship curves (linear scale) for the leaves of three grass species in chalk grassland (from WILLIAMSON 1976)*

characteristics, rather like dosage response curves, as if the risk of death depends on the number of doses of time received! A somewhat similar series of survivorship curves has been obtained for leaves of *Ammophila arenaria* by HUISKES (Fig. 5). A special advantage of demographic analysis of growth is that unlike conventional growth analysis it can be applied to plants growing naturally and does not require destructive harvests.

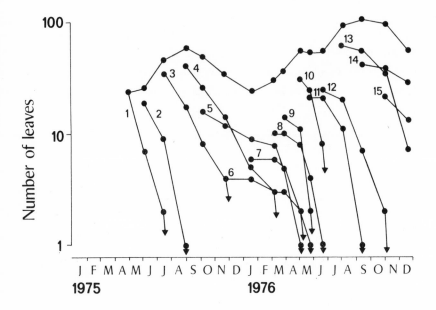

FIG. 5.   *Survivorship curves (logarithmic scale) for the leaves of* Ammophila arenaria *on sand dunes at Newborough Warren, Anglesey, N. Wales. The data show survivorship for cohorts of leaves produced at approximately monthly intervals and (top curve) the number of leaves present per m^2 at various stages in the seasonal cycle (unpublished data of A.H.L. HUISKES)*

3. THE QUANTITATIVE DESCRIPTION OF LIFE CYCLES

Life cycles are composed of phases and usually these have a distinct and defined sequence (Fig. 6). In higher plants the phases are:

1) juvenile – non reproductive – period composed of (a) the maternally supported phase of zygote development to a mature embryo, (b) a phase of embryo dormancy in the seed, (c) a growth phase in which vegetative growth is made before flowering;

2) reproductive phase characteristically rising to a peak or plateau and then declining with senescence;

3) a post-reproductive phase of declining vegetative activity (which may be very short or absent).

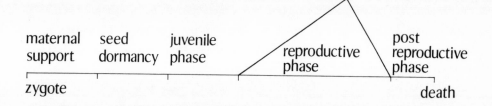

FIG. 6.   *The elements of life cycle structure*

In many plant forms, e.g. trees and annuals, these phases are clearly marked
but in clonal perennials they become confused. In natural environments the
juvenile phases may be very long drawn out. In Tamm's classic studies of the
herbs of meadow and woodland, many individuals of many species remained purely
vegetative over a very long period (Fig. 7). Flowering, when it occurred, was

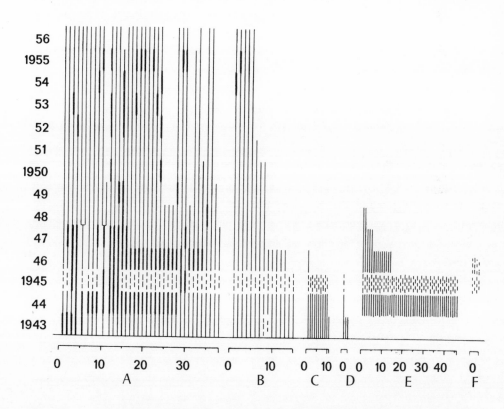

FIG. 7.   *The fate of plants of* Sanicula europaea *in a quadrat repeatedly
mapped for 14 years. Each vertical line represents a plant.
Ψ indicates clonal multiplication, heavy lines indicate flowering,
A, B, C, and D represent different size classes present at the
start of the study, and E and F represent populations of seedlings
recruited in 1944 en 1945 (from* TAMM 1956).

curiously distributed through the population with little synchrony, either between plants (there was nothing like the synchronised mast years of many temperate trees) or between ramets of the same plant. It is as if the chance that a shoot will flower or not may be a purely local reaction to the conditions experienced in the community by that shoot.

In a declining population such as one of *Centaurea jacea* studied by TAMM (1956) the years of decline were wholly without flowering and there was no terminal lethal burst of reproduction but instead a long process of vegetative extinction in local plants. The reproductive period (phase 2) is rarely if ever a plateau but instead a series of occasional flowering episodes, involving a few individuals and set as an occasional feature within a life cycle of predominantly vegetative growth. It is interesting that in some of the populations of *Ranunculus acris* studied at Bangor by Sarukhán no individuals were ever observed to flower, all individuals were short lived, new members were continually recruited – the population was permanently juvenile (SARUKHáN & HARPER 1973).

LINKOLA (1935) pointed out how different was the behaviour of perennial herbs grown as spaced plants in a garden, from that of the same species in nature. In the experimental garden juvenile phases were short and, subsequently, flowering was reliable from year to year. This is generally the experience of gardeners with clonal perennials. In the field the situation is quite different (e.g. LINKOLA 1935; TAMM 1956; SARUKHáN & HARPER 1973 and many others) juvenile phases are long, flowering is unreliable and large numbers of individuals die young; of the few that have long lives, many die without ever flowering at all (vegetative eunuchs). These intimate details of individual life patterns are revealed only by long term mapping of individual shoots: the brilliant studies of Tamm pioneered this type of demography. Above all, this type of study transforms the vision of how most perennial herbs spend their lives and how death strews the scene.

Not only do clonal herbs fail to fit into neat life cycle descriptions and create major difficulties for the mathematical modeller, but what is normally conceived as a typical process of ageing may be reversed. Just as a branch of a suppressed tree may develop strongly and emerge with the full vigour of maturity into a newly created opening in a canopy, a branch of a suppressed clone may develop juvenile vigour as it grows out of dense competitors and invades an open zone; it may then follow through an active period of flowering before it is again suppressed by colonising neighbours. The work of T.A. Rabotnov and his colleagues (e.g. RABOTNOV & SAURINA 1971; see also HARPER & WHITE 1974) has stressed the importance of this phase reversal in clonal plants; in place of a demography based on plant ages they describe life states through which an individual may pass, in which the sequence of development may involve more or less frequent reversals of direction. Such an approach offers a new challenge in the interpretation of community processes. For the prediction of the ecology of a community a knowledge of life states may be of much greater value than a knowledge of plant ages. For example, a grassland farmer can rejuvenate a suppressed population of *Trifolium repens* by appropriate management without introducing new seed and quite regardless of the age of the genetic individuals (genets) of clover in the sward. Juvenility, maturity and senescence are, in a sense, reflections of community condition rather than of real age.

Nevertheless, for the study of adaptation and evolutionary process, the real age of genets is of vital interest, if only because the speed of evolutionary changes is dependent on the time interval from parent zygote to daughter zygote. The only studies that I know that attempt to give real ages to genetic individuals in clonal plants are those of OINONEN (1976a, 1976b) of *Lycopodium annotatum*, *Calamagrostis epigeios*, *Convallaria majus*, and *Pteridium aquilinum*. For the latter species individuals clones could be dated with some accuracy into the Finnish bronze age.

To demographers the length of life is less important in many ways than expectation of life. Where seedling establishment can be mapped, and mortality followed over time, it is sometimes possible to determine survivorship curves and half lives. Most clonal plants (perhaps like all plants) have very short life expectancy and the plants that grow to a reproductive stage are a tiny minority. Two distinct types of life expectancy can be analysed, i.e. depletion curves and true survivorship curves. We can count all the shoots (ramets, or, if there is a genetic marker, genets) present in an area at a point in time and, ignoring new recruits, follow the decay or depletion of the population. Such depletion curves, calculated from Tamm's data show much variation in half life. Thus *Sanicula europaea* had a population half life of >50 years, *Filipendula vulgaris* ca. 18.4 years and a declining population of *Centaurea jacea* 1.9 years (HARPER 1967). Similar calculations from data of RABOTNOV & SAURINA (1971) for *Ranunculus acris* in the meadows of the Oka river basin give half life values of ca. 3.0 years which correspond with values obtained for the same species by Sarukhán (SARUKHáN & HARPER 1973) in a N. Wales pasture.

True survivorship curves are rarer; these involve measuring the fate of populations that form cohorts of known age. SARUKHáN & HARPER (1973) followed the fate of cohorts of seedlings of *Ranunculus repens* and also the fate of ramets (the modular units of clonal growth) from the time of their appearance as buds in leaf axils through the phase of outgrowth of stolons, rooting and establishment as rosettes. Such true survivorship curves are illustrated in Fig. 8a and b and are very different for genets and ramets. Genets have survivorship curves that a zoologist would describe as Deevey Type III, i.e. an initially extremely hazardous juvenile stage with a steadily declining mortality risk. Ramets followed a Deevey Type II survivorship curve, i.e. a mortality risk independent of age. It is interesting to consider whether the extreme juvenile risk of seedlings is just an expression of risks of a weak unprotected stage, making a difficult physiological transition from hetero-and autotrophy and whether the rather constant death risk in ramets may be due to the continued maternal support via the stolon during establishment. A quite

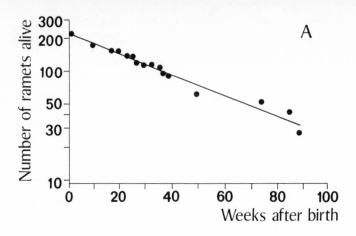

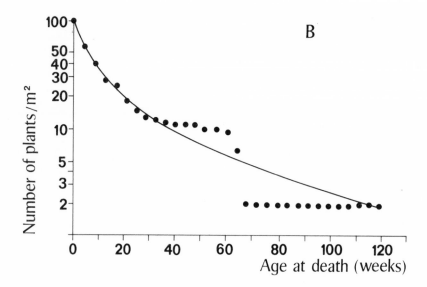

FIG. 8.   *Survivorship curves for* Ranunculus repens: *(a) ramets born at the same time (summer 1969). A ramet was recorded as soon as it was visible as an outgrowth from a leaf axil; (b) seedlings. Data for Fig. 8a and b were obtained from mapped quadrats in a field of permanent grassland at Henfaes, Aber, Gwynedd, N. Wales (from SARUKHáN & HARPER 1973)*

different interpretation is that the high risk that occurs with seedlings is due to the death of new unadapted genetic combinations, an unloading of genetic load. In that case the greater safety of establishment of ramets might reflect the fact that clonal growth repeats modules of already proven genotypes.

In monocarpic plants the act of reproduction is itself a lethal event; the formation of seed involves a cost; metabolites and nutrients are diverted to filling seed instead of to continued growth. In clonal perennials the production of seed is not a lethal activity but the ripening of seed seems to be sub-lethal, slowing down the rate of growth or increasing the risk of death. The setback that occurs in the growth of trees during a mast year is a further measure of the cost of seed production (HARPER 1977). STEBBINS (1950) has pointed to the marked vegetative vigour that is associated with the sterility of inter-specific hybrids; the removal of dead flowers is a well known horticultural practice for increasing both the vigour of garden plants and the extent of their subsequent flowering. In clonal species the proportion of assimilate devoted annually to seed production tends to be much lower than in annuals (HARPER & OGDEN 1970; OGDEN 1974). This is presumably a major reason why seed production is sub-lethal and not lethal. There is a sense in which seed production and continued vegetative growth can be considered as alternative costs in a life cycle strategy. In a comparison of two species of *Agropyron, A. caninum* and *A. repens* (TRIPATHY & HARPER 1973), there appeared to be near perfect correspondence between the number of seeds produced by the non-rhizomatous tussock forming *Agropyron caninum* and the number of seeds plus rhizome buds produced by the rhizomatous *Agropyron repens*. When grown under comparable conditions in pot culture, *Agropyron caninum* produces 258 seeds per plant, and *Agropyron repens* produces 30 seeds and 215 rhizome buds per plant. Such close correspondence suggests that plants possess a limited capacity for producing seeds and rhizome buds and that evolutionary divergence between these two species has involved different partitioning of available resources between these two activities. Even though seed production may represent only a small part of the allocation of resources in the growth of a clonal plant, the life of the clone may be very long and the accumulated seed production of a genet throughout its life may mean that it has very high effective fecundity.

SARUKHáN & HARPER (1973) made a study of three species of *Ranunculus* living in a small area of permanent grassland. These three species were chosen because they were not only closely related but had markedly different life cycles. *Ranunculus bulbosus* produces a corm and each year this is replaced and the old corm dies; there is no clonal multiplication except under very exceptional conditions. Both, increase in numbers within a population and the invasion of new areas, occur solely by an increase in the number of genets. *Ranunculus acris*, in contrast, produces a rhizome which occasionally branches and, because the old rhizome rots away, a clone of disconnected rosettes may eventually be formed. A small fraction (in the studied field circa 10 per cent) of the individual rosettes produced daughter ramets each year. The species has vigorous flowering and produces abundant seed. In contrast *Ranunculus repens* produces stolons and rather short-lived rosettes and has much less vigorous flowering and seed production. Of the order of 90 per cent of the rosettes in a population were replaced each year by daughter ramets. These three species with their very contrasted degree of clonal growth are commonly found intermingled in permanent pastures and raise the question of the real ecological significance of such deep biological differences.

4. QUESTIONS ABOUT PLANT DEMOGRAPHY

There are only partial answers to a number of the questions that need to be answered about the demography of clonal plants:

1) When has it payed (the question is asked in evolutionary terms and has to be answered in terms of increased fitness) to extend vegetative growth at the expense of reproduction by seed? - A partial answer to this question is that the higher the risk of seedling establishment the more often will longevity have enabled genets to capitalize on and exploit the rare chances of establishment.

2) When has it payed a plant that grows clonally to fragment and lose interconnections or to stay interconnected with a persistent rhizome? - One of the advantages of fragmentation is undoubtedly that diseases, particularly viruses, spread rapidly through the parts of an interconnected plant. Fragmentation may guard parts of a genet from a lethal disease that would kill the whole of an interconnected shoot system. Clearly there are also countervailing advantages if the parts of a genet remain interconnected. This may permit correlative inhibition of bud growth permitting regulated control of which buds germinate and it may also allow daughter ramets to be provided with resources from the parent during establishment.

3) When has it payed for a plant to possess a wandering (guerilla) strategy and when to adopt a phalanx strategy? - Clearly, the wandering stoloniferous habit of e.g. *Trifolium repens, Fragaria vesca,* maximises the speed with which a genet can colonize open areas and also maximises the change of inter-specific contact. In contrast a phalanx type of growth e.g. *Glyceria maxima* maximises interference within the species and within the clone but may offer a more powerful defence against invasion by other species.

4) When has it payed to root at nodes or simply be procumbent? - A surprising number of species both annual and perennial maintain a fully procumbent sprawling habit that never or scarcely ever roots nodally. Examples are: *Polygonum aviculare* and most annual species of *Veronica* amongst annuals; *Betula nana* amongst woody perennials. It may be that, in environments in which mineral resources and water are readily available and light is the critical resource in short supply, rooting at nodes is a costly activity with little return; the nodal rooting habit is then seen primarily as associated with a strategy of mineral capture: long internodes and rooted nodes extend access to further nutrient supplies.

5) When has it payed to associate clonal growth with food storage? - Generally thin, short-lived, non-food storing stolons are associated with the guerilla strategy of wandering and invasive growth. Where food storage is involved in a perenating structure it appears that the distance of spread is usually sacrificed and systems of clonal growth involving root tubers, (e.g. *Ficaria verna*), bulbs (e.g. *Narcissus*) and storage rhizomes (e.g. *Iris*) severely restrict the opportunity for exploratory growth into new areas: the compensating advantage is the precocious seasonal growth that comes from stored reserves.

5. THE INTERFACE OF EVOLUTION AND ECOLOGY

To me, one of the most exciting challenges to the plant demographer is the integration of information about ecological behaviour into the context of evolutionary thinking. Hutchinson wrote about "The Ecological Theater and the Evolutionary Play" (HUTCHINSON 1965). The demographer studies the members of the cast in the scenes of a long drawn out play. Ultimately, the explanation of evolutionary phenomena must lie in studies of ecology at the level of the genet, it is at this level that relevant processes of selection occur. Population and community biology are abstractions from the ultimately relevant levels of evolutionary change, i.e. the process by which individual zygotes are successful or not in leaving daughter zygotes.

WILLIAMS (1975), has recently suggested (in a way strongly foreshadowed in a paper by STEBBINS (1958) that the individuals selected from long-lived fecund species represent the few high peaks of fitness resulting from genetic recombination, the few highly fit residues from a vast pool of unfit recombinants (Sisyphean fitness). This suggests that we might regard clonal growth as fixing (as does apomixis) local specialised genetic combinations. It is highly relevant to this argument that the clonal habit is usually tightly linked with outbreeding (dioecy or self-incompatibility). One set of data supporting this view comes from recent studies by TURKINGTON (1975) and CAHN (in preparation). Cahn isolated clones of *Trifolium repens* from various positions in a 1 ha field of old permanent grassland. He showed that these clones differed in aggressiveness towards a standard clone. Turkington followed this by isolating clones from the same field but choosing them from patches dominated by *Lolium perenne, Holcus lanatus, Agrostis tenuis* and *Cynosurus cristatus*. He multiplied the clones in a glasshouse in potting compost and then transplanted them back into the field into all the 16 possible combinations of site of clone origin x site of transplanting. After a season of growth the clones transplanted back into the sites of their origin significantly outyielded those transplanted into other sites in the field (P<0.001). He also transplanted the four bulk populations of clover into sown plots of the four grass species, again in all 16 combinations. The clover transplanted back into the grass species from which it had originally been sampled again outyielded that in the other clover-grass combinations (P<0.001). This strongly suggests a subtle level of micro-adaptation within the populations of *Trifolium repens* in a single small field and it may not be fanciful to see this as Sisyphean fitness. Certainly the wandering guerilla type stoloniferous growth habit of white clover would permit it sensitively to track and remain associated with specialised clones of particular grass neighbours within the sward.

I am very sensitive to the enormous chance of advance in the study of the demography of clonal plants that may be made using sensitive techniques such as enzyme electrophoresis for identifying genets. We are just about at the stage at which demographic study can interact powerfully with genetic analysis of clonal plants. This interaction is absolutely required before we can properly appraise either life cycles or comparative population dynamics or genetic systems. The plant demographer and the geneticist need now to marry and to consummate the marriage.

6. REFERENCES

BAZZAZ, F.A. & J.L. HARPER, 1977 - Demographic analysis of the growth of *Linum usitatissimum*. *New Phytol.*, 78, 193-208.
BELL, A.D., 1976 - Computerized vegetative mobility in rhizomatous plants. *In:* A. LINDENMAYER & G. ROSENBERG (Editors), *Automata, Languages and Development,* North Holland, Amsterdam.
CAHN, M.G. & J.L. HARPER, 1976a - The biology of the leaf mark polymorphism in *Trifolium repens* L. 1. Distribution of phenotypes at a local scale. *Heredity,* 37, 309-325.
CAHN, M.G. & J.L. HARPER, 1976b - The biology of the leaf mark polymorphism in *Trifolium repens* L. 2. Evidence for the selection of leaf mark by rumen fistulated sheep. *Heredity,* 37, 327-333.
DOBZHANSKY, Th., 1973 - Nothing in biology makes sense except in the light of evolution. *Am. Biol. Teach.,* March 1973, 125-129.
HARBERD, D.J., 1963a - Observations on natural clones of *Festuca ovina*. *New Phytol.*, 68, 93-108.
HARBERD, D.J., 1963b - Observations on natural clones of *Trifolium repens* L. *New Phytol.*, 62, 198-204.
HARBERD, D.J., 1967 - Observations on natural clones in *Holcus mollis*. *New Phytol.*, 66, 401-408.
HARBERD, D.J. & M. OWEN, 1969 - Some experimental observations on the clone structure of a natural population of *Festuca rubra*. *New Phytol.*, 68, 93-108.
HALLE, F. & R.A.A. OLDEMAN, 1970 - *Essai sur l'architecture et la dynamique de croissance des arbres tropicaux*. Masson, Paris.
HARPER, J.L., 1967 - A Darwinian approach to plant ecology. *J. Ecol.*, 55, 247-270.
HARPER, J.L. & J. OGDEN, 1970 - The reproductive strategy of higher plants. 1. The concept of strategy with special reference to *Senecio vulgaris*. *J. Ecol.*, 58, 681-698.
HARPER, J.L. & J. WHITE, 1974 - The demography of plants. *Ann. Rev. Ecol. Syst.*, 5, 419-463.
HUTCHINSON, G.E., 1965 - *The ecological theater and the evolutionary play*. Yale Univ. Press, Newhaven, Conn.
LINKOLO, K., 1935 - Über die Dauer und Jahresklassenverhältnisse des Jungenstadiums bei einigen Wiesenstauden. *Acta forest. fenn.,* 42, 1-56.

NOBLE, J., 1976 - The population biology of rhizomatous plants. *Ph. D. Thesis*, University of Wales.

OGDEN, J., 1974 - The reproductive strategy of higher plants. II. The reproduction of *Tussilago farfara* L. *J. Ecol.*, 62, 291-324.

OINONEN, E., 1967a - Sporal regeneration of bracken in Finland in the light of the dimensions and age of its clones. *Acta forest. fenn.*, 83, 3-96.

OINONEN, E., 1967b - The correlation between the size of clones and certain periods of site history. *Acta forest. fenn.*, 83, 1-51.

RABOTNOV, T.A. & N.I. SAURINA, 1971 - The density and age composition of certain populations of *Ranunculus acris* L. and *R. auricomus* L. *Bot. Zh.*, 56, 476-484 (in Russian).

SARUKHáN, J. & J.L. HARPER, 1973 - Studies on plant demography: *Ranunculus repens* L., *R. bulbosus* L. and *R. acris* L. 1. Population flux and survivorship. *J. Ecol.*, 61, 675-716.

STEBBINS, G.L., 1950 - *Variation and evolution in plants*. Columbia University Press, New York.

STEBBINS, G.L., 1958 - Longevity, habitat and release of genetic variability in the higher plants. *Cold Spring Harbour Symp. Quant. Biol.*, 23, 365-378.

TAMM, C.O., 1956 - Further observations on the survival and flowering of some perennial herbs. 1. *Oikos*, 7, 274-292.

TAMM, C.O., 1972a - Survival and flowering of some perennials herbs. II. The behaviour of some orchids on permanent plots. *Oikos*, 23, 23-28.

TAMM, C.O., 1972b - Survival and flowering of perennial herbs. III. The behaviour of *Primula veris* on permanent plots. *Oikos*, 23, 159-166.

THODAY, J.M., 1953 - Components of fitness. *In:* R. BROWN & J.F. DANIELLI (Editors), *Evolution. Symp. Soc. exp. Biol.*, 7, 96-113.

TRIPATHI, R.S. & J.L. HARPER, 1973 - The comparative biology of *Agropyron repens* (L.) Beauv. and *A. caninum* (L.) Beauv. 1. Growth of mixed populations established from tillers and from seeds. *J. Ecol.*, 61, 353-368.

TURKINGTON, R. & J.L. HARPER, 1978 - The growth distribution and neighbour relationship of *Trifolium repens* in a permanent pasture. 2. Inter- and intra-specific contact. *J. Ecol.*, (in press).

TURKINGTON, R., CAHN, M.G. & J.L. HARPER, 1978 - The establishment and growth of *Trifolium repens* in natural and pertubed sites. *J. Ecol.*, (in press).

WILLIAMS, G.S., 1975 - *Sex and evolution*. Princeton Univ. Press, Princeton, New Jersey.

WILLIAMSON, P., 1976 - Above-ground primary production of chalk grassland allowing for leaf death. *J. Ecol.*, 64, 1059-1075.

7. DISCUSSION

MOSSE: I wonder if Professor HARPER would also consider it possible that these very impressive clovers and grasses growing together are competing for nutrients in a different way. It is known that grasses do that very extensively.

HARPER: The Hill experiment was done both in the presence of rhizobia but no mineral nitrogen and with mineral nitrogen but no rhizobial infection. The interactions are highly complex, but the results suggest that the grass-legume interaction is profoundly affected by the rhizobial component. But to what extent we are dealing in this natural environment with a highly polymorphic system with respect to rhizobium we just do not know.

MOSSE: Could I suggest that they are possibly competing more strongly for phosphorus? And that the phosphorus situation is much more important than the nitrogen situation in this case. Because nodulation or fixation will not occur unless there is a certain level of phosphorus.

HARPER: I am entirely in sympathy with this interpretation, because this is a low-phosphate field and in the written text I said at the end that the rhizobial interaction is interesting but in fact the mycorrhizal interaction in relation to phosphate may be a more critical element.

LEVIN (Texas): The data on *Carex arenaria* were very fascinating, especially with regard to the rate of turnover when fertilizer was added. I was wondering whether you had data on long-term fertilizer application pertaining to the question of whether the flux actually diminished, and whether you reestablish a stability that may be related to an early genotype replacement and then a re-equilibration of genotypes after the initial flux.

HARPER: One has to be very careful to get the *Carex* story in perspective, and, as I said, we do not know how many genets were involved. What is certain from the excavations done by NOBLE at the end of these experiments is that very large parts of these populations of *Carex arenaria* do represent a single genet, and that the flux occurs among those parts that almost certainly consist mainly of single genets. So the opportunity for genetic change within them is not there. The underground excavation suggests that in the case of *Carex* one has not got the tight clone intermingling as it occurs in *Trifolium*. The consequence of a fertilizer application is a very much higher flux in both birth rate and death rate. This is a fairly general phenomenon for tillering species. It is certainly true for grasses like *Lolium perenne* that if you put on heavier doses of fertilizer, particularly nitrogen fertilizer, the birth rate of new tillers goes up and the death rate of old tillers goes up as well, and this is again part of a yet more general phenomenon. The death risk in the population has a seasonal rhythm; there are seasons when death is more likely than in other seasons. The death risk is highest when the survivors are growing fastest, and this fits neatly with these *Carex* data. I know of no experiments on perennial systems in northern temperate regions in which the major death risk occurs within the period that is traditionally thought of as the nasty season. I know of no data in which the greater part of the death risk is in the cold winter or in the heat of summer, when it is dry. The death risk found in this sort of demographic study appears to be highest when the survivors are growing fastest, which is of course wholly compatible with the view that the nastiest thing for a shoot to have in its environment is another vigourously growing shoot of the same species.

WOLDENDORP (Arnhem): In my opinion it still does not answer the question why the plants die when they are growing fastest.

HARPER: One can go a little way in beginning to answer it, but not very satisfactorily. The frequency distributions of dry weights of plants within densely planted systems are normally not normal distributions of shoot weights; they are log-normal distributions, and this represents a preponderance of small individuals and relatively few larger ones. The mortality is largely amongst these small individuals. Mortality does not occur in the bigger ones. The birth-death is a terribly dangerous concept, because it represents an arbitrariness in the process of death in which the number of dead parts tends to take dominance over the number of younger parts. But if I take your question further and ask what are the factors of ultimate causation of death, I think there is scarcely any individual plant in nature for which we know the real factors associated with its ultimate death.

WOLDENDORP: Perhaps the plant physiologists among the audience can shed some light on this problem, because I am still astonished that it is just the fast-growing plants which die most easily.

HARPER: Now, be careful, I did not say it was the fastest-growing plants that died, I said the greatest risk of death was in the populations in which the survivors were growing fastest, which is very different.

VAN DOBBEN (Wageningen): The competition for light is of course enormously increased under such conditions.

HARPER: Yes, certainly.

VAN ANDEL (Amsterdam): You talked about TURKINGTON's transplantation experiments with *Trifolium repens,* and I wondered whether you did such an experiment with seeds from the different sites. I ask this with regard to the ecological sense or non-sense of seed production by species that persist mainly by vegetative propagation.

TURKINGTON (Bangor): I did a small part of the experiments with seeds, but not actually in the context you are speaking of. A number of people have shown that the importance of reproduction of white clover by seed in the permanent pasture situation is completely irrelevant. Out of a total of about seven hundred transplanted seeds, I got only three or four that germinated.

HARPER: These sorts of features of what we might call high fitness, that turn up in perennial clonal species, break down very rapidly. It is not easy to select them from the progeny of plants that show these high fitness qualities. What is happening in such a pasture -and I am interested in Professor ALLARD's comment- may be what WILLIAMS calls Sysiphian fitness. His concept, which he called the oyster-elmtree model, is that in species that are reproducing (outbreeding systems), one has the continuous generation of new recombinants on a large scale, among which a few highly successful combinations show high fitness. If they are capable of clonal growth, these few momentary high-fitness states can be maintained as long as those individuals continue clonal growth; but as soon as the process of seed production resulting in new recombinations occurs, there is a breakdown of those special highly recombinant peaks and the process has to start again. New genotypes are, despite what TURKINGTON said, continually injected into this population, but they do not occur in the main sward-dominated areas. All pastures are heterogeneous. There are places where cows defecate, urinate, jump about, pour, lie down, and in which there is patchiness. If you look for

Trifolium repens seedlings via which new genotypes have a chance of entering this population, they will be found on these disturbed sites, particularly on ant hills and mole hills. But within permanent quadrats in which such disturbances do not happen, you do not see seedling establishment.

WATKINSON (Bangor): Could I make a comment in relation to that last point about genetic input into the system? I have looked at some of the data of the *Ranunculus repens* system of which Professor HARPER has been talking, and in that system you also have one or two seedlings going in per year. Only that very low input is needed to maintain very high genetic diversity within the field. I tried a simulation study on this, and found that with one or two genets introduced into the field per year, a clonal diversity of about twenty-five to thirty clones can be maintained within a one-metre square quadrat. There is a tendency to dismiss seedlings input if it is at a very low level, but on the genetic level I think it is very important.

WENT (Nevada): You have not been using the word competition. You say death risks. Do you have any idea of what competition actually is or whether it exists?

HARPER: I avoided the use of the word competition because there is so much danger in this word that is used in different senses by different people. I tend, if there is a risk of argument on the subject of competition, not to use the word competition at all but to talk about interference and to use this blanket word to describe the ways in which one organism may harm or change the environment of a neighbouring organism. Where difficulties arise as soon as we use the word competition is when people describe competition by its end-point. That is to say, you prove competition by showing that X excluded Y or defeated Y or suppressed Y. There is another group that says you can only show competition by either adding individuals to the population and showing that those which were there suffer from the addition, or removing individuals and showing that those that remain benefit from the removal. There is, thirdly, the group who define competition as that which is dependent on the extraction of resources to a mutual demand and who are therefore working on a competitive depletion hypothesis, in which case any possibility of toxic interactions is almost excluded from the word as defined and most commonly used by agronomists, who usually define competition as the consequence of resources in short supply, where the demand is greater. I think that because of these very different usages we get into terrible difficulty if we use the word competition, and that is why I did not use it.

R.W. ALLARD, R.D. MILLER & A.L. KAHLER

The relationship between degree of environmental heterogeneity and genetic polymorphism

1. INTRODUCTION

It has frequently been stated that the extent of genetic variation in populations is related to the degree of heterogeneity of the environment. The reasoning behind this notion was well expressed nearly a quarter century ago by DOBZHANSKY (1955):

"no one genotype is likely to be a paragon of adaptability, superior to all other genotypes in all environments."

Hence:

"the adaptedness of a Mendelian population may ... be advanced if it contains a variety of genotypes suited to different adaptive niches and facets on the environment which the population inhabits .... Granted that genetic variability is an instrumentality whereby Mendelian populations master environmental diversity ... populations which control a greater variety of ecological niches will be more variable than those having a limited hold on the environment."

Thus Dobzhansky proposed genetic diversity as an adaptive strategy by which populations cope with environmental heterogeneity and he postulated that genetic diversity would increase with increased heterogeneity of the environment.

The notion that regulation of genetic variability is a strategy by which Mendelian populations adapt to the spatial and temporal structure of the environment has been one of the most actively investigated areas in contemporary population biology. There is now a vast literature on this topic, a literature that has itself been extensively reviewed. Recent reviews of theoretical aspects include those of CHRISTIANSEN & FELDMAN (1975), FELSENSTEIN (1976), HADELER (1976), KARLIN (1976) and LEVIN (1976); and ANTONOVICS (1971), GOULD & JOHNSON (1972), HEDRICK *et al.* (1976), VALENTINE (1976) and WIENS (1976) have reviewed empirical studies. What is the present status of understanding of genetic variability as an adaptive strategy? HEDRICK *et al.* (1976) stated their conclusions as follows:

"... a substantial amount of circumstantial evidence has accumulated indicating that genetic polymorphisms are related to environmental heterogeneity. There is, however, only a small amount of experimental evidence supporting the hypothesis that environmental heterogeneity is a major factor in maintaining genetic variation."

"Single-locus theory indicates that selection acting differentially in space, coupled with limited migration and/or habitat selection, will maintain a substantial amount of polymorphism .... There have been, however, few documented cases of this in laboratory or other situations where the magnitude and type of selection can be ascertained."

"... many papers will be written before it is clear what proportion of polymorphic loci is maintained or affected by environmental heterogeneity and how environmental differences result in genetic polymorphisms."

In short, the search for relationships between environmental heterogeneity and genetic diversity has not provided clear-cut evidence that such relationships exist.

This ambiguous result is, however, perhaps not surprising when we consider that most studies have treated adaptation in terms of single loci whereas there is an increasing body of evidence from multilocus studies of both natural and experimental populations (e.g. ALLARD *et al.* 1972; CLEGG *et al.* 1972; HAMRICK & ALLARD 1975; ALLARD *et al.* 1977; CLEGG *et al.* in press) that adaptation to specific environmental regimes depends on constellations of genes that act in different stages of the life cycle and affect many different morphological and physiological characteristics. If this is the case the entire genotype as an integrated system of interacting genes, rather than single loci, is the proper framework in which to examine genetic variability as an adaptive strategy. This is the topic considered in this paper. First, we will review the evidence that adaptation does in fact depend on synergistically interacting complexes of genes and show that multilocus Mendelian formulas can be written for ecotypes that occupy specific habitats. Then, we will analyze the question: Do the Mendelian formulas indicate whether there is a relationship between environmental heterogeneity and genetic variability?

2. MULTILOCUS ORGANIZATION

To identify the issues clearly, it is appropriate to discuss the theory of multilocus genetic organization briefly prior to considering experimental evidence that adaptation depends on integrated systems of genes, i.e. whether the genetic structure of populations features coadaptation in the sense of DOBZHANSKY (1955). The simplest model on which we can discuss the genetic basis of coadaptation is one involving just two loci, each with two alleles. Assume that the gametic types $A^{(1)}B^{(1)}$ and $A^{(2)}B^{(2)}$ produce genotypes that are superior in viability and that the alternative gametic types $A^{(1)}B^{(2)}$ and $A^{(2)}B^{(1)}$, produce selectively inferior genotypes. During the life cycle from zygote formation to reproductive maturity viability selection will favor individuals that carry the 11 and 22 combinations of alleles, causing their frequency to increase from early stages in the life cycle to the reproductive stage. If the two loci are unlinked, i.e. they are located on different chromosomes or if they are located 50 or more crossover units apart on the same chromosome, free recombination will occur at gametogenesis and this will reduce the frequency of the favored combinations of alleles. Thus the recombination and segregation that occur during reproduction will undo the work of selection, and the association between the favored alleles will not persist. If, however, the loci are physically linked on the same chromosome, the suppression of recombination due to the linkage will tend to bind the concordant allelic complexes together. Theoretical studies (review in TURNER 1967) have shown that when the crossover value is small enough relative to the intensity of selection, stable nonrandom associations of alleles can develop and persist in the population; and if the linkage is very tight and selection sufficiently strong, the genetic variability will become so organized that only two among four possible gametic types, and only three among 9 possible genotypes, will occur in the population. In other words, the correlation between alleles will be complete ($|r| = 1$), and $|D'|$, the relative gametic phase disequilibrium (linkage disequilibrium) parameter will also take its maximum value of unity.

Theoretical studies have shown that any factor that restricts recombination will have an effect similar to linkage in binding concordant nonalleles together and that positive assortative mating in particular can lead to sharp restriction of recombination (JAIN & ALLARD 1966; WEIR & COCKERHAM 1973). This can be illustrated by considering two individuals with genotypes $A^1A^1B^1B^1$ and $A^2A^2B^2B^2$ in a predominantly selfing population. Due to the predominant selfing these individuals will produce only $A^1A^1B^1B^1$ and $A^2A^2B^2B^2$ progeny generation after generation and the A^1B^1 and A^2B^2 alleles will remain correlated within both lineages, just as if they were linked. But when hybridization occurs between the two lineages, producing the $A^1A^2B^1B^2$ heterozygote, segregation and recombination will occur, the A^1B^2 and A^2B^1 gametic types will be produced, and the association will be broken. When such intercrosses between lineages occur only once in 50 generations or more, as is the case in many plant species, it is obvious that the frequency of heterozygotes will be low and hence that recombination will be severely restricted because effective crossing over occurs only in heterozygotes and not in homozygotes.

It is helpful to have a quantitative measure of the restriction of recombination that is caused by linkage on the one hand and by mating system on the other hand, and for present purposes the rate at which D, the gametic phase disequilibrium parameter, converges to zero for _neutral_ alleles is convenient. For two loci this rate is given by

$$1 - \frac{1}{2}\{\frac{1+\lambda+s}{2} + [(\frac{1+\lambda+s}{2})^2 - 2s\lambda]^{\frac{1}{2}}\},$$

in which λ is the amount of linkage ($0 \leq \lambda \leq 1$) and s is the probability ($0 \leq s \leq 1$) that an individual chosen at random in any generation is the offspring of a single individual in the previous generation (t = 1-s is the probability that it had two parents) (WEIR & COCKERHAM 1973). Note that λ and s enter this expression in the same way and that the magnitude of their effect on rate of decay of D is equal. With random mating (s = 0) and no linkage (λ = 0, or c = 0.5, where c is the crossover value), one half of any disequilibrium is lost in the next mating cycle. With random mating but tight linkage (λ = 0.98 or c = .01) only one percent of the disequilibrium is lost per generation. This is the well-known result that the asymptotic approach of D to zero for neutral alleles under random mating is at the geometric rate of (1-c) per generation so that D in any generation, t, is given by $D_{(t)} = (1-c)^t D_{(0)}$. With no linkage but 98 percent self fertilization, the rate of decay of D is also one percent per generation or 1/50 as large as with random mating and no linkage. And, when tight linkage is combined with heavy selfing, recombination is reduced to the point where little selection is required to hold favorable combinations of alleles together. With complete selfing no recombination will occur and all existing allelic combinations, both favorable and unfavorable, are expected to remain together indefinitely. Populations that reproduce by complete selfing are therefore not favorable for the study of coadaptation because all existing allelic combinations, whether favorable or unfavorable, will be locked together permanently. However, this problem appears to be hypothetical rather than real because, to our knowledge, no plant species is completely self fertilizing.

It is important to note a difference in the effect of restriction of recombination due to linkage and that due to inbreeding. Linkage keeps combinations of alleles involving loci that are physically close on the same chromosome from breaking up; but it does not preserve allelic combinations for loci located on different chromosomes. Inbreeding, in contrast, restricts recombination between all loci, whether on the same or different chromosomes. Theory therefore predicts that as the level of inbreeding increases the entire genotype will become more closely bound together and hence that inbreeding may be a very efficient mechanism for organizing the whole of the gene pool into an integrated system. It also follows that inbreeding populations should be much more favorable than random mating populations for investigating coadaptation experimentally. This is because all loci in inbreeding populations behave as if they are linked in some degree and, consequently, it is unnecessary to search for difficult-to-find closely linked loci upon which to base experiments designed to detect nonrandom associations of alleles.

The introduction of electrophoretic techniques into population genetics have provided a means by which precise multilocus Mendelian formulas can be written for ecotypes that occupy specific habitats. This is now possible because electrophoretic banding patterns can be identified precisely with genotype so that all individuals can be scored unambiguously as a homozygote or heterozygote at each locus. We will now illustrate the results of electrophoretic analysis of population structure on a multilocus basis with

examples taken from our own studies of the Slender Wild Oat, *Avena barbata*
Brot. The mating system of this species, one of approximately 98 percent of
self fertilization, is nearly ideal for studies of coadaptation, because this
amount of inbreeding is expected to protect favorable complexes of alleles
from break up due to segregation while at the same time leaving ample free
genetic variability for evolutionary change in response to spatial and/or
temporal variations in the environment.

3. *AVENA BARBATA*

A. barbata was introduced into California from the Mediterranean basin
approximately 250 years ago. Records indicate that its spread was rapid and
that it soon became a prominent component of grass and oak-savanna communities
in California. *A. barbata* is a tetraploid (2N = 4X = 28) winter annual that
germinates with the first wetting rains in the fall and continues to grow
throughout the winter. It does not survive hard frost; hence it is not
unexpected that it is found only in areas where mean minimum January
temperatures remain about -4°C (Fig. 1). Also, there are limitations to its
drought resistance and it does not occur in areas where mean annual rainfall
is less than 250 mm, also indicated on Fig. 1. These two parameters appear to
identify the major climatic limitations to the distribution of the species in
California.

In 1972 CLEGG & ALLARD reported the results of a survey of allelic
variability at five enzyme loci and two loci governing morphological variants
in 16 populations of *A. barbata* in California. This survey showed that all
populations in southern California and the semiarid grasslands bordering the
central valley are monomorphic for all of the seven loci assayed. Furthermore,
all populations are fixed for the same allele at each of these seven loci. In
other words only a single homozygous genotype occurs in this large
geographical area encompassing about half of the area of California. This
survey also showed that this same genotype is the exclusive one in populations
occupying the most xeric sites in northern California, e.g. sites on steep
slopes with western exposure. On the other hand, populations occupying the
most mesic habitats in northern California (e.g. bottomland sites with deep
dark soil) are monomorphic and fixed for a genotype carrying the opposite set
of alleles at these seven loci. However, the great majority of habitats in
northern California are intermediate between the xeric and mesic extremes and
populations occupying these intermediate habitats were found to be polymorphic
for the seven loci. Furthermore, allelic frequencies in the intermediate
habitats were found to be closely correlated with degree of xerism, i.e. the
frequency of the "xeric" set of alleles increased with increasing xerism of
the habitat.

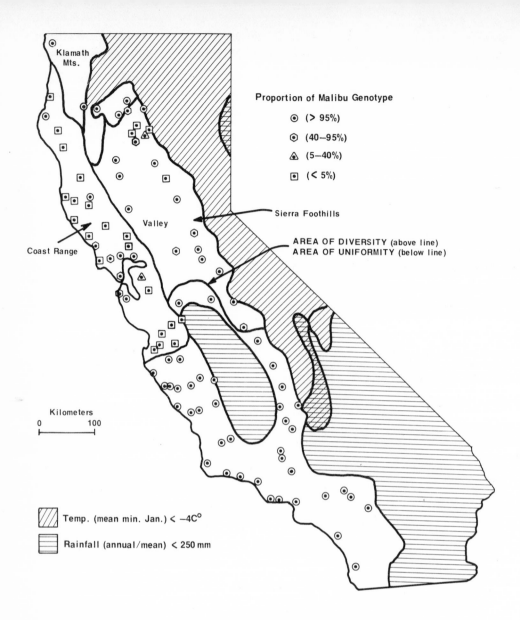

Proportion of Malibu Genotype

⊙ (> 95%)

⊙ (40—95%)

△ (5—40%)

▣ (< 5%)

Klamath Mts.

Sierra Foothills

Valley

AREA OF DIVERSITY (above line)
AREA OF UNIFORMITY (below line)

Coast Range

Kilometers
0   100

▨ Temp. (mean min. Jan.) < −4C°

▤ Rainfall (annual/mean) < 250 mm

FIG. 1.   Avena barbata *collection sites in California, factors limiting distribution, proportion of Malibu type, and genetic regions* (MILLER *in preparation*)

In the topographically diverse coast range of northern California, changes from mesic to xeric habitats frequently take place over very short distances and, to determine whether the remarkable correspondence between allelic frequencies that occurs over most of California is repeated on a micro-geographical scale, detailed studies were made of some populations that span transitions from mesic to xeric vegetational associations in distances of 200 meters or less (HAMRICK & ALLARD 1972; ALLARD *et al.* 1972). The results revealed a consistent pattern of change in allelic frequencies correlated with environment. A typical result was obtained on a hillside transect, designated CSA, located about one mile north east of Calistoga in the Napa Valley. In a distance of less than 200 meters, the habitat on this hillside changes from highly mesic on the valley floor at the bottom to highly xeric on the steep west-facing slope at the top of the hillside. Although the increase in xerism is fairly regular up the hillside, there are some deviations from regularity associated with local topography, such as a small well-watered depression in the hillside. It was found that allelic frequencies tracked the environment almost exactly on this hillside transect: "mesic" alleles decreased steadily from the mesic bottom to the xeric top, except in those local areas where the habitat did not follow the general trend of increasing xerism. Correlation coefficients between alleles at pairs of loci are all of the order of 0.90 and significant statistically. Thus as the "mesic" allele at any locus changed in frequency a similar change in frequency occurred in "mesic" alleles at the other loci, and these changes in allelic frequency reflected changes in degree of xerism with remarkable precision.

To determine the extent to which these loci are inherited as a unit requires that sufficiently large samples be taken from individual populations to permit comparisons of observed enzyme five-locus gametic frequencies with expected frequencies computed as products of observed single-locus frequencies. When such studies were made in polymorphic populations, it was found that two among the $2^5 = 32$ possible 5-locus gametic types were in great excess over expectations based on single-locus allelic frequencies. These were the two gametic types characteristic of xeric and mesic habitats, respectively. There was, of course, a corresponding deficiency among the remaining   5-locus gametic types and it is interesting that this deficiency tended to be greatest for the more extreme recombinants relative to the two favored types. The relative gametic phase disequilibrium parameter (D'), averaged over pairs of loci, usually took more than 50 per cent of its maximum possible value showing that these five loci, while highly correlated in their inheritance, are by no means completely associated. Thus "leakage" in the system provides free genetic variability for quick response to temporal changes in the environment and for opportunistic colonization of unusual habitats. Allozyme frequencies for the same five loci have been examined in Mediterranean populations of A. barbata and the two complexes found in California have not been found. The patterns of coadaptation, correlated with environment, that are found in California have therefore developed since the introduction of this species more than two centuries ago.

In addition to the enzyme loci and the morphological variants, several quantitative characters have been studied both in nature and in common garden experiments (HAMRICK & ALLARD 1972, 1975). These studies show that individuals homozygous for the mesic and xeric complex of enzyme alleles, taken from the same population, differ in time to maturity, stature, tillering capacity, outcrossing rates and other quantitative characters. Thus loci governing these quantitative characters are also components of the gene complexes marked by the enzyme loci. Measurement characters, such as height and time to maturity, are almost certainly governed by many genes of small effect and these genes are presumably distributed over many chromosomes. The xeric and mesic gene complexes therefore include loci located on several chromosomes and most likely on all of the N = 14 chromosomes of *A. barbata*. This provides experimental support for the theoretical prediction that restriction of recombination due to inbreeding helps in holding together favorable associations of alleles of unlinked loci. It also shows that the distribution of *A. barbata* in typical grassland and oak savanna habitats in California can be accounted for in large part on the basis of two ecotypes with Mendelian formulas that can be specified in terms of the enzyme loci. Populations that occupy extreme xeric habitats are made up exclusively of the ecotype whose genetic formula is identified by the multilocus homozygous xeric set of allozymes. Populations occupying extreme mesic habitats have Mendelian formulas marked by the homozygous mesic set of allozymes. Populations that occupy intermediate habitats are polymorphic and large populations are expected to contain all of the $3^5 = 243$ possible geno- types specified by the five enzyme loci. This is because the average level of heterozygosity at any locus is about 4 per cent and this level of heterozygosity provides for sufficient recombination to allow all 32 possible gametic types to appear in each population in each generation. However, the important point is that the "xeric" and "mesic" homozygotes are both found in much higher frequency in polymorphic populations than predicted on the basis of expectations calculated as products of the five single-locus genotypic frequencies. The observed excesses of these two homozygotes in segregating populations provide strong evidence that they represent interacting complexes of genes favored by selection. LEWONTIN (1974) did not take the genetic facts into account in discussing *A. barbata* and he was led to the erroneous conclusion that the selfing that occurs in this species results in "lack of recombination". This in turn led him to conclusions regarding multilocus associations of alleles that are at variance with both theoretical expectations and the observed results.

4. ADDITIONAL PATTERNS OF GENETIC AND GEOGRAPHICAL VARIABILITY

The distribution of genetic variability in *A. barbata* in California has more recently been mapped much more precisely by examining 97 additional populations (Fig. 1) and extending the number of loci assayed from seven to 35. This more extensive study provides additional information concerning the allelic composition and the geographical distribution of the "xeric" complex of alleles and it also shows that seven additional correlated complexes of genes each marked by a specific Mendelian formula for enzyme loci, occur in specific habitats.

Perhaps the most striking feature of the more detailed study is the impressive confirmation it provides for the homozygosity of the xeric complex of enzyme loci in the more arid regions of Southern and Central California: it shows that the xeric genotype is in fact fixed for all of the 35 loci analyzed, i.e. all populations in this region are made up exclusively of the same 35-locus homozygote. This genotype has been designated the Malibu genotype after one of the sites in Southern California in which it occurs.

Analysis of electrophoretic genotypes in the 97 additional populations distributed over California reveal a striking pattern of variation for the Malibu genotype. As shown in Fig. 1, all populations in the southern half of the range of *A. barbata* in California are fixed for this genotype. This area is consequently designated the area of uniformity. The remainder of the range is designated the area of diversity because the frequency of the Malibu genotype varies widely and additional multilocus genotypes appear. The area of diversity can be broken into four main subregions based on geography and frequency of the Malibu genotype: (1) the Sacramento Valley and Sierra foothills in which populations are sometimes polymorphic and the frequency of the Malibu type usually exceeds 40 per cent; (2) two local areas within the valley-foothill region in which Malibu is absent; (3) the Klamath mountains where all populations are fixed or nearly fixed for Malibu; and (4) the coast and the coastal ranges in which a great diversity of types occur.

Is the Malibu genotype highly adapted to xeric conditions as suggested by earlier work? Evidence on this point comes from samples taken from paired locations located less than 100 meters apart within six sites where there was visible ecological differentiation. The paired locations were rated subjectively as to degree of xerism on the basis of exposure, slope, drainage and associated vegetation. Samples were then taken and scored electrophoretically, with the results given in Table 1.

TABLE 1.   *Frequency of the Malibu complex in paired mesic and xeric subpopulations located less than 100 meters apart* (MILLER *unpublished*)

Location	Mesic subsite	Xeric subsite
Balls Ferry*	0.00	0.00
Del Loma	1.00	1.00
Paicines	0.00	1.00
Hooker Creek	0.00	1.00
Weott	0.00	1.00
Bells Station	0.00	0.26

* does not contain Malibu complex

One site (Del Loma) was highly xeric and both subdivisions were fixed for the Malibu genotype. In another site (Balls Ferry) the Malibu genotype did not occur. However, in three of the four remaining sites, the xeric subsites were fixed for the Malibu genotype and in the fourth site (Bells Station), which was judged to be generally mesic, only the more xeric subsite contained the Malibu genotype.

Fig. 2 gives an overlay of the frequency of the Malibu genotype on annual rainfall data for California. This figure shows that the dividing line between the area of uniformity and the area of diversity is the 500 mm rainfall line,

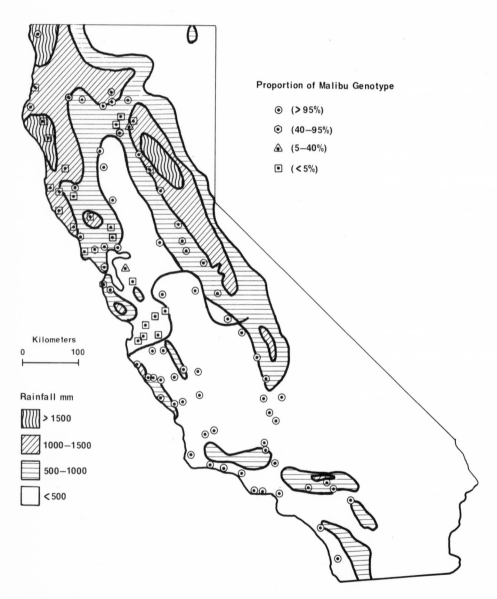

Fig. 2.   *Mean annual rainfall and the distribution of the Malibu genotype in California* (MILLER *in preparation*)

i.e. all polymorphic populations and all non-Malibu populations occur above the 500 mm line. There is one exception: the area to the east of Monterey Bay where non-Malibu types occur in an area with less than 500 mm of rainfall. This is a very foggy area in which fog drip is known to be an important supplement to the light rainfall in the growing of specialty crops. The distribution of the Malibu type thus suggests a model for the relationship between environment and genotype. Malibu is an ecotype highly adapted to dry areas. However, once a moisture threshold is reached, other genotypes become competitive. Local factors can be very important. In the Klamath Mountains, for example, most populations are fixed or nearly fixed for the Malibu geno-type, even though rainfall exceeds 1000 mm. In this area, however, *A. barbata* is restricted to very steep south- and west-facing slopes with shallow or sandy soils, i.e. to habitats that are extremely xeric despite the high rainfall.

Fig. 3 gives the distribution of several additional correlated complexes of alleles that were observed in the area of diversity. In this figure the different allelic complexes are given in code. There are three major groups of genotypes: (1) the Malibu complex, the basic xeric type for which all alleles are coded 1; (2) the mesic complex of the early studies; and (3) the Geyserville complex, which was found in only a single population in the early studies. Two variants occur within each the Malibu and mesic complexes, and three variants in the Geyserville complex, bringing the total number of major complexes to seven. In four locations a complex mixture of genotypes was found. Note that populations with a particular complex of alleles are usually widely dispersed geographically and when two nearby populations have the same complex of alleles, they are usually separated geographically by a population with a different genotype. The distribution of allelic complexes thus suggests a mosaic pattern rather than clinal changes in the environment.

5. ENZYME DIVERSITY IN ISRAEL

Thirty-one populations collected over a wide geographical area have been assayed electrophoretically for seven enzyme systems to determine patterns of enzyme variability in *A. barbata* in Israel (KAHLER *et al.* in preparation). Phenotypic frequencies were scored in nine zones of enzyme activity, probably representing 27 loci. The results show that all loci are polymorphic in Israel and that extensive variability occurs both among populations and within populations. Each population differed from each other population with respect to enzyme genotype and no population was monomorphic for all enzyme loci. In Israel *A. barbata* is therefore more variable genetically than it is in California.

Principal component and multiple regression analyses were used to determine whether there are correlations between environmental parameters and enzyme

B	Is	6Pg	Lap	E_1	E_2	E_3	Cp	Ap	P	Code
GENOTYPIC COMPLEXES										
1	1	1	1	1	1	1	1	1	1	11
1	2	1	1	1	1	1	1	1	1	12
2	2	2	2	2	1	2	1	2	2	21
1	2	1	2	2	1	2	1	2	2	22
2	2	3	1	5	2	2	2	3	1	31
1	2	3	1	3	2	2	2	3	1	32
2	2	2	3	4	2	2	2	3	1	33
						Other Genotypic Complexes				40

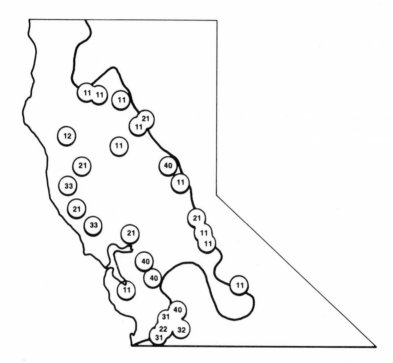

FIG. 3.   *Distribution of* Avena barbata *genotypic complexes in the area of diversity*   (MILLER *in preparation*)

phenotypes in Israel. It was found that temperature related variables and, to a lesser degree, moisture regime variables are correlated with particular enzyme phenotypes. In six out of seven instances temperature related variables were moderately to highly correlated with phenotypic frequencies and in three of seven cases there were moderate associations with moisture regime variables. Temperature and moisture are clearly important factors in adaptation in both

Israel and California, although temperature appears more directly implicated in Israel and moisture more implicated in California.

When an overlay was made of phenotypic frequencies on map locations in Israel, it was found that populations with high frequencies of particular phenotypes were widely dispersed geographically. Also, when phenotypic frequencies were similar in nearby populations, a population with a very different phenotypic frequency was usually located in between. Thus, as in California, isozyme frequency variations appear to reflect a mosaic pattern rather than clinal changes in environmental variability. The importance of this observation is that clinal patterns of genetic variability can arise through drift of neutral alleles (KARLIN & RICKTER-DYN 1976) whereas mosaic patterns are difficult to explain except on the basis of selection.

6. ENVIRONMENTAL HETEROGENEITY AND GENETIC VARIABILITY

We are now in a position to address the question we set out to analyze: Do multilocus Mendelian formulas indicate whether a relationship exists between environmental heterogeneity and genetic variability? To answer this question we must establish: (1) that populations occupying different habitats in an area differ genetically; and (2) that the observed genetic differences are related to variations in the environment of these habitats. With respect to the first of these issues, the multilocus Mendelian formulas make it clear that the extent of genetic variability differs widely from area to area for A. barbata. The entire southern half of California is an area of genetic uniformity in which all populations contain the same 35-locus homozygous enzyme genotype. Israel, in contrast, is an area of genetic variability: all of the enzyme loci assayed are variable and all of the populations surveyed are polymorphic for some loci. The northern half of California differs from both southern California and Israel with respect to genetic variability. In this area some populations are monomorphic and some are polymorphic; further, the polymorphic populations differ widely in degree of polymorphism.

This brings us to the second issue above, i.e. whether these observed differences in genetic variability can be related in a meaningful way to variations in the environment. Unfortunately, we do not know which aspects of the environment are pertinent in most cases. However, the above data provide information concerning the way A. barbata perceives the environments of various areas. In Israel each of the 31 populations surveyed has its unique multilocus enzyme genotype and the populations also differ substantially from each other with respect to within-population genetic variability. The observed genetic differences among the populations indicate that the environment differs from place to place in Israel, i.e. that A. barbata perceives the environment as spatially heterogeneous, or coarse grained, in the sense of LEVINS & MacARTHUR (1966). The genetic variability within all populations indicates that they also perceive the local environment as coarse grained, either spatially on a microscale or temporally. Correlations with temperature and moisture suggest that these are two of the environmental factors implicated. However, too little is known about environmental heterogeneity in Israel to relate it to the observed genetic variability in any meaningful way other than to say that *A. barbata* perceives the environment as heterogeneous on both macro- and microgeographical scales and that the adaptive strategy adopted is one of genetic variability.

The southern half of California is the opposite case from Israel with respect to extent of genetic variability. All of the 35 loci assayed were monomorphic for the same set of alleles and these loci therefore perceive the environment as fine grained. This region, taken by itself, therefore also provides little information concerning the relationship between environmental heterogeneity and genetic variability other than indicating that genetic uniformity can be a workable adaptative strategy even in an area as large, topographically diverse, and environmentally heterogeneous as the southern half of California.

The northern half of California is an area of genetic diversity in which populations from different places are often strikingly different from each other genetically. Several coadapted allelic complexes occur in this area, indicating that *A. barbata* perceives the environment of northern California as spatially heterogeneous on a macrogeographical scale.

Let us now focus on one pattern that is particularly useful in relating genetic variability to variations in the environment. This is the pattern of monomorphism and polymorphism associated with the xeric (Malibu) and mesic complexes of alleles of the grassland and grass-oak savanna habitats, which are the typical habitats of *A. barbata* in California.

In southern California all populations are fixed for the xeric complex but in the northern half of California most populations are polymorphic. However, some populations in the north are monomorphic. Populations that occupy xeric sites, such as those located on steep slopes with thin rocky soils, are mono-morphic for the xeric Malibu complex of southern California. Populations that occupy mesic sites, such as well-watered meadows with deep soils, are mono-morphic for an opposite set of alleles, the mesic set. The monomorphic populations thus occupy identifiable habitats which they perceive to be environmentally homogeneous. The question to be asked is whether the sites occupied by polymorphic populations, which the populations perceive as coarse grained, are visibly more heterogeneous environmentally than sites occupied by monomorphic populations. Evidence has already been cited that this is the case and a recent detailed mapping of the CSA hillside (HAMRICK, unpublished) provides striking support that the habitats occupied by polymorphic populations are in fact conspicuously heterogeneous on a microgeographical scale.

The CSA hillside is transitional between two vegetation types. The bottom of the hillside is a well-watered grassland area with deep dark soil. The top of the hillside, less than 200 meters distant, is a steep west-facing slope with rocky light-colored soil occupied by mixed patches of brush and grass. Total change in altitude is about 30 meters. *A. barbata* occurs in 23 patches on the hillside (Fig. 4), each of which was classified into one of five

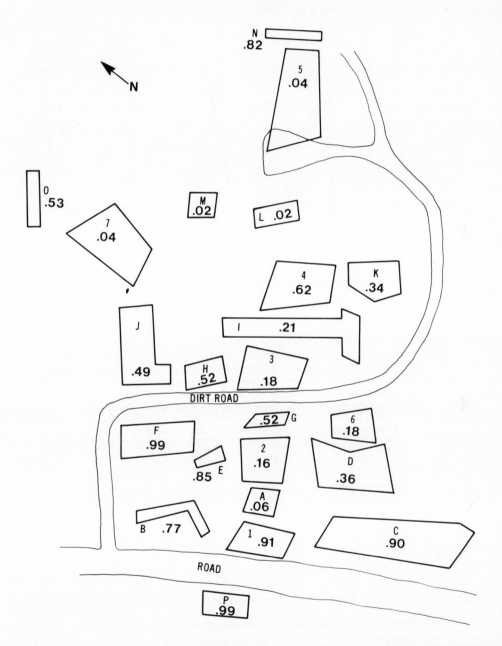

FIG. 4. *Distribution of* Avena barbata *in 23 locations on the CSA transect. Decimal fractions give genetic identity values (Nei) for each patch with the "mesic" complex of alleles. The environment classification of each patch is given in Table 2 (HAMRICK unpublished)*

environmental categories (mesic, intermediate mesic, intermediate, intermediate
xeric, xeric) on the basis of its topographical, soil, and vegetational
features. Two of the patches (2 and 4) were subdivided into 6 subsites, each
of which was classified environmentally (Fig. 5). This environmental
classification showed that, although xerism increased from the bottom to the
top of the hillside, the environment of the hillside is a very complex mosaic
of micro-habitats. A genetic classification of the hillside was obtained by

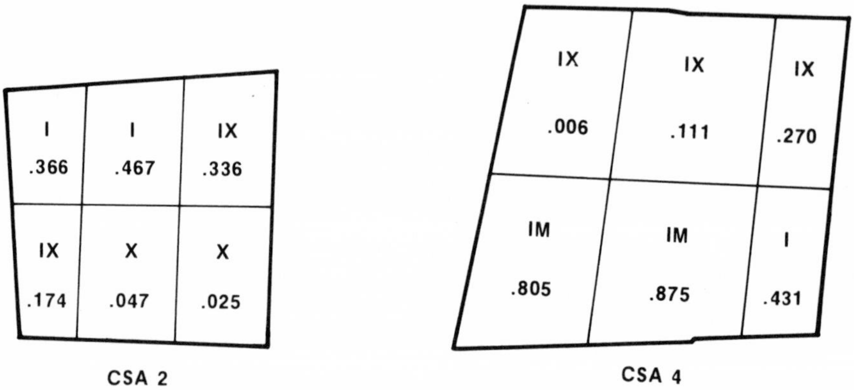

I	I	IX
.366	.467	.336
IX	X	X
.174	.047	.025

CSA 2

IX	IX	IX
.006	.111	.270
IM	IM	I
.805	.875	.431

CSA 4

FIG. 5.   *Subdivisions of the CSA2 and CSA4 sites. Letters give the environmental
classification: X, xeric; IX, intermediate xeric; I, intermediate;
IM, intermediate mesic. Decimal fractions give genetic identity values
(Nei) for each subdivision with the "mesic" complex of alleles
(HAMRICK unpublished)*

assaying samples of *A. barbata* taken from each of the 23 patches for allozyme
genotype. The results, given in Fig. 4 and Table 2, show that the frequency of
mesic and xeric alleles differ sharply from place to place on the hillside and
also that the agreement between the environmental and the genetic
classifications is very close. Agreement was also very close between the
environmental and genetic classifications for the subsites of CSA2 and CSA4,
as shown in Fig. 5.

A polymorphic index, defined as

$$\text{P.I.} = \frac{1}{m} \sum_{i=1}^{m} \sum_{j=1}^{n_i} p_{ij}(1-p_{ij}),$$

where m is the number of loci, n is the number of alleles per locus, p_{ij} is
the frequency of the jth allele at the ith locus, was calculated as a measure
of genetic variability. This index is equivalent to the probability of
heterozygosity at a locus, assuming Hardy-Weinberg equilibrium; thus with two

TABLE 2.   *Genetic identity values (Nei) with the "mesic" genotype for each*
          of the 23 patches on the CSA transect (HAMRICK unpublished)

Mesic	Intermediate mesic	Intermediate	Intermediate xeric	Xeric
F (.99)	B (.77)	O (.53)	D (.36)	6 (.18)
P (.99)	4 (.62)	H (.53)	K (.34)	3 (.18)
I (.91)		G (.52)	I (.21)	2 (.16)
C (.90)		J (.49)		A (.06)
E (.85)				7 (.04)
N (.82)				5 (.04)
				M (.02)
				L (.02)

alleles at each locus, as in the present case, this polymorphic index can
take values between zero and 0.5. Actual P.I. values were found to be zero or
near zero for the environmentally uniform mesic and xeric sites at the bottom
and top of the hillside, respectively. Along the transect P.I. values were
found to be correlated with the degree of heterogeneity of the local
environment and they reached values of 0.46, which is near the maximum
possible value of 0.50, in the environmentally most heterogeneous locations
on the hillside, such as CSA4.

These results thus provide strong evidence that *A. barbata* perceives the
environment of this hillside as spatially heterogeneous and that the extent
of genetic variability is very closely correlated with the degree of spatial
heterogeneity of the environment. It should be noted that the balance between
mesic and xeric factors of the environment fluctuates with the differences in
amount and distribution of rainfall that occur from year to year in
California. Thus, in extremely dry years, the most mesic patches no longer
appear uniform environmentally but become visibly heterogeneous. Similarly,
in wet years more xeric patches also become visibly more heterogeneous. It is
therefore evident that environmental heterogeneity has a temporal as well as
a spatial component.

7. ADAPTIVE STRATEGIES IN *AVENA BARBATA*

Theoretically the most efficient strategy for coping with heterogeneous
environments is the development of an "all-purpose" genotype with sufficient
phenotypic plasticity to perform well in each of the environments that is
likely to be encountered. The observed monomorphism for electrophoretically
detectable variants in extreme xeric and mesic habitats indicates that genetic

uniformity is in fact one of the adaptive strategies that has been adopted by
A. barbata.

In this connection it should be noted that *A. barbata* is a diploidized
tetraploid, i.e. that regular bivalent formation occurs between the 7 pairs
of chromosomes of each of the two ancestral genomes, giving 14 chromosome
pairs of meiosis. This has important implications concerning genetic
variability that are illustrated in Fig. 6 in terms of the genetics of the
enzyme 6-phosphogluconate dehydrogenase (6-PDGH). Two types of families are

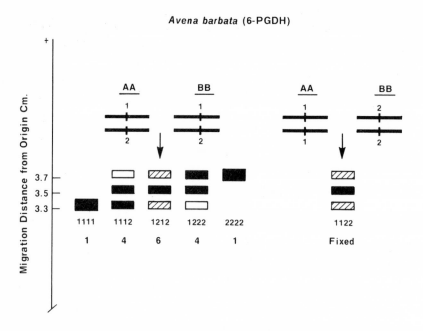

FIG. 6.   *Diagramatic representation of "fixed heterozygosity" for
6-PGDH in* Avena barbata. *Left: duplicated 6-PDGH locus is
heterozygous for alleles 1 and 2 in both genomes, giving
a two-locus segregation (1/16, 4/16, 6/16, 4/16, 1/16).
Right: Genome A is homozygous for allele 1 and Genome B
is homozygous for allele 2, leading to formation of a
fixed intercistronic "hybrid" enzyme* (KAHLER *et al. in
preparation*)

found when plants with one of the phenotypes of this enzyme (designated
$A_1A_2B_1B_2$ and $A_1A_1B_2B_2$ in Fig. 6) are self pollinated and progeny tested: some
families segregate in a 1/16:4/16:6/16:4/16:1/16 ratio whereas other families
include only individuals like the maternal parent. The explanation is the
$A_1A_2B_1B_2$ maternal individuals are heterozygous in both genomes, and hence
segregation in their progeny follows a two-locus pattern, while $A_1A_1B_2B_2$
maternal individuals are homozygous in both genomes, and there is no

segregation. $A_1A_1B_2B_2$ individuals, which are homozygous for different alleles
in their two ancestral genomes, thus form a hybrid enzyme of intercistronic
origin and they are fixed for all three forms of the enzyme. There is substantial
evidence to indicate that biochemical diversity due to this "fixed heterozygosity"
is high in *A. barbata*. The key point is that populations that have been judged
to be devoid of genetic heterogeneity on the basis of enzyme phenotypes, such
as the populations in the area of enzyme locus uniformity, may in fact be
little less diverse biochemically than polymorphic populations. Hence caution
is called for in interpreting the results of electrophoretic surveys of genetic
variability especially in polyploids. Further, increasing evidence that
evolution by gene duplication is a common phenomenon indicates that this caveat
should also be heeded in studies involving diploids, particularly when the
formal genetics of the enzyme variants have not been worked out, as has
usually been the case.

   There is another reason for suspicion that populations monomorphic for
electrophoretically detectable variants are not entirely devoid of genetic
variability: studies of measurement characters made long before electrophoretic
methods were used in population genetics show that there is extensive variability
for such characters both within and between populations (ALLARD unpublished).
In these studies seeds collected from random individuals in nature were sown
in replicated common garden experiments and the resulting progenies were
measured for continuously varying characters, such as flowering time and
height. Significant differences were found in progeny means which indicates
that the populations are genetically variable respecting quantitative
characters. In addition responses were obtained when plus and minus selection
was practiced within progenies derived from single plants. Response to such
selection provides clear evidence that at least some of the loci governing
each measurement character are heterozygous in the natural populations. These
common garden experiments also showed that the mean value for each population
differed significantly from the mean of each other population, and hence that
each population differs genetically from each other population respecting
measurement characters. The adaptive strategy adopted with respect to the
measurement characters is therefore one of genetic variability. This
illustrates the point that different adaptive strategies may be adopted for
different loci and that no single class of loci, such as the enzyme variants,
or those governing measurement characters, gives a complete picture of the
relationship between environmental heterogeneity and genetic variability.

   The area of diversity in northern California illustrates another adaptive
strategy that has been adopted by *A. barbata*. In the southern half of
California all 35 enzyme loci are monomorphic and in northern California many
of these are also monomorphic in all populations. However, the remaining
enzyme loci are polymorphic in many populations in northern California. Thus,
on one side of environmental threshold associated with the 500 mm rainfall

67

line, the adaptive strategy is one of genetic uniformity for all enzyme loci while, on the other side of this line, the strategy is a mixed one featuring genetic uniformity for some loci and genetic variability for other loci. In Israel all of the enzyme loci are variable. Hence, with respect to the enzyme loci, Israel represents the extreme case in which the adaptive strategy is one of genetic variability for all enzyme loci.

The area of diversity in California illustrates still another adaptive strategy that has been adopted by *A. barbata*, namely the strategy of evolving different multilocus complexes of genes for the colonization of habitats that differ from the standard grassland and oak savanna habitats in which the species typically occurs in California. Seven major coadapted complexes have been identified thus far and almost certainly additional complexes will be found as the genetic mapping of *A. barbata* proceeds in California. In Israel every population of *A. barbata* appears to be genetically different. Thus the evolution of ecotypes adapted to specific habitats, each marked by a particular set of enzyme alleles, has evidently proceeded further in Israel, where *A. barbata* is endemic, than in California where it is a recent introduction.

8. CONCLUSIONS

The studies reviewed indicate that adjustment of genetic variability is a strategy by which populations of *A. barbata* cope with variations in the environment. However, the results also show that relationships between environmental heterogeneity and genetic variability are complex and that identification of patterns of relationship requires information concerning the genetic system on a multilocus basis together with detailed coordinate information concerning the responses of populations to the environment.

9. ACKNOWLEDGEMENT

This work was supported in part by National Science Foundation Grant BMS 73-01113-A01.

10. REFERENCES

ALLARD, R.W., BABBEL, G.R., CLEGG, M.T. & A.L. KAHLER, 1972 - Evidence for coadaptation in *Avena barbata*. *Proc. natn. Acad. Sci.*, 69, 3043-3048.
ALLARD, R.W., KAHLER, A.L. & M.T. CLEGG, 1977 - Estimation of mating cycle components of selection in plants. *In:* F.B. CHRISTIANSEN & T. FENCHEL (Editors), *Measuring selection in natural populations*, Springer-Verlag, Heidelberg.
ANTONOVICS, J., 1971 - The effects of a heterogeneous environment on the genetics of natural populations. *Amer. Scient.*, 59, 593-599.

CHRISTIANSEN, F.B. & M.W. FELDMAN, 1975 - Subdivided populations: a review of one- and two-locus deterministic theory. *Theor. Popul. Biol.*, 7, 13-38.

CLEGG, M.T. & R.W. ALLARD, 1972 - Patterns of genetic differentiation in the slender wild oat species *Avena barbata*. *Proc. natn. Acad. Sci.*, 69, 1820-1824.

CLEGG, M.T., ALLARD, R.W. & A.L. KAHLER, 1972 - Is the gene the unit of selection? Evidence from two experimental plant populations. *Proc. natn. Acad. Sci.*, 69, 2474-2478.

CLEGG, M.T., KAHLER, A.L. & R.W. ALLARD - Estimation of life cycle components of selection in an experimental plant population. *Genetics*, (in press).

DOBZHANSKY, Th., 1955 - A review of some fundamental concepts and problems of population genetics. *Cold Spring Harbor Symp. in Quant. Biol.*, 20, 1-15.

FELSENSTEIN, J., 1976 - The theoretical population genetics of variable selection and migration. *A. Rev. Genet.*, 10, 253-280.

GOULD, S.J. & R.F. JOHNSON, 1972 - Geographic variation. *A. Rev. Ecol. Syst.*, 3, 457-498.

HADELER, K.P., 1976 - Travelling population fronts. *In:* S. KARLIN & E. NEVO (Editors), *Population genetics and ecology,* Academic Press, New York, p. 585-592.

HAMRICK, J.L. & R.W. ALLARD, 1972 - Microgeographical variation in allozyme frequencies in *Avena barbata*. *Proc. natn. Acad. Sci.*, 69, 2000-2004.

HAMRICK, J.L. & R.W. ALLARD, 1975 - Correlations between quantitative characters and enzyme genotypes in *Avena barbata*. *Evolution, Lancaster, Pa.*, 29, 438-442.

HEDRICK, P.W., GIVEVAN, M.E. & E.P. EWING, 1976 - Genetic polymorphism in heterogeneous environments. *A. Rev. Ecol. Syst.*, 7, 1-32.

JAIN, S.K. & R.W. ALLARD, 1966 - The effects of linkage, epistasis and inbreeding on population changes under selection. *Genetics, Princeton*, 53, 633-659.

KAHLER, A.L., ALLARD, R.W., KRAZKOWA, M., NEVO, E. & C.F. WEHRHAHN - Isozyme phenotype-environment associations in natural populations of the slender wild oat (*Avena barbata*) in Israel. (in preparation).

KARLIN, S., 1976 - Population subdivision and selection migration interaction. *In:* S. KARLIN & E. NEVO (Editors), *Population genetics and ecology,* Academic Press, New York, p. 617-657.

KARLIN, S. & N. RICKTER-DYN, 1976 - Some theoretical analyses of migration selection interaction in a cline: a generalized two range environment. *In:* S. KARLIN & E. NEVO (Editors), *Population genetics and ecology,* Academic Press, New York, p. 659-706.

LEVIN, S.A., 1976 - Population dynamic models in heterogeneous environments. *A. Rev. Ecol. Syst.*, 7, 287-310.

LEVINS, R. & M. MACARTHUR, 1966 - The maintenance of genetic polymorphism in a spatially heterogeneous environment: variation on a theme by Howard Levine. *Amer. Natur.*, 100, 585-589.

LEWONTIN, R.C., 1974 - *The genetic basis of evolutionary change*. Columbia Univ. Press, New York.

MILLER, R.D. - Genetic variability in the slender wild oat *Avena barbata* in California. (*Ph. D. dissertation,* in preparation).

TURNER, J.R.G., 1967 - On supergenes. I. The evolution of supergenes. *Amer. Natur.*, 101, 195-221.

VALENTINE, J.W., 1976 - Genetic strategies of adaptation. *In:* F.J. AYALA (Editor), *Molecular evolution*, Sinauer, Sunderland, Massachusetts, p.78-94.

WIENS, JOHN A., 1976 - Population responses to patchy environments. *A Rev. Ecol. Syst.*, 7, 81-120.

WEIR, B.S. & C.C. COCKERHAM, 1973 - Mixed self and random mating at two loci. *Genet. Res.*, 21, 247-262.

11. DISCUSSION

VAN DELDEN (Groningen): Have you investigated whether there is variation in the amount of selfing in the different parts of your populations? Looking at the large-scale variation pattern, you find one homozygous genotype in the south and in the xeric environments. I can imagine that in the polymorphic regions the amount of outbreeding is higher, causing a higher degree of polymorphism. This is also suggested by the data, derived from the small-scale variation, on genetic identities. These values were very low for the xeric environments, and could have been caused by genetic drift. In that case the differentiation between the populations from these xeric patches could have been brought about by drift instead of local differentiation due to selection in different environments.

ALLARD: Thank you very much for asking that question. I left out something I did not intend to leave out. When you are doing population genetics the first thing you have to do is mating systems, and that should include *Drosophila* and other species. The reason is that the mating system is rarely random mating and of course here we absolutely have to know what the mating system is, so we always do the mating system first. With respect to mating system, *Avena barbata* is about 98 per cent selfed and about 2 per cent outcrossed on the average, but this varies a great deal. In the more xeric habitats it is a tenth of one per cent outcrossing and in the more mesic habitats it can go up to about 10 per cent outcrossing, so there is a great variation. This brings up another thing I forgot to mention. I talked about electrophoretic genotypes in relation to habitat. We have also studied these with respect to metric characters such as the size of the roots (and incidentally the xeric type has a much larger root system than the mesic type), and with respect to plant stature, maturity data, and a whole series of other things, including mating systems. One of the things associated with the xeric electrophoretic genotype, the handle we can use to get hold of this, is that the mating system is one with a smaller amount of outcrossing than is the mesic type. This is part of this whole complex of things.

VAN DELDEN: Do you think this has a genetic basis or is it environmentally induced?

ALLARD: No, it is genetic, because what we have done is take this back to a common garden experiment. While the mating system is obviously not the same as in nature, because we take it in a different environment, the difference we observe in nature remains in the common garden experiment. So in this experiment we can show that there is a genetic component to this. These xeric and mesic complexes have a whole series of things associated with them The electrophoretic variance is the handle that allows us to identify a whole series of measurement character differences that we were not able to proceed with before. They were there but were rather indefinite, and it is very hard to pick them up unless you have some framework to look at. Concerning your question whether the differences could be due to drift, the answer is that in no way this could be due to drift.

ANONYMOUS (1): Is there some physical or perhaps functional reason for the deviation from random mating of the gene complexes in barley populations? Are there some chromosomal complexes or genetic mechanisms that distort the random distribution of gametic types?

ALLARD: No, there is no problem of that kind. The whole thing is due to the mating system operating on the population level. If you take heterozygotes from the populations and self them and you do the standard sort of genetic analysis, you get perfectly good 1:2:1 ratios, and so on. There are no disturbances in meiosis, no meiotic drive, and no other genetic disturbances. This is purely associated with the mating system. We looked at this very carefully, and there is no reason to think that other things are involved.

ANONYMOUS (1): You used NEI's formula for genetic similarity, but there are genotypic associations, could HEDRICK's genotypic identity be more relevant biologically?

ALLARD: We have computer programs for these calculations and we routinely calculated not only NEI's genetic-distance measure but also HEDRICK's one, which is on a genotypic basis rather than on a gene-frequency basis. We also did the one that ANTHONY EDWARDS worked out. The answer is that they all tell you the same thing, the values are very close. The reason that I used NEI's measure is that most people who work with animals use it.

ANONYMOUS (2): Would you care to comment on whether or not there is a positive relationship between niche width and genetic variability? It seems to me that *Avena* is a weed and occurs in all kinds of habitats, so it has to be broad-niched.

ALLARD: First of all, *Avena barbata* is really not very weedy. *Avena fatua* or *Avena sterilis* is a real weed, but *Avena barbata* is not.

ANONYMOUS (2): But it is still broad-niched, because there is such a variety of habitats that it occurs in.

ALLARD: After GRIMES' discussion yesterday I have some hesitation about commenting very much on this. Let me compare *Avena barbata* with *Avena fatua*; that might possibly give you some indications. *Avena fatua* is a hexaploid and you migh expect that it would have more buffering, more biological diversity. It is also more heavily inbreeding than *Avena barbata*. It is also much more of a generalist; it occurs in a very wide range of habitats. The interesting thing about it is that it is also more polymorphic, despite the lower outcrossing rate, despite its having more phenotypic plasticity, at least you might guess it might have in terms of the biochemical diversity. But it is much more polymorphic than *Avena barbata*; much less phenomorphic. So again you have to be very careful about making broad generalizations. Each case seems to be pretty much unique in itself. There are some patterns that show up, but you really ought to know quite a bit about the situations before you say very much about such things as niche width, at least in genetic terms.

ANONYMOUS (3): How can you be sure, since California is a very large state, that the history of the species is not important. You might, given the way it has been colonized, have had very discrete founder events, because the situation in Israel seems to be really difficult to interpret.

ALLARD: A great deal is known about the introduction of *Avena barbata* and many other species of the Mediterranean basin, because during the exploration period there were historians along. Very shortly thereafter, botanists came. So the mission period is very well documented. There were massive, multiple introductions of *Avena barbata*, because there was a great deal of trade back and forth, particularly with Spain, but also with other parts of the Mediterranean. I might say that we have got hold of samples of other places than just Israel, and the complexes we see here do not occur in the Mediterranean basin, at least we have not found them in very extensive sampling. So what has happened in California has been an evolution to suit the particular circumstances of our super-Mediterranean climate. We have a more Mediterranean climate than the Mediterranean.

LEVIN (Texas): The idea of fixed heterozygosity, as you pointed out, is a very intriguing possibility which offers the plant all kinds of biochemical plasticity. I was wondering if you considered looking at fixed heterozygosity over habitats, to see whether there is a relationship in that respect?

ALLARD: That is one of the things we are looking at now in *Avena barbata* but also in another group of species: the *Festucas*. Here we have diploids, tetraploids, and hexaploids. All we can say so far is that *Festuca microstachys*, which is very widely distributed and has a great variety of habitats, is a hexaploid. The species that is almost certainly the diploid ancestor is, however, very narrowly distributed; it occurs only in a few specialized habitats in California. Whether in fact the fixed heterozygosity, which is very much greater in *Festuca microstachys*, has anything to do with it, is a question. But at least the result here, and in many other species too, suggests that fixed heterozygosity does have something to do with broadening the ability to handle broader niches.

LEVIN: In view of your interpretation of selection for the differences that you pointed out, have you calculated selective coefficients?

ALLARD: We have done this a number of times. We have a huge paper in preparation right now, for which we took the whole life cycle apart. We took viability selection, that is, selection from the time of zygote formation to reproductive maturity, as one part of the life cycle. As another part we took from reproductive maturity to zygote formation. That is a very complicated part of the life cycle and it turns out that the selection in it goes in different intensities and at different directions for the same locus in different parts of the life cycle. Thus, there is one set of alleles favoured in viability selection and the opposite set is favoured in gamete formation.

So this is a very complicated area, as TIMOTHY PROUT has pointed out for animal populations.

LEVIN: What kind of values do you get in terms of selection coefficients?

ALLARD: There are 20 to 30 to 40 per cent differences between allele one *versus* allele two, except that what happens at one locus depends very much on what happens at the second locus and what happens at the third locus and so on. The single locus data can be very misleading.

ANONYMOUS (4): Have you tried to map your populations, or one part of your populations, in order to see if the partition of your different genotypes is random from one year to the other?

ALLARD: That is another thing I intended to mention and forgot. I talked mostly about spatial heterogeneity, but it is very clear that there is also temporal heterogeneity. For example, this year in California we are having a drought that is probably as bad as the one in Britain last year, and we had one the year before. It is very clear that the CSA hillside, for example, was very xeric last year and this year. We have mapped this over ten years now and there are fluctuations in the genotypes associated with the amount of rainfall. A moisture parameter keeps coming out and there are fluctuations over time. But there is an average for any particular little part of the habitat. The populations fluctuate around that particular average, depending on what the environment was the previous year.

VALDERON (Montpellier): Is there any evidence concerning the physiological significance of the enzymes you are working with in the electrophoretic studies?

ALLARD: We regard the eletrophoretic variance as very nice marker genes and I am not very interested in getting into the argument of specifically what they do, particularly the systems that we work with. If we worked with such enzymes as alcohol dehydrogenase or amylase, that might be another matter. But for the enzymes we work with we have no clue and furthermore we are not about to put out the amount of work it would mean to try to get a clue. Partly because we suspect we would fail, but mostly because we want them as markers to study mating systems and migration and the standard parameters of population genetics. I would be very happy if somebody else would take this on, but we are not about to do it.

The genetic implications of ecological adaptations in plants

1. INTRODUCTION

During the past decade, great strides have been made in empirical and theoretical plant population ecology. The empirical studies have been highlighted by work on the effect of intraspecific competition on plant size and numbers (HARPER & WHITE 1970, 1974), mortality and fecundity schedules (WHITE & HARPER 1970; HARPER & WHITE 1974; SARUKHAN & HARPER 1973; SARUKHAN 1974; HARPER & WHITE 1970), the spatial dynamics of succession and colonization (YARRANTON & MORRISON 1974), safe site specificity (HARPER 1961, 1965; SHELDON 1974), pollen and seed dispersal (LEVIN & KERSTER 1974), and the dynamics and longevity of the seed pool (ROBERTS 1972; HARRINGTON 1972). Within the realm of theory, special attention has been given to optimizing life history strategies with special regard to seed dormancy (COHEN 1967, 1968), and dispersal (GADGIL 1971; ROFF 1975), reproductive schedules and developmental switching (COHEN 1971, 1976; BRADSHAW 1974; TAYLOR et al. 1974; LEVINS 1968; SCHAFFER 1974a, b), and longevity (GADGIL & BOSSERT 1970; GADGIL & SOLBRIG 1972; SCHAFFER & GADGIL 1975; KAWANO 1975).

The descriptions and predictions of life history tactics and responses of plant species adapted to different environments largely have been considered without regard to the genetic consequences which accrue from the adaptations. Some aspects of demographic attributes which have received theoretical genetic treatment are population growth with density-dependent regulation (CHARLESWORTH & GIESEL 1972; CHARLESWORTH 1971, 1973), age-specific selection (ANDERSON 1971; KING & ANDERSON 1971), age-specific fecundity (GIESEL 1974; DEMETRIUS 1975), and migration (JAIN & BRADSHAW 1966; ANTONOVICS 1968; GILLESPIE 1974, 1975, 1976; BULLOCK 1976; NAGYLAKI 1976). With the exeption of migration, many of the models employed do not take into account the unique properties of plants, nor do they address many interesting questions relevant to plants.

Our purpose is to discuss some genetic implications of demographic properties of plant populations. Emphasis will be placed upon the genetic consequences of realistic fecundity distributions, reproductive schedules, patterns of differentiated plant subdivisions, and seed-pool characteristics. The cameos which emerge will be painted with broad strokes because of time restrictions and the limited depth to which these problems thus far have been pursued.

2. THE GENETIC CONSEQUENCES OF FECUNDITY DISTRIBUTIONS IN PLANTS

2.1. INTRODUCTORY REMARKS

Recent studies on the population ecology of plants have shown that in natural and experimental populations only a small proportion of the seedling crop survive to reproductive maturity, and that a size hierarchy will be established among the seedlings early in the season which may persist and be magnified as the population matures (WHITE & HARPER 1970; ROSS & HARPER 1972). As a consequence, populations typically are characterized by biomass and fecundity distributions which are L-shaped, wherein a very small proportion of the population may make a substantial and grossly disproportionate contribution per capita to the yield or seed output of the population (KOYAMA & KIRA 1956; RISSER 1969; SARUKHAN 1974; LEVERICH 1977). Indeed, in some populations, the highest fecundity classes may be more frequent than plants of medium fecundity. Conversely, a very large fraction of the population may make only small contributions to yield or seed output. The difference in seed production between large and small plants within a population may differ by several orders of magnitude especially in species with great developmental flexibility. Even the difference between mean seed set and seed set in the most reproductive individuals may be enormous (SALISBURY 1976). The L-shaped performance distribution may arise from competitive interactions between cohorts and the statistical distribution of site quality, as determined in part by the genetic make-up of the population.

The L-shaped fecundity distribution in plant populations is in stark contrast to the poisson distribution which is a cornerstone of population genetic theory (KARLIN & McGREGOR 1968). What are the genetic consequences of an excess of small plants and an excess of large ones relative to a poisson distribution? Whether we talk of a response to directional selection, decay in variability, or other parameters, it seems likely that the effect would be great. The specific consequences would depend on the importance of the genotype in dictating the observed fecundity hierarchy. We sought answers to the afore-mentioned questions using a simulation model which takes into account micro-environmental hospitality.

2.2. THE MODEL

We began with a population composed of 250 annual plants. They were assumed to compete at the seedling and juvenile stage resulting in the establishment of a size hierarchy which would persist through the remainder of the season. The relative performance of young plants was assumed to be controlled by genetic differences, and genotype-independent factors such as site hospitality. Seedlings were assigned relative fitness values and site hospitality coefficients (based upon a random number between 0.0 and 1.0). The extent to which the genotype contributed to seedling vigor, which we will refer to as

C_{vigor} also was entered into the model so that we might look at different levels of penetrance or plastic response. The vigor of a seedling is established as follows:

seedling vigor = (genotype fitness).(C_{vigor}) + (site hospitality).$(1-C_{vigor})$

The ultimate performance of the plants was established as seedlings by assigning a potential fecundity value to each in order of the vigor of the seedlings, the most vigorous seedling having the highest fecundity potential. The fecundity upon which seedlings thereby were placed is illustrated in Fig. 1.

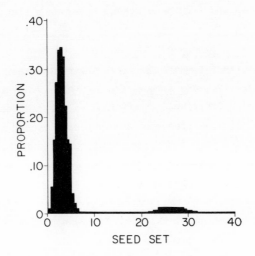

FIG. 1.   *The distribution of fecundities under the L-shaped distribution*

This distribution was obtained by superimposing two poisson distributions with different means, $\lambda 1$ and $\lambda 2$. We will refer to it as L-shaped, as its shape suggests. For most simulations the mean values chosen were $\lambda 1 = 7.15$ and $\lambda 2 = 35.75$ thus affording a 5 fold difference between the two means and an overall mean of 10. In most simulations 90 per cent of the plants lies within the first distribution. We will contrast a simple poisson distribution with a mean of 10 with our distribution.

The distribution we have chosen to contrast with the standard poisson is not meant to represent a specific study or species. Rather it represents the type of schedule we may expect to find, and contains the essence of the small empirical base in the literature.

2.3. RESULTS

The most important question that arises is how the fecundity distribution affects the rate of response to selection. To examine this, simulations were performed with selection for one allele at a locus, starting with that allele at a frequency of 0.1, and following the population until it reached a frequency of 0.9. The starting point of 0.1 was chosen for the sake of expediency for any lower value would have resulted too often in the chance loss of the favoured allele, or excessively extended response times, both especially in the case of a recessive allele. The minority allele was introduced in the form of homozygous seed.

The effect of an L-shaped fecundity schedule versus a simple poisson on the response to selection is quite striking. If we assume that a recessive homozygote has a relative fitness of 1.0 and that the dominant homozygote and heterozygote have relative fitnesses of 0.75, and that C_{vigor} is .5, the mean sojourn time for a recessive gene from $q = .10$ to $q = .90$ averages about 75 generations for a poisson distribution compared to about 45 generations for and L-shaped one (Fig. 2). The presence of manifestly fecund plants whose success is in part related to a favorable genotype is responsible for the accelerated response to selection. Moreover the response to selection of a recessive gene in the .1 to .9 frequency window, given an L-shaped distribution, is similar to the response features favored of a dominant gene given a poisson distribution. For a dominant gene, the rate of advance also is faster with the L-shaped distribution than with the poisson.

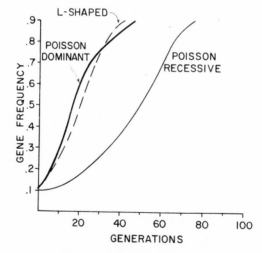

FIG. 2.   Rate of response to selection as a
function of the fecundity schedule.
The "double poisson" refers to the
L-shaped schedule. Genotype fitnesses:
$W_{AA} = .75$, $W_{Aa} = .75$, $W_{aa} = 1.0$

The rate of evolution with an L-shaped fecundity schedule is a function of the ratio of the mean values of the two distributions comprising the L. In the initial model, the ratio of means was 1:5, and evolution under the L-shaped model was about twice the pace of the poisson. Maintaining an overall mean of 10 seeds as in the first simulation but a lambda ratio of 1:2.5, the sojourn time for a recessive allele was about 1.2 times faster than with a poisson. On the other hand, if the fecundity of the high performing plants is very much greater than the remainder of the population, yielding for example a 1:10 ratio, the rate of response to selection is about 3.2 times greater than with a poisson distribution.

Whereas L-shaped fecundity distribution permits a more rapid response to selection than a poisson, the former also is more likely to foster the extinction of a gene when it is rare and less likely to protect polymorphism when two or more alleles are common. Once again the greater variation in fecundity which accrues from the L-shape is the cause. Consider a population of 10 plants with two alleles in equal frequency, both equally fit, and with an L-shaped fecundity distribution. Based upon 100 runs, the loss of polymorphism occurs in an average of 14.2 generations. In contrast, a similar population with 7 plants and a poisson distribution retains polymorphism for an average of 19.5 generations, and a population with 6 plants retains polymorphism for an average of 13.3 generations. Thus, the decay of variability in populations with "L" fecundity schedules is similar to that in much smaller populations with poisson schedules. With regard to the loss of a rare favored recessive genotype, (relative fitness of AA and Aa = .75) we found that in runs in which q = .1, a population size of 250, relative fitnesses of the dominant homozygote and heterozygote of .50, C_{vigor} is .5, mean ratio 1:5, the recessive allele was lost in 80 per cent of the trials when an l-shaped distribution was used versus 65 per cent when a poisson was used.

Our discussion thus far has been couched in the context of a plant population which is exhibiting a plastic or developmental response to density-stress from conspecifics, other species or the environment as a whole. However, plants may exhibit a mortality response as the environment deteriorates (HARPER & WHITE 1974). The suppressed fraction of the population may fail to produce seed. The effect would be represented by an L-shaped distribution with $\lambda 1$ being zero. Since an increase in the ratio between the means of the two distributions comprising "L" results in a more rapid response to selection, removing the less productive plants from the parental pool (assuming size is genotype-dependent) will permit a large variable population to experience much more rapid evolution than a similar population with a poisson fecundity schedule. Note that we are not placing a larger fraction of the population in the high fecundity category; that proportion is assumed to remain constant. If we were simply to increase the proportion of the population in the high fecundity distribution from 10 per cent (as in our initial runs) to higher percentages, the special advantage which accrues to superior genotypes is diluted and the rate of evolution declines.

A mortality response to exigencies of the environment lowers the genetically effective size of the population thus making it more prone to genetic drift. Consider a situation referred to earlier, namely a population of 10 plants whose fecundity distribution is L-shaped, whose lambda ratio is 1:5, and whose 2 alleles were in equal proportions. We estimate that polymorphism would be maintained for an average of 14.2 generations. Eliminating the contribution of plants from the low fecundity distribution, polymorphism will be lost in an average of 3.1 generations. The relationship between rate of gene frequency change in response to a selection model and the decay of genetic variability in a neutral model is not qualitatively altered by a mortality response as opposed to a plastic response. The greater the potential rate of evolution by selection the greater also is the potential for evolution by random drift relative to population genetic theory. Most plants show plastic and mortality responses to density-stress which means that they could respond to selection quite rapidly should variation be present in population, but their populations are not well adapted to protect polymorphism (Fig. 3).

In view of the striking effect of an L-shaped fecundity distribution on the response to selection, it is of interest to determine the effect of this distribution on equilibrium frequencies in a population faced with the immigration of a deleterious gene. This problem was examined with a simulation of a population with 250 plants, poisson and L-shaped fecundity schedules, and immigration rates of 1 per cent, 2 per cent and 4 per cent, a 50 per cent contribution of the genotype to seedling vigor, and a range of selection coefficients. Immigration is in the form of homozygous seed. Dominant and

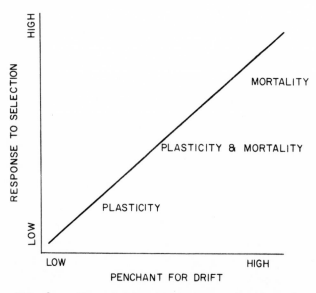

FIG. 3. *The relationship between plastic and mortality responses to stress and the evolutionary potential of populations*

recessive cases were examined. The population is assumed to be monomorphic at the start. The equilibrium frequencies of the alien allele are lower with an L-shaped fecundity distribution than with a poisson. Our estimates of poisson equilibria are in close accord with those obtained by analytical methods. The difference is most extreme when the coefficient of selection against the immigrant is low. For example, consider the case of a dominant gene whose immigration rate is .01 (Fig. 4). When the selective disadvantage is only 1 per cent, the

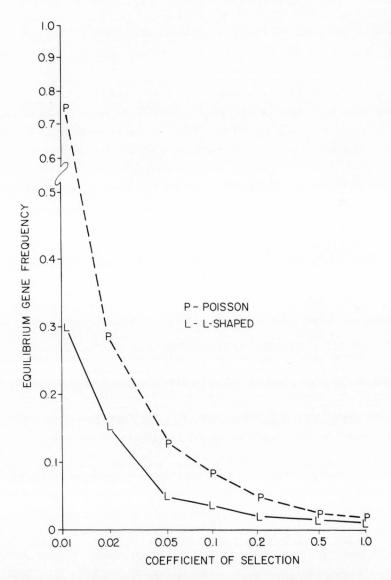

FIG. 4.   *Gene frequency equilibria for a dominant gene under different levels of selection showing the effect of different fecundity distributions. The term "double poisson" refers to the L-shaped distribution. The immigration rate is 1 per cent. The coefficient of selection against the dominant homozygote and heterozygote are the same*

equilibrium level with an L-shaped distribution is only .30 as compared to .75 when a poisson is used. Selection coefficient notwithstanding, the equilibrium alien gene frequency with the poisson is more than twice that with the L. This is generally true for higher levels of immigration. Thus the fecundity schedule may have a profound effect on the impact of deleterious gene flow. When the alien gene is recessive, there is only about a 30 per cent difference at different immigration levels.

3. GENETIC IMPLICATIONS OF DIVERSE REPRODUCTIVE SCHEDULES

The reproductive schedules of plants are extremely diverse. Some plants flower the first year of their life and then die, others flower periodically after a pre-reproductive period of two to several years, and others only flower once after achieving considerable age (HARPER & WHITE 1974). Correlatively, the generation times of different species may vary by at least an order of magnitude. Most theoretical and experimental population geneticists have been concerned with the rate of response to selection per generation, and as such have ignored the genetic implications of diverse reproductive schedules in terms of a chronological time table. Only plant breeders have addressed the matter of progress per unit time, but their interests have centered on annual and short-lived perennials (FALCONER 1960). Molecular geneticists have suggested that the rate of evolution is independent of generation time (NEI 1975). Regardless of the validity of their arguments, they have not considered characters which are only intermittently exposed to selection, namely, juvenile or reproductive characters. Surely the rate of evolution of juvenile and floral characters per unit time in an annual could be different than from a century plant assuming equivalent selection pressures.

We have constructed a series of life histories depicted in Fig. 5. Assuming a population of infinite size and an initial frequency of .2, semidominance (relative fitness: AA = 1.0, Aa = .75, aa = .50), a L-shaped fecundity distribution and a 50 per cent environmental impact on seedling vigor (C_{vigor}), we determined the mean number of years for a gene to move from p = .2 to .8 as a function of absolute time. The relationship between these variables is shown in Fig. 6. The correlation mean between generation time and rate of response is .98. Changing the selection coefficients alters the slope but not the compelling relationship between generation time and response to selection.

Having emphasized directional selection, consider next the implications of different reproductive tactics in perennials when long-term selection is cyclic or random. Long-lived populations with delayed reproduction would be poorly equipped to respond to a change in selection regime, and thus would be more likely to retain genetic polymorphism than would short-lived populations which reach reproductive maturity in one or a few years. The genetic structure of long-lived populations would be tuned to their long-term environmental

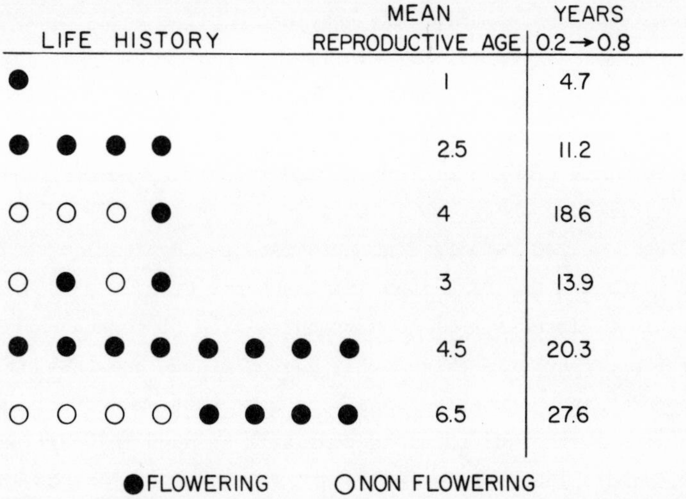

LIFE HISTORY	MEAN REPRODUCTIVE AGE	YEARS 0.2 → 0.8
●	1	4.7
● ● ● ●	2.5	11.2
○ ○ ○ ●	4	18.6
○ ● ○ ●	3	13.9
● ● ● ● ● ● ● ●	4.5	20.3
○ ○ ○ ○ ● ● ● ●	6.5	27.6

●FLOWERING   ○NON FLOWERING

FIG. 5.   *The effect of different life histories on rates of change in gene frequency*

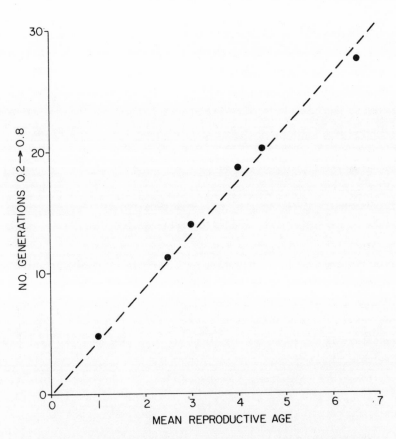

FIG. 6.   *The relationship between rate of evolution and mean reproductive age*

experience, whereas the genetic structure of short-lived populations would reflect recent experiences which may have had a pronounced impact upon their genetic structure.

Thus far the potential to respond to selection has been within the context of 100 per cent sexual reproduction. However, hundreds of species are facultative apomicts capable of reproducing in part by agamospermy, or by stolons or rhizomes. Agamospermy is more common in fugitive species, whereas vegetative reproduction is more common in late successional or climax species (GUSTAFFSON 1946; VAN DER PIJL 1969; SALISBURY 1976; MYGREN 1954). Apomixis is important in a discussion of generation related phenomena, since this mode of reproduction effectively extends the age of plants and thus their generation time. Agamospermy and vegetative expansion have the same effect, although the spatial pattern of the individual in time will be much more diffuse in the case of the former. Thus species which are facultative apomicts should respond more slowly to selection than those which are strictly sexual. Moreover, we expect apomixis to reduce the response to selection because it reduces the effective population size and thus the amount of additive genetic variance in individual fitness (FISCHER 1930).

4. GENETIC IMPLICATIONS OF SEED POOL DYNAMICS

Thus far rates of evolution have been considered with regard to life history features of perennial plants. Annuals also should be considered since they too have age-structured populations by virtue of their seed pools. The annual habit in plants is accompanied by specialized physiological mechanisms which permit seeds to remain dormant in the soil for a few years to decades (MAYER & POLJAKOFF-MAYBER 1975; ROBERTS 1972; KOLLER 1969). The nearly universal imposition of germination regulating mechanisms prevents the reproductive potential of a population from being gambled except when the probability of survival is maximal. The longevity of seeds tends to be a positive function of environmental unpredictability and is correlated with other adaptations to cope with unpredictability. The seeds of desert ephemerals and fugitive species are relatively long-lived (50 to 100 years) whereas the seeds of mid-successional annuals in mesic habitats often are short-lived (10-20 years) (TOOLE & BROWN 1946; KIVALAAN & BANDURSKI 1973; HARRINGTON 1972). The buried seed population has a constant death risk (HARPER & WHITE 1974).

The seed bank is based upon contributions made over several years, during which time selection may have favored alternate alleles at a particular locus related to temperate tolerance, pubescence, etc. The seed pool therefore provides the population with a memory of its recent history and is tuned to the general experience of the population over several years rather than just the past 1 or 2. A seedling crop is drawn from this potentially diverse seed pool.

The optimal germination tactics for a seed pool have been studied by COHEN (1966, 1967, 1968). The question now is not what is best for a population in terms of its persistence, but what are some genetic consequences of different seed pool histories. What constraints do seed pools impose on evolution at a single locus unrelated to dormancy? A. TEMPLETON (unpublished) has considered the effects of directional and disruptive selection in time taking into account the impact of the environment on population size. I will summarize some of his findings which are based upon analytical solutions.

By virtue of its history, the seed pool retards the rate of evolution when fitness is constant, balancing or directional selection notwithstanding. There is a simple linear relationship between the change in gene frequency per year and the average number of years a seed has spent in the seed pool prior to germination. Consider a case of directional selection in a population of annuals. If the mean time that plants spent in the seed pool prior to germination were 10 years, then the number of years required for selection to alter the frequency of a gene from one value to another would be 10 times greater than if there were no seed pool. If the average time in the seed pool were 3 years, than the rate of evolution would be 3 times lower than if there were no seed pool. Consider a case of balancing selection at a single locus where the heterozygote was favored and both homozygotes had relative fitness of .99. The number of years required to move a gene from one frequency to the equilibrium value (.5) also would be a simple function of time plants spent in the seed pool (Fig. 7).

The extent to which the seed pool retards the response to selection not only is dependent upon the age structure of the seed pool, but also is dependent upon the growth rate of the population. The discussion thus far assumed that population size is constant. Should populations pass through a period of growth, the seed pool would be weighted in favor of recent seeds since they would be increasingly abundant and constitute a higher proportion of the pool than if the population was not expanding. The mean age of the seeds in the pool would decline. On the other hand, if populations were contracting, the seed yield per year would also contract so that the pool would be weighted in favor of the older seeds relative to what would be the case with a stable population size. The mean age of seeds in the pool would increase. Accordingly, the rate of evolution will be faster in expanding populations with seed pools than in stable populations. Correlatively the rate of evolution will be slower in contracting populations than in stable ones.

The seed pool allows the existence of a covariance between fitness and absolute seed production during years in which the environment is hospitable and plants generally are large. The genotype favored in that environment will produce more seeds and thus have a greater impact on the seed pool than an alternate genotype favored in years when conditions for survival are marginal. With a seed pool not only the relative fitness is important but the absolute

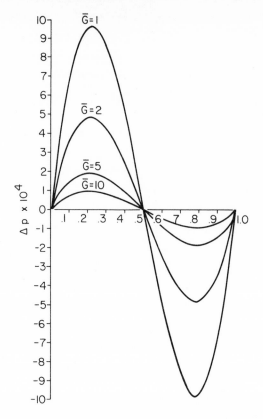

FIG. 7.   *Rate of change in gene frequency toward an
equilibrium value of .5 as a function of
mean numbers of years seeds spend in the
soil prior to germination. The plants are
assumed to be annuals*

performance of the best genotype in year 1 vs. the best in year 2. Differential
contribution of seeds to the seed pool both within and between years, but
especially the latter, may have the same consequences as directional selection.
Let us assume that in good years the relative fitness of 3 genotypes are AA =
1.00, Aa = 0.75, and aa = 0.50. Let us also assume that in marginal years the
fitnesses are AA = 0.50, Aa = 0.75, and aa = 1.00. Population size is constant.
In the good years the seed production of the favored homozygote (AA) is 100,
whereas in the marginal years the seed production of the favored homozygote
(aa) is only 50. If the environment alternates between good and marginal years
in a regular cyclic fashion the A gene eventually will be fixed even though
environments and relative fitnesses are symmetrical. Random fluctuations in the
environment will have the same effect. Accordingly selection will favor genes
that do the best when the environment is most hospitable. The greater the mean
age of seeds in the pool the more pronounced is the filtering effect of
selection in a fluctuating environment.

5. THE EFFECT OF PATCH SIZE AND SHAPE ON MIGRATION-SELECTION EQUILIBRIA

5.1. INTRODUCTORY REMARKS

There has been much interest in the joint action of migration and selection on the level of genetic variability in subdivided populations occupying spatially heterogeneous environments. Two and multi-niche models have been constructed which relate to organisms in general (CHRISTENSEN 1975; KARLIN 1976; FLEMING 1975; GILLESPIE 1974, 1975, 1976; HEDRICK *et al*. 1976) or plants specifically (JAIN & BRADSHAW 1966; ANTONOVICS 1968; NAGYLAKI 1976; LEVIN & KERSTER 1975; GLEAVES 1973). Unfortunately, most treatments make assumptions about the mating system, the spatial distribution of patch types, and dispersal distributions which are contrary to what one actually finds in plant populations. Plant populations are not panmictic, incoming genes are not randomly distributed throughout a population, and environmental patches are two-dimensional arrays with specific dimensions.

Using computer simulation, we chose to answer the following questions: What are the equilibrium gene frequencies in subpopulations adapted to different environments when patch size and shape are variables? To what extent do the equilibria differ with different gene dispersal schedules?

5.2. THE MODEL

The model involved a population of 4096 plants occupying a site with patch type A and B in equal proportions. Only a single locus with two alleles (A and a) was considered possible. The relative fitnesses of genotypes in patch A are AA = 1.0, Aa = 0.75, aa = 0.50. The relative fitnesses of genotypes on patch B are AA = 0.50, Aa = 0.75, aa = 1.0. The size and shape of patches varies as prescribed with at least 4 plants per patch. The plants were assumed to occupy a 64 x 64 grid of uniformly spaced safe sites. Every plant was one map unit from its nearest neighbors.

The initial population is composed of AA plants in the A patches and aa plants in the B patches. Patches alternate in space in a checkerboard fashion. The breeding structure of the population was defined by the dispersal of pollen. Dispersal followed random, leptokurtic, or nearest-neighbor patterns. No seed dispersal is assumed. Seeds occupy the same site as the seed parent. The stepping stone schedule involves the movement of seeds to one of the four or eight nearest sites at the cardinal compass points. For the leptokurtic schedule, pollen dispersal distances assumed the distribution shown in Fig. 8. The direction of broadcast was designated by a random number generator. Each plant produced 18 pollen grains, a number sufficient to insure that each plant almost certainly would receive 1 pollen grain. Pollen falling outside the grid was lost. Plants were assumed to be self-incompatible. The number of ovules per plant is assumed to be very large so that the number of seed produced is equivalent to the pollen receipt.

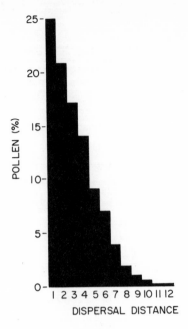

FIG. 8.   *Distribution of pollen dispersal*
distances under the leptokurtic
pollen schedule

Competitive selection occurs when more than one seed is deposited per site.
The surviving seed genotype is a function of frequency of the genotypes and
the relative fitness of each. The frequency of the genotypes after selection
is calculated. Each genotype then is assigned a region on a scale from 0 to 1
equivalent to its frequency. The genotype of the surviving seed is determined
by the genotype domain into which a randomly drawn number between 0 and 1 falls.

Since the model was of a finite population, there were considerable edge
effects. For the examination of most of the results, therefore, the grid was
divided into a center and an edge. This enables the center results to be
generalized to the center of any sized population, and also gives an indication
of the magnitude of the edge effects. The boundary between the edge and center
and the edge was placed at a patch boundary, such that the areas of the center
and the edge were as similar as possible. In the case of 32 x 32 patches, the
boundary had to run through the middle of the patches, but this introduces no
bias. In those cases where the patch arrangement was at random, the random
process was constrained to ensure equal representation of both patch types
both in the center and round the edge.

5.3. RESULTS

Consider first the proportion of pollen entering a patch whose source was from
the other patch type. We will refer to this pollen as alien pollen. The alien

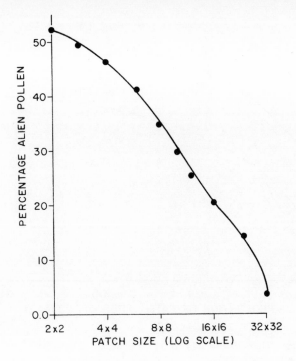

FIG. 9.   *The relationship between patch size and the percentage pollen received from the alternate patch type*

pollen influx is a function of patch size, large patches receiving less of it than small ones (Fig. 9). The proportion of alien pollen decreased from about 50 per cent with 3 x 3 patches to only 5 per cent with 32 x 32 patches. Note that for the latter a boundary of several rows was omitted. The overall pattern would not change appreciably were the entire grid treated. If we shift from a checkerboard to a random distribution of patches (frequency of A patches still = B patches) the level of alien pollen receipt declines for each patch size.

The influx of alien pollen also is affected by the patch shape. Given sub-populations of 64 plants, the proportion of alien pollen varies from 55 per cent for 1 x 64 patches to 26 per cent for 8 x 8 patches. Regardless of patch size or shape, the levels of alien pollen receipt is a curvilinear function of the amount of contact between opposite patch types (Fig. 10).

The mean frequency of the maladapted gene in each patch type is a function of patch size. Equilibrium frequencies range from .38 for 2 x 2 patches arranged in a checkerboard fashion to 0.11 for 32 x 32 patches. Moving to a random patch distribution results in lower equilibria values. Equilibrium gene frequencies also are greatly influenced by patch shape. With a patch size of 64 plants, equilibrium frequencies of the maladapted gene vary from .16 with 8 x 8 patches to 0.25 with 4 x 16 patches to 0.34 with 2 x 32 patches and 0.38 with 1 x 64 patches. These values are based upon checkerboard patch patterns (Fig. 11).

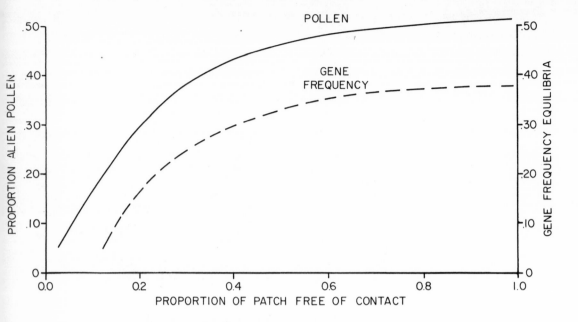

FIG. 10. *Alien pollen influx and gene frequency equilibria as functions of level of patch contact*

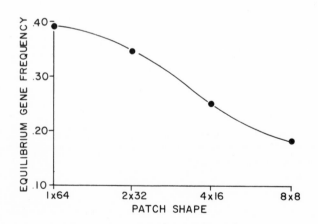

FIG. 11. *Equilibrium gene frequency as a function of patch shape*

At equilibrium the maladapted gene is distributed in a non-random fashion within patches. This is most evident where patch size is large (Fig. 12). This gene tends to be higher in frequency near the edge of the patch than near the interior and in patches larger than 8 x 8 rarely reaches the center. This pattern results from narrow pollen dispersal between and within patches. The former dictates the initial deposition sites, and the latter the penetration rate. Once in a patch an alien gene will move over short distances which are random with regard to direction. This provides ample opportunity for its selective elimination beyond the area of continual immigration.

X = AA   ⊠ = Aa   ● = aa

FIG. 12.   *The distribution of genotypes within patches*

91

Thus far equilibrium frequencies have been considered within the context of a given leptokurtic dispersal schedule. It is of interest to determine the consequences of various schedules including that assumed in most population genetic models. We will compare the equilibrium levels of maladapted genes under the following pollen dispersal schedules: (a) random, (b) leptokurtic with a mean distance of 2.5 and S.D. of 2.8 units as used previously, (c) leptokurtic with a mean distance of 1.25 and S.D. of 1.4 units, (d) leptokurtic with a mean of 5.2 and S.D. of 5.6 units, (e) 4 nearest-neighbors, (f) 8 nearest-neighbors. The leptokurtic distributions have the same kurtosis. The nearest-neighbor distributions provide for equal dispersal of pollen for each of the nearest-neighbors in question.

The effect of the six dispersal schedules on equilibrium frequencies in populations with different patch sizes is shown in Fig. 13. The equilibria

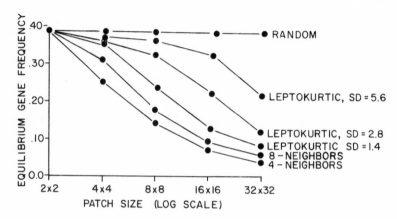

FIG. 13.   Equilibrium gene frequencies in patches of different sizes as a function of pollen dispersal schedule

vary widely especially in populations with large patch size. With random mating, the maladapted gene frequency averages 0.38, patch size notwithstanding. If we consider the 32 x 32 patch population, the frequency of the maladapted gene equilibrates at about 0.04 with pollen dispersal to the 4 nearest-neighbors, and at 0.07 with pollen dispersal to the eight nearest-neighbors. With narrow, moderate and broad leptokurtic dispersal, frequencies are 0.08, 0.11, and 0.21, respectively. The more restricted is pollen flow, the lower the equilibrium value. Correlatively, the more restricted is pollen flow the more highly adapted are subpopulations to their respective environments.

Self-fertilization is a method for insulating plants from the flow of deleterious genes from other populations. Thus we would expect selfing to result in lower equilibrium levels for the maladapted genes in our models.

Consider the consequences of 50 per cent selfing and 50 per cent outcrossing relative to our standard leptokurtic (2.5 mean) schedule. Surprisingly 50 per cent selfing only had a moderate effect on equilibrium frequencies, regardless of patch size (Fig. 14). However, predominantly self-pollinating populations contain a lower level of the maladapted gene than those discussed thus far.

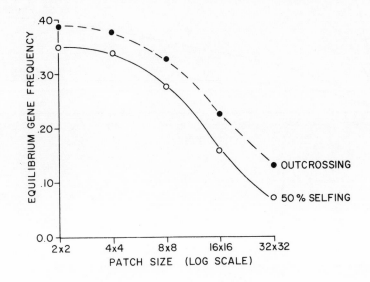

FIG. 14.   *The effect of 50 per cent self-fertilization on equilibrium gene frequencies in different patch sizes*

For example, in a population with 90 per cent selfing and patch size of 8 x 8, the equilibrium gene frequency is 0.18 versus 0.27 with 50 per cent selfing, and 0.35 with obligate outbreeding.

In view of the effect of different dispersal schedules and selfing on equilibrium gene frequencies, it is of interest to determine the time to equilibrium. With panmixia we would expect subpopulations to reach equilibria in one generation. With our standard leptokurtic distribution, equilibria are reached in an average of 12 generations when patches are 8 x 8 or larger. For populations with 2 x 2 patches equilibria are achieved in an average of 5 generations and with 4 x 4 patches in an average of 7 generations. Selfing significantly retards the time to equilibrium. Consider a population with 8 x 8 patches. The time to equilibrium with 50 per cent selfing averages about 17 generations; with 90 per cent selfing the average is about 30 generations (Fig. 15).

Thus far we have considered patches to be contiguous. However, it is possible that distinctive patches may be separated by small uninhabitable areas.

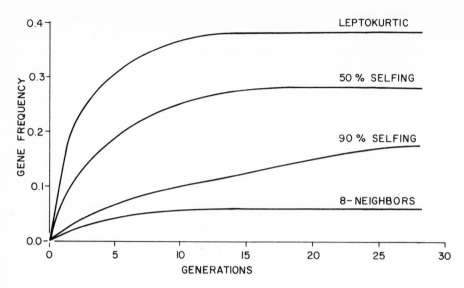

FIG. 15.   *The approach to equilibrium as a function of mating scheme in a population with 8 x 8 patches*

We explored the effect of habitat discontinuity by introducing a 2-position (unit) gap between patches. As one might anticipate, habitat discontinuity and restricted gene flow (in the form of our standard leptokurtic schedule) confers an excellent opportunity for interpatch differentiation. In populations where patch size is large, the frequency of the maladapted gene may be nearly 50 per cent of that in the absence of the gap (Fig. 16). The impact of the gap is even larger with

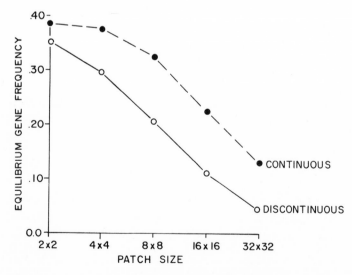

FIG. 16.   *The effect of discontinuity between patches on equilibrium gene frequencies*

94

more restricted dispersal schedules since the size of the gap is proportionally greater relative to the dispersal limits. With nearest-neighbor dispersal, the gap cannot be crossed and subpopulations remain monomorphic.

In summary, the level of polymorphism within subpopulations occupying heterogeneous environments is a complex function of patch size, shape, spatial distribution of patch types, and gene dispersal schedule. Simply to note that an environment is heterogeneous and that gene flow is restricted fails to provide a meaningful insight into gene frequency heterogeneity between subpopulations, the correlation between patch type and gene frequencies, or the extent to which the adaptedness of populations is reduced by a selection-migration equilibrium. We need to know about the organization and size of patch types if we wish to appreciate the potential and actual level of differentiation which may ensue in the face of gene flow.

6. CONCLUSIONS

In 1972 BRADSHAW wrote an essay on some evolutionary consequences of being a plant. He emphasized the implications of the strength and diversity of selection which operate on closely adjacent populations, the restriction of gene flow between neighboring populations, the correlation of habitats between parents and offspring, the flexibility of the breeding system, and the nature and organization of genetic variation within species. He concluded that "to understand what is actually happening in plant species, we need to assume very different premises from many of those exercising the minds of many population geneticists". As we have attempted to convey in this presentation, we vigorously endorse this position. Moreover, we urge the exploration of the evolutionary implications of demographic features which typically have resided in the domain of ecologists. Fecundity schedules, reproductive schedules, seed pool properties and the pattern of differentiated population subdivisions all are variables which have manifest genetic consequences, which in turn may free or restrict the evolutionary potential of populations. Moreover, the aforementioned demographic properties either have not been considered by population geneticists or their actual expression is contrary to many assumptions used in genetic models. The theory and illustrations presented here only represent a small sample of the kinds of problems that we need to address if we hope to understand what is actually happening in plant species, why it is happening, and what it means.

7. REFERENCES

ANDERSON, W.W., 1971 - Genetic equilibrium and population growth under density-regulated selection. *Am. Nat.*, 105, 489-498.
ANTONOVICS, J., 1968 - Evolution in closely adjacent populations. VI. Manifold effects of gene flow. *Heredity, London,* 23, 507-524.

BRADSHAW, A.D., 1972 - Some evolutionary consequences of being a plant. *Evol. Biol.*, 5, 25-47.

BRADSHAW, A.D., 1974 - Environment and phenotypic plasticity. *Brookhaven Symp. Biol.*, 25, 75-94.

BULLOCK, S.H., 1976 - Consequences of limited seed dispersal within a simulated annual population. *Oecologia*, 24, 247-257.

CHARLESWORTH, B., 1971 - Selection in density-regulated populations. *Ecology*, 52, 469-475.

CHARLESWORTH, B., 1973 - Selection in populations with overlapping generations. V. Natural selection and life histories. *Am. Nat.*, 107, 303-311.

CHARLESWORTH, B. & J.T. GIESEL, 1972 - Selection in populations with over-lapping generations. II. Relations between gene frequency and demographic variables. *Am. Nat.*, 106, 388-401.

CHRISTIANSEN, F.B., 1975 - Hard and soft selection in a subdivided population. *Am. Nat.*, 109, 11-16.

COHEN, D., 1966 - Optimizing reproduction in a randomly varying environment. *J. theoret. Biol.*, 12, 119-129.

COHEN, D., 1967 - Optimizing reproduction in a randomly varying environment when a correlation may exist between the conditions at the time a choice has to be made and the subsequent outcome. *J. theoret. Biol.*, 16, 1-14.

COHEN, D., 1968 - A general model of optimal reproduction in a randomly varying environment. *J. Ecol.*, 56, 219-228.

COHEN, D., 1971 - Maximizing final yield when growth is limited by time or by limiting resources. *J. theoret. Biol.*, 229-307.

COHEN, O., 1976 - The optimal timing of reproduction. *Am. Nat.*, 110, 801-807.

DEMETRIUS, L., 1975 - Reproductive strategies and natural selection. *Am. Nat.*, 109, 243-249.

FALCONER, D.S., 1960 - *Introduction to quantitative genetics*. Oliver & Boyd, London.

FISHER, R.A., 1930 - *The genetical theory of natural selection*. Clarendon Press, Oxford.

FLEMING, W.H., 1975 - A selection migration model in population genetics. *J. math. Biol.*, 2, 219-233.

GADGIL, M., 1971 - Dispersal: population consequences and evolution. *Ecology*, 52, 253-261.

GADGIL, M. & W. BOSSERT, 1970 - Life history consequences of natural selection. *Am. Nat.*, 104, 1-24.

GADGIL, M. & O.T. SOLBRIG, 1972 - The concept of r- and K-selection: evidence from wild flowers and some theoretical considerations. *Am. Nat.*, 106, 14-31.

GIESEL, J.R., 1974 - Fitness and polymorphism for net fecundity distribution in iteroparous populations. *Am. Nat.*, 108, 321-331.

GILLESPIE, J.H., 1974 - Polymorphism in patchy environments. *Am. nat.*, 108, 145-151.

GILLESPIE, J.H., 1975 - The role of migration in the genetic structure of populations in temporally and spatially varying environments. I. Conditions for polymorphism. *Am. Nat.*, 109, 127-135.

GILLESPIE, J.H., 1976 - The role of migration in the genetic structure of populations in temporally and spatially varying environments. II. Island models. *Theoret. pop. Biol.*, 10, 227-238.

GLEAVES, J.T., 1973 - Gene flow mediated by wind borne pollen. *Heredity, London*, 31, 355-366.

GUSTAFFSON, A., 1946 - Apomixis in higher plants. *Acta Univ. lund*, 39, 1-370.

HARPER, J.L. & J. WHITE, 1970 - The dynamics of plant populations. *Prov. Adv. Study Inst. Dynamics Numbers Popul.*, (Oosterbeek, 1970), 41-63.

HARPER, J.L. & J. WHITE, 1974 - The demography of plants. *A. Rev. Ecol. Syst.*, 5, 419-463.

HARRINGTON, J.F., 1972 - Seed storage and longevity. *In:* T.T. KOZLOWSKI (Editor), *Seed biology*, Vol. 3, Academic Press, New York, p. 145-245.

HEDRICK, P.W., GINEVAN, M.E. & E.P. EWING, 1976 - Genetic polymorphism in heterogeneous environments. *A. Rev. Syst.*, 7, 1-32.

JAIN, S.K. & A.D. BRADSHAW, 1966 - Evolutionary divergence among adjacent plant populations. I. Evidence and its theoretical analysis. *Heredity, London*, 21, 407-441.

KARLIN, S., 1976 - Population subdivision and selection-migration interaction. *In:* A. KARLIN & E. NEVO (Editors), *Proc. Int. Conf. Pop. Genet. Ecol.*, Academic Press, New York.

KARLIN, S. & J. McGREGOR, 1968 - The role of the poisson progeny distribution in population genetic models. *Math. Bio Sci.*, 2, 11-17.

KAWANO, S., 1975 - The productive and reproductive biology of flowering plants. II. The concept of life history strategy in plants. *J. Coll. Liberal Arts, Toyama Univ.*, 8, 51-86.

KING, C.E. & W.W. ANDERSON, 1971 - Age-specific selection. II. The interaction between r and K during population growth. *Am. Nat.*, 105, 137-156.

KIVALAAN, A. & R.S. BANDURSKI, 1973 - The ninety-year period for Dr. Beals seed viability experiment. *Amer. J. Bot.*, 60, 140-145.

KOLLER, D., 1969 - The physiology of dormancy and survival of plants in desert environments. *Symp. Soc. Expt. Biol.*, 23, 449-469.

KOYAMA, H. & T. KIRA, 1956 - Intraspecific competition among higher plants. VIII. Frequency distribution of individual weight as affected by the interaction between plants. *J. Inst. Polytech. Osaka cy. Univ.* (D), 7, 73-94.

LEVERICH, J., 1977 - Demographic studies of a population of *Phlox drummondii*. *Ph. D. Thesis.* University of Texas, Austin.

LEVIN, D.A. & H.W. KERSTER, 1974 - Gene flow in seed plants. *Evol. Biol.*, 7, 139-220.

LEVIN, D.A. & H.W. KERSTER, 1975 - The effect of gene dispersal on the dynamics of gene substitution in plants. *Heredity, London,* 35, 317-336.

LEVINS, R., 1968 - *Evolution in changing environments*. Princeton Univ. Press, Princeton.

MAYER, A.M. & A. POLJAKOFF-MAYBER, 1975 - *The germination of seeds*. Pergamon Press, London.

NAGYLAKI, T., 1976 - Dispersion-selection balance in localized plant populations. *Heredity, London,* 37, 59-67.

NEI, M., 1975 - *Molecular population genetics and evolution*. North Holland, Amsterdam.

NYGREN, A., 1954 - Apomixis in angiosperms. II. *Bot. Rev.*, 20, 577-649.

PIJL, L. VAN DER, 1969 - *Principles of dispersal in higher plants*. Springer-Verlag, New York.

RISSER, P.G., 1969 - Competitive relationships among herbaceous grassland species. *Bot. Rev.*, 35, 251-284.

ROBERTS, E.H., 1972 - *Viability of seeds*. Chapman & Hall, London.

ROFF, D.A., 1975 - Population stability and the evolution of dispersal in a heterogeneous environment. *Oecologia,* 19, 217-237.

ROSS, M.D. & J.L. HARPER, 1972 - Occupation of biological space during seedling establishment. *J. Ecol.*, 60, 77-88.

SALISBURY, E.J., 1942 - *The reproductive capacity of plants*. Bell & Sons, London.

SALISBURY, E., 1976 - A note on shade tolerance and vegetative propagation of woodland species. *Proc. R. Soc. B.,* 192, 257-258.

SARUKHAN, J., 1974 - Studies on plant demography. *Ranunculus repens* L., *R. bulbosus* L., and *R. acris* L. II. Reproductive strategies and seed population dynamics. *J. Ecol.*, 62, 151-177.

SARUKHAN, J. & J.L. HARPER, 1973 - Studies on plant demography. *Ranunculus repens* L., *R. bulbosus* L., and *R. acris* L. I. Population flux and survivorship. *J. Ecol.*, 61, 675-716.

SCHAFFER, W.M., 1974a- The evolution of optimal reproductive strategies: the effect of age structure. *Ecology,* 55, 291-303.

SCHAFFER, W.M., 1974b- Optimal reproductive effort in fluctuating environments. *Am. Nat.,* 108, 783-790.

SCHAFFER, W.M. & M.D. GADGIL, 1975 - Selection for optimal life histories in plants. *In:* M. CODY & J. DIAMOND (Editors), *Ecology and evolution of communities*, Harvard Univ. Press, Cambridge, p. 142-157.

SHELDON, J.C., 1974 - The behavior of seeds in soil. III. The influence of seed morphology and the behavior of seedlings on the establishment of plants from surface-lying seeds. *J. Ecol.*, 62, 47-66.

TAYLOR, H.M., GOURBY, R.S., LAWRENCE, C.E. & R.S. KAPLAN, 1974 - Natural selection of life history attributes: an analytical approach. *Theoret. pop. Biol.*, 5, 104-122.

TOOLE, E.H. & E. BROWN, 1946 - Final results of the Duvel buried seed
     experiment. *J. agric. Res.*, 72, 201-210.
WHITE, J. & J.L. HARPER, 1970 - Correlated changes in plant size and numbers
     in plant populations. *J. Ecol.*, 58, 467-485.
YARRANTON, G.A. & R.G. MORRISON, 1974 - Spatial dynamics of a primary
     succession: nucleation. *J. Ecol.*, 62, 417-428.

8. DISCUSSION

WATKINSON (Norwich): I accept what you said about seed dormancy and the
unpredictability of the environment, but if one is going to make meaningful
comparisons between environments and between species in different environments,
how would you suggest that one quantifies the predictability of the environment,
because I see that this is a major stumbling block preventing the empirical
workers in the field from getting much further in this area.

LEVIN: I don't think I am in a position to tell you how to quantify the
environment. What I am trying to do is simply show that seed pools have an
effect, have a genetic effect, and that it makes a difference in genetic
currency in terms of the kinds of seed-pool dynamics you have. Now, exactly
how one wants to describe environmental predictability or demonstrate it or
study it, is something which is really beyond an answer from me. I don't think
I really want to respond to that.

HARPER (Bangor): Due to technical difficulties this question was very
badly recorded and cannot be reproduced in detail. Professor Harper asked
why a double poisson was used to imitate the L-shaped curve. His second
question concerned the relationship between the genetic and non-genetic
components in the fecundity frequency distribution.

LEVIN: In response to your last comment: we simply were interested in
looking at a combination of genotype and environmental effects and we just
chose 50 per cent for heuristic purposes just to include some environment and
some genetic effects. We gave equal weight to both, since in many cases we do
not know what the causal factors are. I certainly agree that it will be more
interesting to use real values than the ones we have chosen just because they
relate to the nature of the problem, and the same answer holds for our
L-shaped fecundity distribution. It was simply, let us say, for heuristic
purposes, that we chose that particular distribution. JOHN WILSON can express
exactly why he wanted to use a double poisson to generate the L, although we
did want a little hump out on the tail of the L. We wanted a little bit of
a hump for those special plants out at the end rather than just having a flat
L, and the fact that you suggested that we understated the case and that the
difference would be even greater, makes me feel very fine, because what I am
trying to do is just show that plants are different from the simple theory
that has been applied to them. What we really have to know is what plants are
doing, in order to fully appreciate plant evolution and ecology and the

interface between them.

ANONYMOUS (1): Don't you think that the main effect of the seed bank is to diminish the genetic drift because it increases the size of the population first and also avoids selection for one part of it?

LEVIN: The seed pool acts as a buffering device. You may have environmental fluctuations up and down, but the seed pool will tend to stabilize the populations within a gene frequency range and reduce the probability of genetic drift. So when we look at populations of annual plants and we see populations going from 10,000 to 100, this is not an adequate picture of the true population dynamics in a sense that if seeds are live plants, the populations size may be very constant. The number of individuals that are reproducing in any given year fluctuates considerably, but the actual population size is maintained and the seed pool does act as a buffer but also acts as a drag, because it provides a memory. I think that after a bad year the seed pool will be more important, and in Texas in some of our annuals we have good years and bad years where in the bad year maybe only 10 per cent of the plants were there, compared to what we saw the year before, in which case the next year, if you draw at random from the seed pool, the older seeds will make a relatively larger contribution. So I am not sure that you need catastrophic events. I think you can have good and bad years, where there is always a seed pool that remains intact. I think you can draw from the past in the sense that, like migration seed pools from the past provide heterogeneity as well as all those things that migration would do to maintain genetic polymorphism. But it does retard the response to selection as does migration.

BELL (Bangor): Following up the point that Professor HARPER made on the importance of distance between neighbours, did you take into account in your grid systems, that if you are considering four neighbours and then considering eight neighbours that actually different distances are involved?

LEVIN: Well, the distances were determined in advance, and the distance is reflected in the mating systems, so to speak. If you are crossing with plants that are very close to you, you could be crossing with a brother and a sister, since we are dealing with annuals. So the spatial pattern affects the breeding system, which affects the outcome, as I showed you. So in a sense distance was considered. I mean it is built in, but not in the way that your are suggesting. But it is reflected in the nature of the output, whether we are looking at four nearest neighbours or eight nearest neighbours or whatever. You get different values, and this is due to differences in the restriction of the movement of pollen.

BELL: But the subtle difference that four of the eight are further away than the other four does not make any difference?

LEVIN: I cannot say it does not make any difference. I have to think about exactly what we might expect to find, other than what I have just said that the farther away the pollen is moving the more open the breeding system is and the less likely you are to be mating with relatives. In terms of the plastic response to competition as a function of distance, we could not build that into our model, since we had decided that there should be some simple pattern of safe sites; therefore, there is no way we could get a distance effect, an interference effect over space. We have considered it, but I am not sure how you could bring that in. If you had empty spaces you could do that, but with all spaces filled, I am not sure you could handle it.

ANONYMOUS (2): Did you say that in the cyclic environment your model predicts that the genotypes that perform better in good years are favoured?

LEVIN: Yes, if you simply go back and forth from a good year to a bad year and if you had different genotypes favoured in the two years. If in the good year the best genotype produces 100 seeds and if in a bad year the best genotype only produces 50 seeds, you have differential contribution to the seed pool between years; and by going back and forth, back and forth, you gradually weight the seed pool by the population in terms of the genotypes that do best in the favourable years. So even though you might have a cyclical environmental pattern with a seed pool, the result is the same as directional selection; you end up fixing the genes or the genotype that does best in the best environment.

ANONYMOUS (2): What happened when you had several bad years in succession, what happens in the population?

LEVIN: The seed production would be lower; you would have your seed pool to draw upon in the event of good years, but if you had several bad years you would simply have a lower contribution of seeds in those years weighted in terms of the relative fitness of the three genotypes in those year. I am not sure I am answering your question.

ANONYMOUS (2): It seems to me that the frequency in the seed pool of seeds produced from genotypes that are better performers in good years would be higher. Is that right?

LEVIN: Yes. One good year could greatly outweigh several bad years, according to my model, if the good ones were producing fewer seeds in good years. If the best genotype in poor years was producing few seeds relative to the best genotype in good years, you might have several bad years and one good year and that would compensate for it, or even more than compensate.

Interpretation of small-scale patterns in the distribution of plant species in space and time

1. INTRODUCTION

The vegetation patterns which can be detected within areas of several square metres or less hold a special fascination for the plant ecologist and present us with particular opportunities and problems. When working on a small scale we are obliged to grapple with the mechanisms whereby neighbouring plants interact with each other and we must also define the conditions which often permit species of different biology to co-exist in close proximity. However, the challenge to those who are conducting small-scale studies is not merely to elucidate local patterns but to place the data from such investigations in useful relation to the results obtained by ecologists working on quite different vegetation and in other geographical areas. The rudiments of the conceptual framework required to attain this broader objective are now apparent and in this paper five aspects of the mechanism controlling the structure and species composition of vegetation will be examined.

2. THE THREE PRIMARY STRATEGIES*

It has been asserted (GRIME 1974, 1977) that there are three main extremes of evolutionary specialization in autotrophic plants. Each may be identified by reference to a number of characteristics (Table 1) including morphological features, resource allocation, phenology and response to stress. The ruderal strategy has evolved in severely disturbed but potentially productive habitats, the competitive strategy prevails in productive, relatively undisturbed vegetation, whilst the stress-tolerant strategy is associated with continuously unproductive habitat conditions. The three strategies are, of course, extremes; it would appear that plants have evolved in relation to each of the possible equilibria between competition, stress and disturbance in the triangular model illustrated in Fig. 1.

---

* In the paragraphs which follow reference is made to the terms "strategy", "competition", "stress" and "disturbance", all of which are capable of a variety of meanings to biologists. In the appendix, these words have been defined in order to make clear the sense in which they have been used in this paper.

TABLE 1. *Some characteristics of competitive, stress-tolerant and ruderal plants*

Characteristic	Competitive	Stress-tolerant	Ruderal
1. Morphology of shoot	High dense canopy of leaves Extensive lateral spread above and below ground	Extremely wide range of growth forms	Relatively small stature, limited lateral spread
2. Leaf form	Robust, often mesomorphic	Often small or leathery, or needle-like	Various, often mesomorphic
3. Litter	Copious, often persistent	Sparce, sometimes persistent	Sparce, not usually persistent
4. Maximum potential relative growth-rate	Rapid	Slow	Rapid
5. Life-forms	Perennial herbs, shrubs and trees	Lichens, perennial herbs, shrubs and trees (often very long-lived)	Annual herbs
6. Longevity of leaves	Relatively short	Long	Short
7. Phenology of leaf production	Well-defined peaks of leaf production coinciding with period(s) of maximum potential productivity	Evergreens, with weakly defined patterns of leaf production	Short period of leaf production in period of high potential productivity
8. Phenology of flowering	Flowers produced after (or, more rarely, before) periods of maximum potential productivity	No general relationship between time of flowering and season	Flowers produced at the end of temporarily favourable period
9. Proportion of annual production devoted to seeds	Small	Small	Large
10. Response to stress	Rapid morphogenetic responses (root-shoot ratio, leaf area, root surface area) maximising vegetative growth	Morphogenetic responses slow and small in magnitude	Rapid curtailment of vegetative growth, diversion of resources into flowering
11. Photo-synthesis	Strongly seasonal, coinciding with long continuous period of vegetative growth	Opportunistic, often uncoupled from vegetative growth	Opportunistic, coinciding with vegetative growth
12. Storage of photo-synthate, mineral nutrients	Most photosynthate and mineral nutrients are rapidly incorporated into vegetative structure but a proportion is stored and forms the capital for expansion of growth in the following growing season	Storage systems in leaves, stems and/or roots	Confined to seeds
13. Palatability	Various	Low	Various

2.1. RUDERALS

Of the three primary strategies which have been proposed, the ruderal strategy is the most compact and easy to recognize. The most consistent feature of species or populations adapted to persistently and severely disturbed habitats is the tendency for the life-cycle to be that of the ephemeral, a specialization clearly adapted to exploit environments intermittently favourable for rapid plant growth. A related characteristic of many ruderals is the capacity for high rates of dry matter production (BAKER 1965; GRIME & HUNT 1975), a feature which appears to facilitate rapid completion of the life-cycle and maximizes seed production.

2.2. COMPETITORS

The basic attributes of the competitive strategy are those which permit a high rate of acquisition of resources (light, water, mineral nutrients and space) in crowded environments. Competitive plants are most abundant in circumstances allowing the rapid development of a large biomass of plant material, i.e. where there is high productivity and minimal damage to the vegetation. Under these conditions, species which are best equipped to tap the surplus of resources above and below ground and to maximise production prevail. The competitors include a wide range of perennial herbs, shrubs and trees and are distinguished by a number of characteristics, two of which are of particular importance. The first consists of the potential to produce a dense canopy of leaves and a large root surface area, over the period of the year in which conditions are most favourable to high productivity. The second is high phenotypic plasticity both in the apportionment of photosynthate between root and shoot and in the structure and spatial arrangement of individual leaves and roots, a characteristic which involves a high rate of reinvestment of captured resources in growth and in respiration.

An objection which may be foreseen to this rather specific description of the competitive strategy arises from the fact that competition occurs with respect to several different resources. Hence, it might be supposed that the ability to compete for one resource varies independently from the ability to compete for each of the others. However, as pointed out by MAHMOUD & GRIME (1976), rapid production of a large biomass of shoot material, a pre-requisite for effective above-ground competition, is dependent upon high rates of uptake of both mineral nutrients and water, characteristics which are themselves dependent upon a considerable expenditure of photosynthate on root development.

It seems more likely, therefore, that the abilities to compete for light, mineral nutrients, water and space are inter-dependent to such an extent that they exhibit parallel responses to natural selection.

103

2.3. STRESS-TOLERATORS

The stress-tolerant strategy comprises an extremely diverse assortment of plants which on first inspection appear to be far too varied in life-form (from lichens to certain forest trees) and in ecology (from arctic shrubs to cacti) to be included in the same category. However, morphological and taxonomic diversity among the stress-tolerators belies the conformity of life-history and physiology whereby these plants are adapted to survive in chronically unproductive conditions. This is to suggest that although the various types of stress-tolerators differ through the possession of mechanisms adapted to the specific forms of stress operative in their habitats, all exhibit a suite of characteristics common to plants which are capable of surviving conditions in which the level of production is consistently low. These features, which have been reviewed previously in relation to particular types of stress-tolerators (GRIME 1977), include inherently slow rates of growth, the evergreen habit, long-lived organs, sequestration and slow turnover of carbon, mineral nutrients and water, low phenotypic plasticity, shy flowering, and the presence of mechanisms which allow the vegetative plant an opportunistic exploitation of temporarily favourable conditions.

In environments such as tundra where both biomass and production remain extremely low the evidence that stress-tolerance constitutes the major adaptive strategy is incontrovertible. However, a potential source of confusion arises from the fact that many of the plant characteristics associated with stress-tolerance (Table 1) are typical of the trees and shrubs which form the terminal stages of vegetation succession in many types of forest and scrub. This observation suggests that there is a need to re-examine the tacit assumption in most ecological theory that increasing vegetation density is associated with a progressive increase in the relative importance of competition. As explained under the next heading, where the development of a dense plant biomass in an initially productive environment eventually causes severe resource depletion, survival in the later stages of the succession may depend less upon high competitive ability than upon the capacity to persist under conditions in which only low annual increments of growth are possible.

3. DOMINANCE

In experiments and in studies of natural vegetation examples of the impact of plant size are numerous (e.g. BOYSEN-JENSEN 1929; WATT 1955; BLACK 1958) and a consistent feature of plant succession is the appearance of plants of increasing stature. Moreover, at each stage of succession the major components of the plant biomass are usually the species with the largest life-forms. As pointed out by GRUBB (1976) this suggests that another dimension - that of dominance - must be incorporated into any model of the mechanism controlling the character and species composition of vegetation.

104

Dominance arises from a positive feed-back between (a) the mechanisms whereby the dominant plant achieves a size larger than that of its neighbours and (b) the deleterious effect which large plants may exert upon the fitness of smaller neighbours. The mechanism of (a) varies according to which strategy is favoured by the habitat conditions. In contrast, (b) does not vary substantially according to habitat or strategy and consists principally of various forms of plant-induced stress such as those due to shading, deposition of leaf litter and depletion of the levels of mineral nutrients and water in the soil.

Because of the variable nature of (a), different types of dominance may be recognized. The "ruderal dominants" consist of the larger summer annuals e.g. *Impatiens glandulifera* and include the majority of cereal crops, many of which form dense colonies which temporarily suppress other early colonists of fertile disturbed habitats. These plants owe their dominant status to the synchronous germination of dense populations of relatively large seeds followed by the rapid exploitation of untenanted space above and below ground. "Competitive dominants" include many of the most conspicuous perennial herbs, shrubs and trees of the early stages of succession in productive habitats. In these species, attainment of a comparatively large stature involves the ability of sustain high rates of uptake of resources from densely-crowded vegetation. By contrast, "stress-tolerant dominants" (perennial herbs, shrubs and trees of unproductive habitats or of the late successional stages of initially productive habitats) appear to achieve dominance through the maintenance of slow rates of growth under limiting conditions over long periods.

4. SUCCESSION

4.1. VEGETATION SUCCESSION IN PRODUCTIVE HABITATS

In localities where the soils and climate are conducive to high productivity there is a strong tendency for undisturbed vegetation eventually to become dominated by trees and since the pioneer studies of CLEMENTS (1916) it has been recognized that clearance of woodland is followed, in the absence of further major disturbance, by a characteristic process of recolonization in which annuals, perennial herbs, shrubs and trees successively dominate the vegetation.

Succession in productive habitats is a complex phenomenon reflecting the relative dispersal efficiencies of plants and progressive modification of soil and micro-climate by the developing vegetation. There can be little doubt also that competitive-dominance plays a major part in the process by which the annuals are displaced by perennial herbs. However, the role of competitive-dominance in the later stages of succession is more complex and requires careful analysis.

It is tempting to explain the sequence perennial herbs → shrubs → trees

simply in terms of the longer period required for the development of the
elevated leaf canopies of woody species. The comparatively slow growth-rates
of shrubs and trees are an inevitable consequence of the expenditure of
photosynthate upon supporting structures (trunks, branches, etc.) at the
expense of leaf and root surface areas. However, the possibility may be
considered that the process of succession is further attenuated by the
competitive effects of perennial herbs upon seedlings and saplings of woody
species. This is to suggest that despite their elevated leaf canopies, tree
seedlings and saplings may experience severe effects from herbaceous species
such as *Pteridium aquilinum* and *Chamaenerion angustifolium* which because of
their rapid growth-rates and capacity for lateral spread above and below
ground, and copious litter production, are better equipped, at least in the
short-term, to exercise competitive dominance. As an extension of this
hypothesis, we may suppose that the tendency of shrubs to dominate the
intermediate phase of vegetation succession is related to the branching form
of the shoot which, in contrast to that of most trees, allows, at an early
stage of growth, lateral spread of the leaf canopy and monopolization of the
aerial environment.

The woody species which appear early in the course of vegetation succession
in productive habitats resemble in several respects the competitive herbs
which they usurp. Their competitive characteristics include rapid rates of dry
matter production (i.e. in comparison with other trees), continuous stem
extension and leaf production during the growing season and rapid phenotypic
adjustments in leaf area and shoot morphology in response to shade (BORDEAU &
LAVERICK 1958; GRIME 1966; LOACH 1970). These features are particularly
conspicuous among the deciduous trees and shrubs such as *Rhus glabra*,
Ailanthus altissima, Betula populifolia, Populus tremuloides and *Liriodendron
tulipifera* which occur at the early phase in reafforestation of abandoned
pastures and arable fields on the eastern seaboard of North America. Species
which appear to play an equivalent role in Europe include *Sambucus nigra*, and
species of *Salix* and *Populus*.

An additional characteristic of the competitive trees and shrubs is their
intolerance of deep shade and it is not surprising, therefore, that as the
tree canopy closes, seedling establishment in these species diminishes and
many of the smaller saplings become severely etiolated and die. By this stage
in vegetation succession it is usually evident that changes have also occurred
in the composition of the herb layer. Although the competitive herbs are still
represented, most of them are extremely reduced in vigour and tend to be
replaced by evergreen, shade-tolerant herbs (AL-MUFTI *et al*. 1977).

As the larger of the competitive trees reach maturity most of the shrubs
succumb to shade or reach the end of their life-span and become senescent and a new
element, consisting of the shade-tolerant trees, becomes prominent. Seedlings
of many of the shade-tolerant trees have been present in relatively small

numbers, since an early stage of the succession, but because of their slow growth-rates they have been outstripped by the competitive trees and shrubs and have remained inconspicuous.

Although the shade-tolerant trees of initially productive habitats are slow-growing they include certain species which have a very long life-span, attain a large size, and eventually, in the final stages of succession, become the dominant component of the forest. Over a large area of Central Europe, beech (*Fagus sylvatica*) plays this role, whilst, in Australasia evergreens belonging to the genus *Nothofagus* are the dominant trees in the final phase of succession in productive habitats.

A problem of great theoretical interest and practical significance concerns the mechanism whereby these climax trees attain their dominant status. Some authorities such as WALTER (1973) have attributed success to their competitive ability. However, the characteristics of these trees dictate that they should be classified as stress-tolerators* rather than competitors. Consideration of the circumstances prevailing in the later stages of succession suggests that stress is, in fact, often likely to be a major selective force in climax forest. A consequence of the development of mature forest is the imposition of severe stress by the vegetation itself. This arises not only as a result of shading by a dense, continuous leaf canopy, but also because a high proportion of the mineral nutrients present in the habitat are sequestered in the living and dead parts of the plant biomass.

4.2. VEGETATION SUCCESSION IN UNPRODUCTIVE HABITATS

Where the productivity of the habitat is low, the role of ruderal plants and competitors in succession is much contracted and stress-tolerant herbs, shrubs and trees become relatively important at an earlier stage. The growth form and identity of the climax species varies according to the nature and intensity of the stresses occurring in the habitat. Where the degree of stress is moderate, such as in semi-arid regions or on shallow nutrient-deficient soils, the climax vegetation is often composed of sclerophyllous shrubs and small trees, the canopy of which is sufficiently discontinuous to allow the presence of an understorey of stress-tolerant herbs. Under more severe stress, e.g. arctic and alpine habitats, ruderals and competitive species may be totally excluded and here the succession may simply involve colonization by lichens, certain bryophytes, small herbs and dwarf shrubs.

---

* Many climax trees possess characteristics intermediate between those of the competitor and the stress-tolerator and may be described as "stress-tolerant competitors".

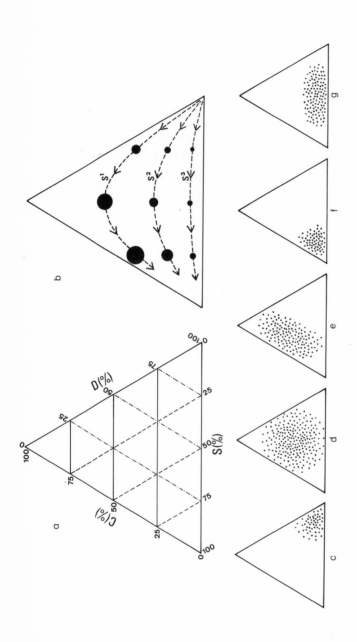

FIG. 1. a. A model describing the various equilibria between competition (C), stress (S) and disturbance (D) in vegetation.

b. The paths of secondary successions under conditions of high (S^1), moderate (S^2) and low (S^3) potential productivity. The relative sizes of the plant biomass at various stages of succession are indicated by the circles.

c.–g. The strategic range of selected life-forms and taxa:
c. annual herbs,   d. perennial herbs and ferns,   e. trees and shrubs,
f. lichens,   g. bryophytes

From the preceding sections it is evident that a major factor determining the relative importance of the three primary strategies in vegetation succession is the potential productivity of the habitat. In Fig. 1 a model has been drawn which summarizes the influence of potential productivity upon the path of vegetation succession.

The model consists of an equilateral triangle in which variation in the relative importance of competition, stress and disturbance as determinants of the vegetation is indicated by three sets of contours. At their respective corners of the triangle, competitors, stress-tolerators and ruderals become the exclusive constituents of the vegetation. Indicated within the triangle are areas corresponding to the strategic ranges of selected life-forms and taxa. Perennial herbs and ferns include the widest range of strategies whilst annual herbs are mainly restricted to severely disturbed habitats. Trees and shrubs comprise competitors and stress-tolerators. Although the lichens are confined to the stress-tolerant corner of the model, bryophytes are more wide-ranging with the centre of the distribution between the ruderals and stress-tolerators.

The curves, S^1, S^2 and S^3 describe, respectively, the paths of succession in conditions of high, moderate and low potential productivity whilst the circles superimposed on the curves represent the relative size of the plant biomass at each stage of the succession. The course of these curves indicates also the probable sequence of life-forms in each succession. In the most productive habitat, the course of succession (S^1) is characterized by a middle phase of intense competition in which first competitive herbs and then competitive, shade-intolerant, shrubs and trees dominate the vegetation. This is followed by a terminal phase in which stress-tolerance becomes progressively more important as shading and nutrient stress coincide with the development of a large plant biomass dominated by large, long-lived forest trees. Where succession occurs in less productive habitats (S^2) the appearance of highly competitive species is prevented by the earlier onset of resource depletion and the stress-tolerant phase is associated with dominance by smaller slow-growing trees and shrubs in various vegetation types of lower biomass than those occurring in S^1. In an unproductive habitat, the plant biomass remains low and the path of succession (S^3) moves directly from the ruderal to the stress-tolerant phase.

The curves of S^1-S^3 refer to succession under conditions of fixed potential productivity, a situation which is unlikely to occur in nature. In most habitats, the process is associated with a progressive gain or loss in potential productivity. Increase in productivity, for example, may arise from nitrogen fixation or input of agricultural fertilizer. Conversely, losses of mineral nutrients may result from leaching of the soil by rainfall. All these processes will modify the course of succession. In addition, succession in natural habitats is often subject to reversal by intermittent disturbance. Where disturbance involves cropping and loss of mineral nutrients (e.g. clear felling of woodland) not only may succession be temporarily reversed, but its subsequent course is likely to be deflected towards the area at the base of the triangle corresponding to successions of lower potential productivity.

5. CONTROL OF SPECIES DENSITY

5.1. INTRODUCTORY REMARKS

As progress is made in understanding the primary mechanism of vegetation, the possibility arises that some generalizations can be attempted with regard to the processes which control the number of plant species per unit area of vegetation. Species density is, of course, a function of sample size and of the number of strata (e.g. tree, shrub, herb and bryophyte layers) and taxonomic groups examined and in the following analysis, attention is confined to species densities measures in square quadrats of one square metre and refers exclusively to herbaceous species.

5.2. EFFECTS OF DOMINANCE

In the preceding discussion of dominance (see section 3) reference has been made to the tendency of larger plants to suppress the growth of smaller neighbours. In derelict grasslands and in perennial communities dominated by tall herbs this phenomenon has been closely studied (WATT 1955; SMITH *et al.* 1971; GRIME 1973).

Where herbaceous vegetation is established under fertile conditions and is not subject to damage (mowing, grazing, burning, trampling, etc.) the taller, more robust species expand and smaller or slower-growing plants are suppressed. If the growth of the larger species remains unchecked, a process of exclusion occurs and this may eventually result in the vegetation approaching a state of monoculture. It seems reasonable to conclude that the dramatic reduction in species densities in meadows and pastures observed over the last thirty years in Europe (e.g. THURSTON 1969) is to a large extent the result of an increasing intensity of competitive dominance brought about by stimulating the yield of the more robust and productive species and genotypes through the application of high rates of mineral fertilizers.

Effects of dominance upon species density are not confined to highly productive environments. Even where herbaceous vegetation is established on shallow infertile soils, a gradual trend towards monoculture is frequently observed providing that conditions allow the plant-biomass to remain relatively undisturbed. Here a well-known example is the widespread fall in species density which has been associated with reductions in grazing intensity by rabbits and sheep in unfertilized calcareous grasslands of the British Isles (e.g. WATT 1957; DUFFEY *et al.* 1974). In this example, the disappearance of the smaller low-growing herbs is correlated with the expansion of grasses such as *Brachypodium pinnatum, Festuca rubra* and *Dactylis glomerata*. As in the case of the dominants associated with more fertile habitats (e.g. *Urtica dioica, Epilobium hirsutum, Chamaenerion angustifolium*) these species develop a dense leaf canopy over the period June-August and produce a high density of persistent litter which appears to prevent the incursion of species with phenologies complementary to that of the dominants.

5.3. EFFECTS OF STRESS AND DISTURBANCE

Stress and disturbance exert parallel effects upon species density in herbaceous vegetation. In both, the impact changes according to intensity and this effect can be described in a simple model (GRIME 1973). The result of moderate intensities of either stress or disturbance is to increase species density by reducing the vigour of the dominants, thus allowing subsidiary species to co-exist with them. At the most extreme intensities of stress and/ or disturbance, however, species densities decline as conditions are created to which only a very small number of species are sufficiently adapted to survive.

5.4. QUANTITATIVE DEFINITION OF THE MECHANISM CONTROLLING SPECIES DENSITY IN HERBACEOUS VEGETATION

It has been argued (AL-MUFTI *et al.* 1977) that certain of the phenomena which influence species density in herbaceous vegetation (dominance, debilitation of dominants, incursion of subsidiary species, reduction of species density by extremes of stress and/or disturbance) are characteristic of particular ranges in shoot biomass and litter. Arising from this suggestion is the "hump-backed" relationship (Fig. 2) in which the potential species density is plotted against the annual maximum in standing crop + litter (both measured at the time of the maximum in the former). The values inserted in Fig. 2 are based upon measurements from the Sheffield area only and are therefore tentative. An interesting feature is a corridor of potentially high species density in the range 350-750 g/m^2. Above 750 g/m^2, diversification is restricted by dominance whilst below 350 g/m^2 species density is limited by the small number of species capable of surviving the severity of environmental stress and/or damage experienced in such extreme habitats. Within the corridor, however, vegetation composition is determined by an equilibrium between the conflicting forces of natural selection associated with moderate intensities of competition, stress (e.g. shortages of light, water, mineral nutrients or sub-optimal temperatures) and disturbance (e.g. trampling, mowing and grazing), and a habitat is created in which co-existence is possible between a wide range of species and ecotypes. The mechanisms whereby species of different habitat requirements may co-exist

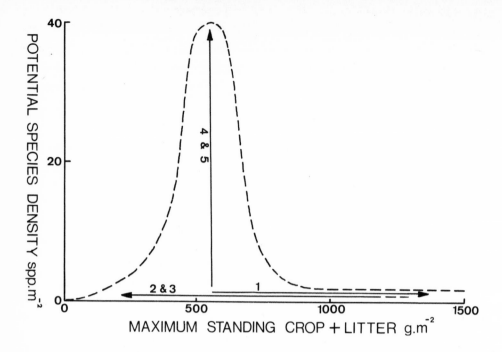

FIG. 2. *Scheme relating the processes (1-5) which control species density*
in herbaceous vegetation.
1. dominance, 2. stress, 3. disturbance, 4. niche-differentiation,
5. arrival and establishment of suitable species

within "corridor environments" may be conveniently analysed in terms of spatial
and temporal niches.

5.5. SPATIAL NICHES

Even in apparently uniform terrain, most vegetation samples include a complex
mosaic of micro-habitats. These arise from such factors as edaphic variation,
interactions between microtopography and climate, selective predation and the
redistribution of nutrients by animals. To these must be added spatial
differences in environment arising from the activity of the plants themselves.
These include variation in nutrient availability, water supply, degree of
shading, accumulation of litter and organic toxins and modifications of the
soil microflora.

5.6. TEMPORAL NICHES

5.6.1. Seasonal niches

Close scrutiny of small samples of vegetation often reveals situations in which
species of contrasted ecology are growing in intimate association and in which
it is difficult to explain their co-existence in terms of spatial heterogeneity

in environment. In many of these cases, it is apparent that the species concerned have different seasonal patterns of shoot expansion and flowering and are adapted to different parts of the annual climatic cycle. In limestone pastures of Derbyshire, for example, there is a tendency for certain grasses to peak in growth earlier than the majority of the forbs, many of which possess long tap-roots and exploit reserves of moisture during periods in which the grasses are subjected to desiccation. This difference in phenology between certain grasses and forbs is at least in part mediated by different responses to temperature and in laboratory experiments it is possible to manipulate the grass/forb ratio by alteration of temperature (MASON & GRIME 1975).

5.6.2. Niches arising from short-term variation

Although phenologies associated with the four seasons are of widespread importance, less conspicuous temporal niches operating on both shorter or longer time scales must be taken into account. Particularly in the British climate, daily, even hourly, fluctuations in radiation, temperature and water potential of the atmosphere are likely to result in a constant shifting in the identity of the constituent species and genotypes for which conditions most nearly approximate to the optimum for photosynthesis and/or growth. Equally deserving of attention are the temporal niches which arise from short-term changes in vegetation structure resulting from grazing or mowing. Field observations and comparative studies (MILTON 1940; MAHMOUD 1973) have shown, for example, that there are profound differences between species and varieties of grasses with respect to their response to defoliation. From experiments such as that illustrated in Plate 1 there is evidence that the main reaction

PLATE 1.   *Response of four species to clipping every two weeks at 5 cm. Left to right:* Lolium perenne *var.S23,* Holcus lanatus, Agrostis tenuis *and* Deschampsia flexuosa. *Photograph taken 7 days after clipping*

of *Lolium perenne* var. S23 to clipping consists of a rapid and almost vertical regrowth of the damaged leaves whereas, under the same treatments, *Agrostis tenuis* responds by producing a large number of small tillers and leaves, many of which are not projected into the clipped stratum but instead form a compact low sward which includes stolons capable of invading bare patches and infiltrating areas of short turf occupied by other species. From these results it is apparent that conditions allowing co-existence of species such as *Lolium perenne* var. S23 and *Agrostis tenuis* are likely to arise where grazing is intermittent and fluctuations in the height of the turf alternately favour erect and prostrate growth-forms.

5.6.3. Niches arising from long-term variation

Floristic diversity may be dependent also upon year-to-year variation in habitat conditions. Cyclical fluctuations in the ratio of *Lolium perenne* to *Trifolium repens* associated with changes in nitrogen status are a familiar example known to grassland ecologists and similar long-term effects correlated with fluctuations in rainfall have been reported (WATT 1960). One of the most detailed and interesting studies of the role of year-to-year variation in maintaining a mixture of species in herbaceous vegetation is that of MULLER & FOERSTER (1974).

6. REGENERATION BY SEED

Features of the seed and its germination play no part in the characterization of the primary strategies attempted in Table 1. This is not to deny the exist-ence of seed and seedling characteristics which maximize the chance of seedling establishment in habitats favourable to later stages in the life history of the plant. However it is evident that many of the forces of natural selection which determine the characteristics of the seed and seedling are quite different from those which fashion the established plant. This dislocation is in part the result of the relatively small size of seedlings, the majority of which are incapable of survival in the conditions tolerated by the established plant. It is also due to the fact that for the maintenance (and even for the expansion) of populations of many perennial species often only low rates of regeneration from seed may be required. In such cases, a species may remain as a viable constituent of vegetation in circumstances where opportunities for seedling establishment are of relatively infrequent occurrence within the habitat. It is hardly surprising, therefore, to find that many species possess mechanisms of seed production, dispersal and dormancy which, together with the germination requirements, facilitate establishment in local, small and temporary gaps in the vegetation cover. The dependence of many species upon gaps for successful regeneration has been established for a considerable period of time and recently a comprehensive review of evidence on this point has been assembled

(GRUBB 1977).

An attempt to classify seedling regeneration patterns would be premature. However, it is useful, at this early stage of synthesis to draw a distinction between species and populations which develop banks of buried seeds in the soil and those which do not.

At least, two major groups of plants may be recognized in which buried seeds play no part in regeneration. The first consists of a miscellany of herbs (including many *Umbelliferae*), shrubs and forest trees. In this group the seeds tend to be comparatively large and provide an initial source of energy and structural materials which is sufficiently large either to buffer the seedling during initial establishment beneath an elevated closed canopy (SALISBURY 1941) or to allow rapid penetration of the shoot through low herbaceous vegetation (GRIME & JEFFREY 1965).

The second group in which buried seeds are relatively unimportant includes many grasses and is composed of species occurring in habitats subjected to highly predictable disturbance by drought (e.g. *Festuca* spp., *Hordeum murinum*, *Bromus sterilis*). In habitats of this type the majority of the detached viable seeds occurs on the soil surface and seasonal sampling (Fig. 3a) usually reveals a strong annual fluctuation in the number of germinable seeds reaching a minimum during the wet season, at which time large crops of seedling colonize the areas of bare ground resulting from the preceding drought.

The best-documented examples of species which accumulate large reservoirs of seeds in the soil are the ruderals of arable fields (BARTON 1961). For these plants there is abundant evidence of the capacity of buried seeds to survive for long periods and to germinate in large numbers when the habitat is disturbed. The advantage of the seed bank is obvious in ruderal species of intermittently open habitats, since long-term survival of their populations depends upon the ability to persist through periods in which the habitat is occupied by a closed cover of perennial species. However, seed banks are not confined to the ruderals of arable land. Seasonal estimations of the germinable seed content of surface soil such as those illustrated in Fig. 3b and 3c reveal that many herbaceous perennials possess seed banks. Studies of the seed physiology of both the annuals and the perennials with seed banks (WESSON & WAREING 1969; KING 1975, 1976; GRIME & JARVIS 1975; THOMPSON *et al*. 1977) indicate that these species have mechanisms which (1) delay germination and facilitate burial in the soil and (2) restrict the germination of buried seeds to microsites in which the insulating effect of canopy, litter or humus has been locally disturbed.

FIG. 3a

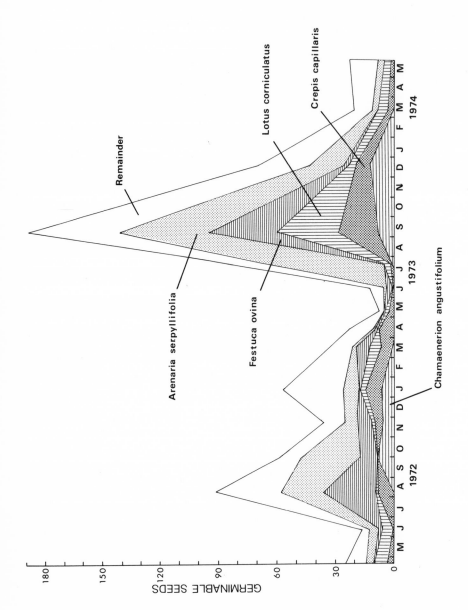

Remainder

Arenaria serpyllifolia

Festuca ovina

Lotus corniculatus

Crepis capillaris

Chamaenerion angustifolium

GERMINABLE SEEDS

180

150

120

90

60

30

0

M J J A S O N D J F M A M J J A S O N D J F M A M J F M A M

1972

1973

1974

For legend see p. 118

FIG. 3b

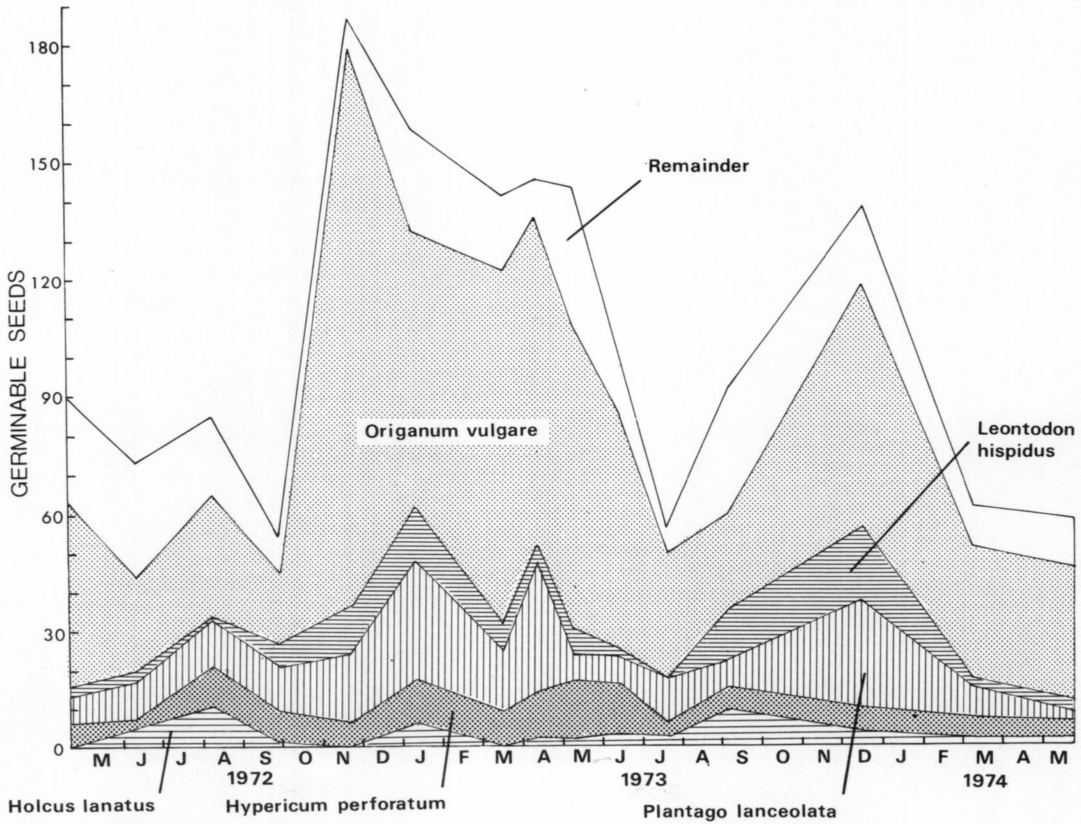
GERMINABLE SEEDS
0
30
60
90
120
150
180
Remainder
Origanum vulgare
Leontodon hispidus
M J J A S O N D J F M A M J J A S O N D J F M A M
1972
1973
1974
Holcus lanatus
Hypericum perforatum
Plantago lanceolata

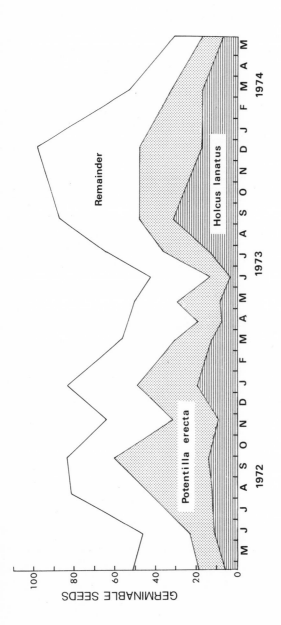

FIG. 3 (a–c). *Estimations of seasonal variation in the numbers of detached germinable seeds present in three herbaceous communities in North Derbyshire. The vertical axis in each figure is based upon the number of seedlings germinating during a standardized laboratory test on samples of soil (total volume 900 cm³, surface area 450 cm²) prepared by thorough mixing of sub-samples removed from the soil surface (0–3 cm) at random positions within the site (M.S. SPRAY & J.P. GRIME, unpublished data).*

a. Sparse discontinuous vegetation on south-facing slope of limestone quarry heap subject to summer drought (Tideswelldale).

b. Tall herb community with continuous vegetation cover but subject to occasional burning: no fires occurred during the period of the investigation (Tideswelldale).

c. Derelict limestone grassland on north-facing slope with continuous vegetation cover (Lathkilldale)

7. ACKNOWLEDGEMENTS

I should like to record my thanks to Dr. J.G. Hodgson, Mr. C. Sydes, Mr. S.B. Furness, Mr. K. Thompson and Dr. R. Law for most helpful discussions during the preparation of this paper. This work has been supported by the Natural Environment Research Council.

8. APPENDIX: DEFINITIONS

PRIMARY STRATEGIES may be defined as similar or analogous groupings of genetic characteristics which recur widely among species or populations and cause them to exhibit similarities in ecology.

COMPETITION between plants is the tendency of neighbours to utilize the same quantum of light, molecule of water, ion of a mineral nutrient or volume of space (GRIME 1973b). Competition is concerned with the acquisition of resources from crowded environments. Competition is merely one component of the struggle for existence and is but one of the three major routes to dominance.

STRESS consists of the external constraints which limit the rate of dry matter production of all or part of the vegetation, e.g. shortages of light, water, mineral nutrients and suboptimal temperatures. These shortages may be an inherent characteristic of the environment or they may be induced or intensified by the vegetation itself. The most widespread forms of plant-induced stress arise from shading and reduction in the levels of mineral nutrients in the soil resulting from their accumulation in the plant biomass. It is worth noting that this use of "stress" to describe an external constraint on dry matter production differs from that of many plant physiologists (e.g. LEVITT 1975) who have used the word to describe the physiological state of the plant.

DISTURBANCE consists of the mechanisms which limit the plant biomass by causing its partial or total destruction. Disturbance may affect either the living or dead components of the vegetation or both and arises from the activities of herbivores, pathogens, man (trampling, mowing, ploughing, etc.) and from phenomena such as wind-damage, frosting, desiccation, soil erosion, and fire. Certain environmental factors (e.g. drought, frost) are capable of inducing both stress and disturbance and often both effects occur in the same habitat. Whether factors such as drought result in disturbance or stress depends not so much upon their severity as upon the constancy of their occurrence. Strongly disruptive effects of climate upon vegetation coincide with marked seasonal changes in factors such as moisture supply.

9. REFERENCES

AL-MUFTI, M.M., WALL, C.L., FURNESS, S.B., GRIME, J.P. & S.R. BAND, 1977 - A quantitative analysis of shoot phenology and dominance in herbaceous vegetation. *J. Ecol.*, 65 (in press).

BAKER, H.G., 1965 - Characteristics and modes of origin of weeds. *In:* H.G. BAKER & G.L. STEBBINS (Editors), *The genetics of colonising species*, Acad. Press, New York, p. 588.

BARTON, L.V., 1961 - *Seed preservation and longevity*. Hill, London, 216 p.

BLACK, J.M., 1958 - Competition between plants of different initial seed sizes in swards of subterranean clover (*Trifolium subterraneum* L.) with particular reference to leaf area and the light micro climate. *Aust. J. agric. Res.*, 9, 299-312.

BORDEAU, P.F. & M.L. LAVERICK, 1958 - Tolerance and photosynthetic adaptability to light intensity in White pine, Red pine, Hemlock and *Ailanthus* seedlings. *Forest Sci.*, 4, 196-207.

BOYSEN-JENSEN, P., 1929 - Studier over Skovtracerres Forhold til Lyset. *Dansk. Skovforen. Tidsskr.*, 14, 5-31.

CLEMENTS, F.E., 1916 - *Plant succession. An analysis of the development of vegetation*. Carnegie Inst., Washington, 512 p.

DUFFEY, E., MORRIS, E.G., SHEAIL, J., WARD, L.K., WELLS, D.A. & T.C.E. WELLS, 1974 - *Grassland ecology and wildlife management*. Chapman & Hall, London, 281 p.

GRIME, J.P., 1966 - Shade avoidance and shade tolerance in flowering plants. *In:* R. BAINBRIDGE, G.C. EVANS & O. RACKHAM (Editors), *Light as an ecological factor*, Blackwell Scientific Publications, Oxford, p. 281-301.

GRIME, J.P., 1973a - Competitive exclusion in herbaceous vegetation. *Nature*, 242, 344-347.

GRIME, J.P., 1973b - Competition and diversity in herbaceous vegetation - a reply. *Nature*, 244, 310-311.

GRIME, J.P., 1974 - Vegetation classification by reference to strategies. *Nature*, 250, 26-31.

GRIME, J.P., 1977 - Evidence for the existence of three primary strategies in plants and its relevance to ecological and evolutionary theory. *Amer. Natur.*, 107 (in press).

GRIME, J.P. & R. HUNT, 1975 - Relative growth-rate: its range and adaptive significance in a local flora. *J. Ecol.*, 63, 393-422.

GRIME, J.P. & B.C. JARVIS, 1975 - Shade avoidance and shade tolerance in flowering plants. II. Effects of light on the germination of species of contrasted ecology. *In:* G.C. EVANS, R. BAINBRIDGE & O. RACKHAM (Editors), *Light as an ecological factor II*. (Bes Symposium no. 16). Blackwell Scientific Publications, Oxford, p. 525-532.

GRIME, J.P. & D.W. JEFFREY, 1965 - Seedling establishment in vertical gradients of sunlight. *J. Ecol.*, 53, 621-642.

GRUBB, P.J., 1976 - A theoretical background to the conservation of ecologically distinct groups of annuals and biennials in the chalk grassland ecosystem. *Biol. Conserv.*, 10, 53-76.

GRUBB, P.J., 1977 - The maintenance of species-richness in plant communities: the importance of the regeneration niche. *Biol. Rev.*, 52, 107-145.

KING, T.J., 1975 - Inhibition of seed germination under leaf canopies in *Arenaria serpyllifolia, Veronica arvensis* and *Cerastium holosteoides*. *New Phytol.*, 75, 87-90.

KING, T.J., 1976 - The viable seed contents of anthill and pasture soil. *New Phytol.*, 77, 143-147.

LEVITT, J., 1975 - *Responses of plants to environmental stresses*. Academic Press, New York, 697 p.

LOACH, K., 1970 - Shade tolerance in tree seedlings. II. Growth analysis of plants raised under artificial shade. *New Phytol.*, 69, 273-286.

MAHMOUD, A., 1973 - A laboratory approach to ecological studies of the grasses: *Arrhenatherum elatius* (L.) Beauv. ex. J. and C. Presl., *Agrostis tenuis* Sibth. and *Festuca ovina* L. *Ph. D. thesis*, University of Sheffield.

MAHMOUD, A. & J.P. GRIME, 1976 - An analysis of competitive ability in three perennial grasses. *New Phytol.*, 77, 431-435.

MASON, G. & J.P. GRIME, 1975 - *Annual Report, Unit of Comparative Plant Ecology (NERC)*, University of Sheffield, p. 10-11.

MILTON, W.E.J., 1940 - The effect of manuring, grazing and cutting on the yield, botanical and chemical composition of natural hill pastures. *J. Ecol.*, 28, 328-356.

MÜLLER, G. & E. FOERSTER, 1974 - Entwicklung von Weideansaaten im Überflutungs-bereich des Rheines bei Kleve. *Z. Acker- und Pflanzenbau*, 140, 161-174.

SALISBURY, E.J., 1941 - *Reproductive capacity of plants*. Bell, London, 244 p.

SMITH, C.J., ELSTON, J. & A.H. BUNTING, 1971 - The effects of cutting and fertilizer treatments on the yield and botanical composition of chalk turf. *J. Br. Grassld. Soc.*, 26, 213-220.

THOMPSON, K., GRIME, J.P. & G. MASON, 1977 - Seed germination in response to diurnal fluctuations of temperature. *Nature*, 267, 147-149.

THURSTON, J.M., 1969 - The effect of liming and fertilizers on the botanical composition of permanent grassland, and on the yield of hay. *In:* I.H. RORISON (Editor), *Ecological aspects of the mineral nutrition of plants*. (BES Symposium no. 9), Blackwell Scientific Publications, Oxford, p. 3-10.

WALTER, H., 1973 - *Vegetation of the earth in relation to climate and the ecophysiological conditions*. English Universities Press, London, 237 p.

WATT, A.S., 1955 - Bracken versus heather, a study in plant sociology. *J. Ecol.*, 43, 490-506.

WATT, A.S., 1957 - The effect of excluding rabbits from grassland B (*Mesobrometum*) in Breckland. *J. Ecol.*, 45, 861-878.

WATT, A.S., 1960 - Population changes in acidophilous grass-heath in Breckland, 1936-1957. *J. Ecol.*, 48, 605-629.

WESSON, G. & P.F. WAREING, 1969 - The induction of light sensitivity in weed seeds by burial. *J. exp. Bot.*, 20, 414-425.

10. DISCUSSION

BAZZAZ (Illinois): I have a question about your productivity corridor in relation to diversity. Some of the most diverse systems have a much higher productivity than your corridor. How, for example, do you explain the high diversity in some North American prairies that produce 1,200 up to 1,500 grams per square metre.

GRIME: I think there is great danger for most of the larger areas, because one is then including a great deal of diversity of environment. I think there is a lot to be said for considering species density at a fairly small unit if our main concern is with the interactions between species, I would not for a moment deny that one might have very high diversity in a place in which the average standing crop and litter was very high. But that might simply be because in one corner there is a very improductive area which contains most of the species.

BAZZAZ: Okay. But the other end of the scale: sometimes in some desert habitats when there is a good rainy year you get tremendous amounts of new species and the productivity would be very low. So you still have high diversity but low productivity on the other sides of the corridor too.

GRIME: Our experience is that one just about gets into this corridor in that sort of circumstances using the measurements that we have been applying.

HARPER (Bangor): I think there are risks of semantics and misunderstandings about the terms competition and stress. One must aim for operational terms so that you can recognize situations in the field and define, preferably empirically, what you are dealing with. You can get into an awful mess in the case of competition, because of the different uses. You can define an intensely competitive situation as one in which one species eliminates another rapidly, and one can in fact define intensity of competition by the differential, the rate at which A gives way to B. But there is a second view, which I suppose is more strictly Darwinian, which would say that you recognize a very intense competition as one in which the two components are so balanced that the rate at which one wins against the other is exceedingly slow. I think these represent two extreme situations, but both of them will be referred to as representing intense competition. I would like to hear from Dr. GRIME his operational definition of competition and the same applies to the term stress.

GRIME: In the case of stress I would simply offer the definition given in this paper: it simply consists of the external factors which restrict production. This seems to be a very straightforward definition.

HARPER: It was straightforward to me until you talked about a species with high productivity in a stressed environment.

GRIME: *Brachypodium pinnatum* is a plant one finds in habitats of intermediate productivity; that is to say, there is a degree of stress. The plants which survive there may well be reasonably efficient to capturing resources, but their survival also involves withstanding degrees of stress, that is, conditions of rather limited resource availability.

HARPER: That is a switch, because now stress becomes "conditions of low resource availability". You say in a stressed environment one must define stress as being recognized by low productivity. Now in the case of *Brachypodium pinnatum* one cannot say it has high productivity in a stressed environment.

GRIME: I have not suggested that it has a high productivity, merely that it gets slightly above the corridor (800-900 grams per square metre). Could I go back to competition? What I am suggesting is that plants which are the most efficient competitors for light are also the most efficient competitors for water and mineral nutrients. And the reason why I believe there is this correlation is that one cannot have efficient competition for light unless one has a fairly large continuous input of water and mineral nutrients, because inherent in high competitive ability is the ability to display high phenotypic plasticity in the mobilization of the photosynthates and in the growth responses.

HARPER: We must surely have an operational means for defining competition in the field.

GRIME: Before we can do that we need a lot more information, information which physiologists are capable of providing.

MINDERMAN (Arnhem): Did I understand you correctly that the ability of trees to give rise to much litter, classifies trees as competitive species?

GRIME: No, competition is primarily involved in the capture of resources. The imposition of stress through becoming rather large and therefore being capable of imposing stress, is a separate component of dominance.

BARKMAN (Wijster): I think when people talk about stress they mean external pressures which affect our well being. I doubt whether stress in plants can be defined as those factors which reduce primary production, because I wonder whether maximum primary production is a sign of maximum well being. Many chalk grassland plants will grow more vigourously if you abolish nitrogen and phosphorus shortage, but they will no longer flower. We can ask whether this is really a stressless situation. I think the way you define stress is a bit misleading. Your low disturbance-low stress situations cannot persist for a long time. If you deprive places of their vegetation and there is little disturbance after that, the vegetation will grow into a close stand and the result will be a lot of competition and a lot of stress. Your high disturbance-high stress situation is equally impossible except temporarily.

GRIME: Yes, it refers to the immediate situation, not to long-term implications.

BARKMAN: Why do you restrict the r-type to annual herbs? Would you think that a tree like *Sambucus nigra,* which grows fast on much disturbed stations where there is low stress in not very extreme habitats, is a typical r-species?

GRIME: It is not an r as an arable weed. We certainly do not find it in persistently disturbed environments. We find it established as a tree in an environment which was severely disturbed 5 to 10 years ago. We are in a rather different phase of vegetation development by that time. The impact of disturbance is less important.

VAN DER LAAN (Oostvoorne): Why did not you mention in your list of characteristics of the various groups of plants the response to disturbance? Could you give a brief characterization?

GRIME: Obviously, ruderal plants have the best chance of survival in the long term under severe constant disturbance, simply because their life cycle is short enough to fit either between successive disturbances or between disturbance and the re-establishment of perennial vegetation, and the main way in which this seems to be done, apart from a short life history, is in the development of large reservoirs of seeds. The only circumstances in which we do not find the large reservoir of seeds are where you have very regular disturbance imposed by climate or where the germination physiology is timed apparently to coincide with the annual opening in the habitat by drought. When you come to the effect of disturbance on the competitive plants and the stress tolerators, clearly they are going to be very vulnerable to disturbance simply because they want to achieve a reproductive age by the time they may well be destroyed. The only other point about disturbance is that obviously many plants which as mature established specimens can survive in crowded environments, are unable to survive these environments in the seedling phase. And I believe that a very important contribution in the last few years has been GRUBB's paper in *Biological Reviews* on the regeneration niche, where he points out that the plants have just the same difficulties in getting into

this world as the animals do and very often we see perennial or even stress-
tolerant plants undergoing the brief almost ruderal phase in the first week
or two after germination and they appear to depend very often upon a gap.
So there is a particular role of disturbance in the reproductive biology of
almost all plants and therefore I would say that we have only to use
characteristics of the established plants to draw out distinctions between
the major adaptive strategies. There is a kind of uncoupling between the
reproductive biology and the germinative biology and the biology of strategic
development in the later stage.

Coexistence of plant species by niche differentiation

1. INTRODUCTION

For many years biologists have been interested in the frequently observed phenomenon that a large number of species can continue to occur within a relatively small area for a long time. It is only during the last ten years that workers from other disciplines have also been actively engaged in this question, which has been much to the benefit of the developed theories.

GRUBB (1977) very recently wrote a comprehensive review on the mechanisms that may cause species-richness in plant communities, so we can confine ourselves to the following very simple scheme. In trying to explain how species can coexist two main groups of suppositions can be distinguished: (1) suppositions associated with a heterogeneous environment, and (2) those based on a homogeneous environment.

For the heterogeneous environment a distinction can be made between theories which consider: species that cannot grow in each other's micro-habitat, and species that can grow in each other's micro-habitat but, nevertheless exclude each other because each species has the greatest competitive ability in its own micro-habitat. In essence these "mixtures" consist of many micro-mono-cultures, therefore the term "apparent coexistence" is preferred here. This type of coexistence has attracted much attention already (see HARPER 1961; THOMAS & DALE 1976; HICKMAN 1977; GRUBB 1977; WERNER & PLATT 1977).

For the homogeneous environment, some functional niche differentiation in the sense of ELTON (1927) must be supposed to occur. In this case the theories depend on species being limited by: one or more factors in different periods, and/or different factors simultaneously. In this sort of environment mixtures are involved in which "true coexistence" may occur.

Within the scope of this paper, we shall only treat this functional niche differentiation within a homogeneous environment. In our opinion an ecological stabilizing mechanism must underly the natural species-richness, which makes that an incidental decrease of the population density is followed up by an increase in reproductive rate. It is very unlikely that species-richness is maintained by accidental events, like climatic fluctuations, which cause changes in the competitive relations between the populations.

In experimental research on niche differentiation it may be very advantageous to grow both monocultures and mixtures under conditions which are exactly the same except for the competitive situation, since this permits determination of the Relative Yield Total (RYT). This parameter, introduced by DE WIT & VAN DEN BERGH (1965) may be used to indicate the degree of niche differentiation.

2. THE RYT CONCEPT

2.1. SPACING

To show the usefulness of the RYT first a spacing experiment will be described. Fig. 1 shows that the individuals of the <u>same</u> plant species begin to interact as soon as yield per unit area is no longer proportional to plant density.

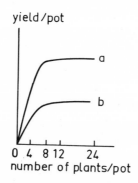

FIG. 1. *Spacing curves of species a and b*

The dependence of yield on plant density is linear up till 6 plants of species a per pot, which means that the per-plant weights are constant. There is a diminishing effect of density on yield per pot as density increases per-plant weights decrease due to competition. With 8 plants per pot the maximum yield is approached and, at higher densities, any doubling of the number of plants leads to a halving of per-plant weights. The same holds for species b; the only difference is that the per-plant weights of species b are smaller.

2.2. REPLACEMENT SERIES

The interference of individuals of <u>different</u> plant species is often studied in replacement series. In these, the yields of the species in the mixtures are compared with their yields in monoculture, on the understanding that the total sowing density of the mixed and monocultures are equal. In these replacement diagrams the yields of the separate species can be plotted against their planting frequencies.

Assuming that the species do not interfere with each other, the curves in a replacement diagram of a mixture experiment of species a and b at a total density of 24 plants per pot (Fig. 2) will be the same as those of the spacing

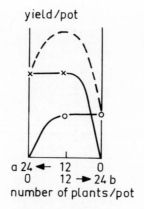

FIG. 2.   *Replacement diagram of species a and b*
without interspecific interference
(see text)
     X   *yields of species a per pot*
     O   *yields of species b per pot*
  – – – *total yields per pot*

experiment of Fig. 1. In this case competition occurs only within the species and the total yield of the 12/12 mixture approaches the sum of the yields of both monocultures. Although the species grow close enough to affect each other, there is no interference, because they grow in entirely different niches.

2.3. RELATIVE YIELDS AND RELATIVE YIELD TOTAL (RYT)

For a better indication of the degree of overlap of the niches of the species, the absolute yields can be converted into dimensionless relative yields. The relative yield (r) of a species is the quotient of its yield in the mixture (O) and of its yield in the monoculture (M).

In Fig. 3a the relative yields of the 1/1 mixture are both approaching 1 and the sum of the relative yields (Relative Yield Total = RYT) approaches 2. The growth processes leading to these curves in this diagram are completely independent. Going from diagram a to d, however, the growth processes represented by the curves become more and more dependent on the interference between the species. In diagram b both curves are still convex, but to a less extent than in the preceding case. In diagram c the convex curve is partly compensated for by a concave curve and finally in diagram d, the one curve is convex to the same extent as the other is concave. In this order the niches in which the species grow overlap more and more; this is also shown by the RYT value

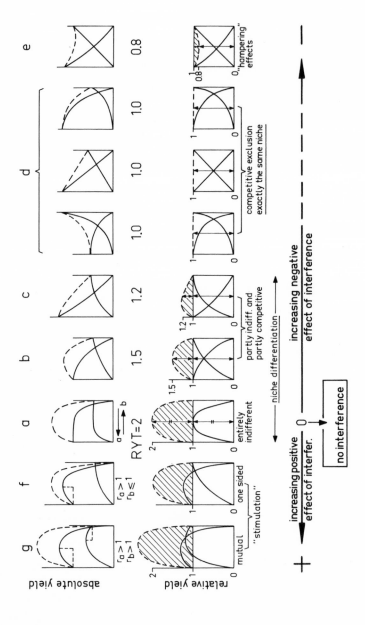

FIG. 3. *Ways of interference (for explanation see text)*

128

decreasing from 2 to 1. When RYT = 1 the species occupy exactly the same niche (diagram d). Ultimately the less agressive species will be crowded out by the more aggressive species in the mixture. Where RYT > 1, on the other hand, the species are occupying partly different niches and coexistence may occur.

When the relative yield of species a (r_a) is greater than 1, there is a stimulating effect of species b on a (Fig. 3f, for example supporting plant and climber, host plant and parasite, legume and non-legume, etc.) and when r_b > 1 also, the stimulating effect is mutual (Fig. 3g). In all these cases the RYT values may be even greater than 2 (RAININKO 1968).

Finally RYT < 1 (Fig. 3e) may indicate hampering effects. This low RYT value may be caused by excretion of toxic substances or by a disease by which the carrier species is not damaged, but the neighbouring species is affected (SANDFAER 1970).

3. CONDITIONS FOR EQUILIBRIUM

The curves in the replacement diagrams of many competition experiments may be described by the hyperbolic functions generated by the model of DE WIT (1960):

$$O_a = \frac{k_{ab} \cdot z_a}{k_{ab} \cdot z_a + z_b} M_a \text{ and } O_b = \frac{k_{ba} \cdot z_b}{k_{ba} \cdot z_b + z_a} M_b$$

O_a and M_a represent the yields of species a in mixture and monoculture respectively, z_a and z_b the relative planting frequencies, and k_{ab} and k_{ba} the relative crowding coefficients of species a with respect to species b and *vice versa*. In the model the k-values determine the degree of curvature of the curves: k > 1 results in a convex curve, k = 1 gives a straight line and k < 1 a concave curve. If k_{ab} is the reciprocal of k_{ba} and therefore their product is equal to 1, the species are supposed to occupy the same niche ("crowd for the same space" in DE WIT's terms), whereas with niche differentiation the product of the k-values is expected to be greater than 1 (which also implies a RYT > 1). RYT > 1 indicates niche differentiation between two species, but does not guarantee that the species will attain a stable equilibrium.

To study the changes in botanical composition of mixtures through time the relative reproductive rate α has been introduced, i.e. the quotient of the proportions in which the species are represented in the mixture at time (t+1) and time t.

$$\alpha_{ab} = \frac{(O_a/O_b)_{t+1}}{(O_a/O_b)_t} \text{ or } \log (O_a/O_b)_{t+1} = \log (O_a/O_b)_t + \log \alpha_{ab}$$

By plotting the log's of these proportions for different mixtures in a ratio diagram (Fig. 4), it is shown in some experiments that α varies with planting frequency.

According to the model, if $k_{ab} \times k_{ba}$ > 1 α indeed depends on planting frequency (DE WIT 1960). With increasing frequencies of species a, α decreases, which means that species a is becoming less aggressive with respect to species b.

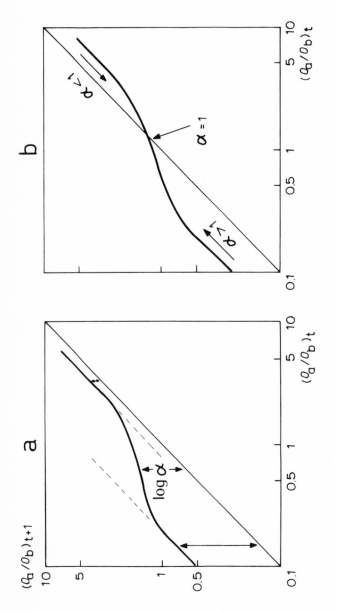

FIG. 4.   *Ratio diagrams in which the ratio of the yields of species a and b in a mixture*
at time (t+1) is plotted against this ratio at time t.
(a) species a replaces species b at all frequencies;
(b) at high frequencies of species b, species a replaces species b; at high
*   frequencies of species a, species b replaces species a; a stable equilibrium*
*   is reached at the intersection with the diagonal (α = 1)*

However, in the example of Fig. 4a, α remains greater than 1, hence finally species b will be crowded out by species a, in spite of the frequency dependence of α.

In Fig. 4b also the aggressiveness of species a with respect to species b decreases with increasing frequency of species a in the mixture, but in this example the curve intersects the diagonal. This means that to the left of this intersection species a wins and to the right species b wins but that at the intersection the species are in a stable equilibrium (α = 1). BRAAKHEKKE (in preparation) has shown that the quotient of the monoculture yields determines whether the diagonal is intersected and, therefore, whether a stable equilibrium is possible. This condition is formulated by the following equation:

$$k_{ba} > M_a/M_b > {}^1\!/k_{ab}$$

4. A FIELD EXPERIMENT WITH MULTI SPECIES MIXTURES

The seeds of 13 species were collected on an old extensively grazed pasture: 7 grass species and 6 herbs. 3.5 m^2 plots were sown as monocultures as well as mixtures according to the replacement principle, with a total density of 1600 viable seeds/m^2. An area of 90 x 90 cm^2 was harvested in each plot once a year and the botanical composition analysed and dry weight determined.

Table 1 shows the relative yields of the species in the first and the second year in the various mixtures. 100/0 means that the mixture consists of grass species only; 75/25 means that 75% of the number of seeds were from grasses and 25% were from herbs etc.; 0/100 means that the mixture consists of herbs only. Two herbs *Prunella vulgaris* and *Ranunculus repens* disappeared already in the first year, due to very slow establishment. *Lolium perenne* was by far the most aggressive species with very high relative yields.

When looking at the sum of the relative yields, we see that the RYT value of the mixture involving grass species only is about equal to 1, which means that the species occupy the same niche. With the introduction of herbs in the mixtures the RYT values are greater than 1. Obviously, the grasses and the herbs show some niche differentiation. Finally the herbs among themselves show the greatest niche differentiation: RYT = 1.41.

In the second year *Lolium perenne* has almost disappeared, probably because of no fertilizers were given (compare also the total absolute yields of the mixtures at the bottom of Table 1). Instead *Agrostis tenuis* has taken over the dominating position, whereas *Hordeum secalinum* and *Centaurea pratensis* have disappeared entirely. Here, too, the RYT values increase from about 1 to 1.45 when going from pure grass mixtures to pure herb mixtures. In the second year in the mixtures with herbs only the sum of the relative yields of *Chrysanthemum leucanthemum* and *Plantago lanceolata* even on their own is greater than 1, namely 1.26. Hence, these two species show remarkable evidence of niche differentiation and therefore we chose them for further experimentation.

TABLE 1. *Relative yields, RYT and absolute yields of different mixtures in a field experiment in the first and second year (for explanation see text)*

Seed ratio grasses/herbs	1e year					2e year				
	100/0	75/25	50/50	25/75	0/100	100/0	75/25	50/50	25/75	0/100
Lolium perenne	0.68	0.77	0.64	0.39		0.05	0.01	0.02	0.01	
Agrostis tenuis	0.12	0.13	0.18	0.17		0.64	0.38	0.34	0.45	
Anthoxanthum odoratum	0.05	0.05	0.06	0.05		0.18	0.13	0.15	0.11	
Trisetum flavescens	0.04	0.05	0.06	0.05		0.04	0.06	0.04	0.01	
Hordeum secalinum	0.11	0.07	0.11	0.05		–	–	–	–	
Festuca rubra	0.01	0.01	0.02	0.03		0.01	†	0.02	0.04	
Cynosurus cristatus	0.01	0.01	0.01	0.02		0.01	0.02	0.01	†	
Chrysanthemum leucanthemum		0.03	0.06	0.10	0.40		0.29	0.22	0.18	0.82
Plantago lanceolata		0.06	0.13	0.20	0.45		0.22	0.28	0.31	0.44
Rumex acetosa		0.06	0.09	0.25	0.53		0.03	0.06	0.07	0.19
Centaurea pratensis		†	0.01	†	0.03		–	–	–	–
RYT	1.01	1.24	1.37	1.31	1.41	0.93	1.14	1.14	1.18	1.45
Yield (g.d.m./0.8 m^2)	245	301	345	316	434	73	93	87	94	114

5. GROWTH LIMITED BY COMPETITION FOR LIGHT

First of all a possible niche differentiation with regard to light interception has been investigated. Competition experiments between *Chrysanthemum leucanthemum* and *Plantago lanceolata* have been carried out by BERENDSE (in preparation) in a growthroom under optimum nutrient conditions and at two light intensities. Light was regarded to be the only factor which could limit growth of the species.

The RYT values at both light intensities were systematically somewhat greater than 1 (1.05 and 1.08), which points to a slight degree of niche differentiation. The taller *Plantago lanceolata* leaves were erect and the leaves of the shorter *Chrysanthemum leucanthemum* plants were prostrate. Simulations of daily photosynthesis carried out by TRENBATH (1974) showed that the RYT values which may be expected from this favourable leaf arrangement are only about 1.09.

It may be concluded that in grassland vegetation niche differentiation with respect to the light factor is of little importance.

6. GROWTH LIMITED BY COMPETITION FOR MINERALS

The question now arises whether differences in underground behaviour of *Chrysanthemum leucanthemum* and *Plantago lanceolata* may cause niche differentiation. In the field experiment no differences between rooting depths of these species could be observed, hence the occupation of different soil volumes is unlikely. On the other hand it is wellknown that the behaviour of plant species with regard to minerals may vary considerably.

HARPER (1961) already pointed out that two species can coexist only when they are controlled by different factors. This author mentioned the difference in nitrogen source between legumes and grasses. BRADSHAW (1969) argued that not only qualitative but also quantitative differences in mineral requirements can enable species to coexist. In 1970 LEVIN gave this suggestion a theoretical basis.

Recently this possibility has caught the attention of some phytoplankton research workers (STEWART & LEVIN 1973; PETERSEN 1975; TITMAN 1976). As far as we know no studies have been devoted to the possibility of differential limitation by minerals in relation to terrestrial plants. GRUBB (1977) passed it off with only a few words.

It is easy to see that a mixture can produce more than monocultures in the fictitious case that each monoculture does not take up the very nutrient that limits the other monoculture. Consequently RYT > 1 and an equilibrium may be possible. This argument has been worked out in a more sophisticated manner by BRAAKHEKKE (in preparation). He has considered two double quotients: that of the minimum concentration (m) of two minerals 1 and 2 in two species a and b and that of the amounts of these minerals taken up (U). The minimum concentration of nutrient 1 in species a is reached at the moment that growth stops because of deficiency of this nutrient, everything else being not limiting. The uptake rates are not considered explicitly in the model but instead the results of the uptake processes; U_a^1 means the amount of nutrient 1 taken up by species a grown in a mixture. Under conditions of limited supply of these nutrients the RYT can be greater than 1, if the following equation holds:

$$\frac{m_a^1/m_a^2}{m_b^1/m_b^2} > \frac{U_a^1/U_a^2}{U_b^1/U_b^2} > 1$$

In addition another condition must be fulfilled. The ratio of the limiting nutrients in the substrate has to be in between certain limits. These depend on the minimum contents in the plants, the uptake abilities and the planting frequency (BRAAKHEKKE, in preparation).

In an attempt to test whether the *Plantago* (P) and *Chrysanthemum* (C) species refered to earlier fulfil the first condition, two experiments were performed using K and Ca, because it was known that these species often differ to some extent with respect to the contents of these nutrients.

In the first experiment, to determine the minimum K and Ca contents, the two species were separately grown on a series of 10 nutrient solutions with an increasing K and a decreasing Ca concentration (Table 2); the other nutrients

TABLE 2.   *K and Ca concentrations (me/pot) of nutrient solutions in an experiment with* Plantago lanceolata *and* Chrysanthemum leucanthemum. *The other nutrients were added according to ½ Hoagland solution*

Solutions	I	II	III	IV	V	VI	VII	VIII	IX	X
KNO_3	0.10	0.15	0.20	0.25	0.30	0.35	0.40	0.45	0.50	0.55
$Ca(NO_3)_2$	0.50	0.45	0.40	0.35	0.30	0.25	0.20	0.15	0.10	0.05

were abundant. The solutions were not renewed and the plants were harvested when the growth ceased on most of the solutions due to K or Ca deficiency.

Fig. 5 shows the relation between K and Ca concentration in the whole plants. On solution X a minimum concentration of Ca is reached in the plants whereas K is accumulated and on solution I a minimum concentration of K is reached with

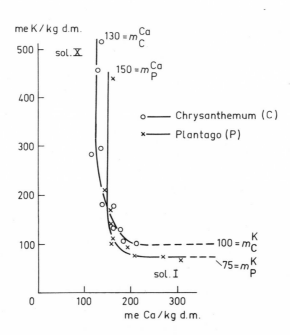

FIG. 5.   *The K concentrations of* Chrysanthemum leucanthemum *and* Plantago lanceolata *plotted against their Ca contents when grown in a culture experiment with 10 combinations of K and Ca*

Ca accumulated to a less extent. The two curves show that the species differ in their minimum contents. These values may be used in the first part of the above-mentioned equation:

$$\frac{m_P^{Ca}/m_P^K}{m_C^{Ca}/m_C^K} = \frac{150/75}{130/100} = 1.5$$

According to this result a RYT value > 1 for *Chrysanthemum leucanthemum* and *Plantago lanceolata* may occur.

In the second experiment, to determine the double quotients of the amounts of these nutrients taken up by the plants growing in mixture, the two species were grown together in replacement series at low K or low Ca levels, and at low levels of both nutrients.

The value of this double quotient varied between 2 and 3, and hence does not lay in the narrow range between 1 and the value of the double quotient of the minimum contents 1.5. This means that the other part of our first condition is not satisfied.

The RYT values of this experiment appeared to vary between 1.00 and 1.23, which suggests some niche differentiation, that is still unexplained. Anyhow these values are too small to explain the result of the field experiment. It has still to be evaluated if other combinations of nutrient shortages may do better.

7. SOME CONCLUDING REMARKS

In spite of the above-mentioned result, the model just discussed may help us to understand equilibria of plant species in environments which are supposed to be homogeneous. For every plant combination the ratio's of the nutrient concentrations in the substrate have to meet specific requirements to allow this kind of equilibrium. When the substrate is changed for example by fertilizing, the frequencies of the species at equilibrium will change, provided the change of the substrate is within certain limits mentioned above. Larger changes in the substrate destroy the possibility of an equilibrium; they are followed by the extinction of one species.

Raising the concentrations in the substrate without changing the ratio between the nutrients, will disturb the equilibrium also, because the limiting effect of the nutrients disappears and other factors such as light become decisive.

To what extent fluctuations of the nutrient concentrations in the substrate may disturb the equilibrium, depends on the amplitude and the period of the fluctuation in relation to the rate of change of the frequency ratio between the species. This offers a way to describe the rather vague concept of environmental dynamics and to indicate how it affects the species richness of the vegetation. Models like this one can deepen our insight into the relation between diversity and stability and into the problem of eutrophication.

8. ACKNOWLEDGEMENTS

The authors are much indebted to Prof. Dr. Ir. C.T. de Wit for the stimulating discussions, to Mr. W.Th. Elberse for his continuous support in so many ways and to Dr. B.R. Trenbath for criticizing the text.

The investigations were supported in part by the Foundation for Fundamental Biological Research (BION), which is subsidized by the Netherlands Organization for the Advancement of Pure Research (ZWO).

9. REFERENCES

BERENDSE, F. - Competition for light between populations of *Plantago lanceolata* L. and *Chrysanthemum leucanthemum* L. (in preparation).

BERGH, J.P. VAN DEN & W.Th. ELBERSE, 1975 - Degree of interference between species in complicated mixtures. XII. *Intern. Bot. Congr., Leningrad, Abstr.*, Vol. 1, p. 138.

BRADSHAW, A.D., 1969 - An ecologist's viewpoint. *In*: I.H. RORISON (Editor), *Ecological aspects of the mineral nutrition of plants*. Blackwell, Oxford, p. 415-427.

ELTON, C.S., 1927 - *Animal ecology*. Sidgwich & Jackson, London, 209 p.

GRUBB, P.J., 1977 - The maintenance of species-richness in plant communities: the importance of the regeneration niche. *Biol. Rev.*, 52, 107-145.

HALL, R.L., 1974 - Analysis of the nature of interference between plants of different species. II. Nutrient relations in a Nandi *Setaria* and Greenleaf *Desmodium* Association with particular reference to potassium. *Austr. J. agric. Res.*, 25, 749-756.

HARPER, J.L., 1961 - Approaches to the study of plant competition. *In*: F.L. MILTHORPE (Editor), *Mechanisms in biological competition*. University Press, Cambridge, p. 1-39.

HICKMAN, J.C., 1977 - Energy allocation and niche differentiation in four co-existing annual species of *Polygonum* in Western North America. *J. Ecol.*, 65, 317-326.

LEVIN, S.A., 1970 - Community equilibria and stability, and an extension of the competitive exclusion principle. *Amer. Natur.*, 104, No. 939, 413-423.

PETERSEN, R., 1975 - The paradox of the plankton: an equilibrium hypothesis. *Amer. Natur.*, 109, No. 965, 35-49.

RAININKO, K., 1968 - The effects of nitrogen fertilization, irrigation and number of harvestings upon leys established with various seed mixtures. *Suom. maatal. Seur. Julk.*, 112, 1-137.

SANDFAER, J., 1970 - An analysis of competition between some Barley varieties. *Danish Atomic Energy Comm., Risö Rep.*, 230, 1-114.

STEWART, T.M. & B.R. LEVIN, 1973 - Partitioning of resources and the outcome of interspecific competition: a model and some general considerations. *Amer. Natur.*, 107, No. 171, 171-198.

THOMAS, A.G. & H.M. DALE, 1976 - Cohabitation of three *Hieracium* species in relation to the spatial heterogeneity in an old pasture. *Can. J. Bot.*, 54, 2517-2529.

TITMAN, D., 1976 - Ecological competition between algae: experimental confirmation of resource-based competition theory. *Science*, 192, No. 4238, 463-465.

TRENBATH, B.R., 1974 - Biomass productivity of mixtures. *Adv. Agron.*, 26, 177-210.

WERNER, P.A. & W.J. PLATT, 1976 - Ecological relationships of co-occurring goldenrods (*Solidago: Compositae*). *Amer. Natur.*, 110, No. 976, 959-971.

WIT, C.T. DE, 1960 - On competition. *Versl. Landbouwk. Onderz.*, 66.8.

WIT, C.T. DE & J.P. VAN DEN BERGH, 1965 - Competition between herbage plants. *Neth. J. agric. Sci.*, 13, 212-221.

10. DISCUSSION

<u>WHITE</u> (Dublin): How does RYT vary with the biomass production? The production of the mixtures with herbs only in the first year is four times higher than in the second year, and still in both years RYT is about equal to 1.4.

<u>VAN DEN BERGH</u>: There is, I suppose, no direct relationship between RYT and biomass production. In general, one could say that with high production levels under optimal conditions, the RYT tends to be equal to 1, because only light is growth limiting and there are few possibilities for niche differentiation in light use. With low production levels, say less than 6 ton $ha^{-1}year^{-1}$, the possibilities of niche differentiation seem to be far greater, because many factors may limit different species in their growth. All the yields in this experiment, including those of the first year, were less than 6 ton $ha^{-1}year^{-1}$.

<u>HARPER</u> (Bangor): One interesting consequence of this model counting for diversity within an otherwise homogeneous environment is that the various species that are persisting together will each of them be suffering from deficiency of a different nutrient component. This might enable us to detect this happening in the field by studying deficiency symptoms in wild plants. This would be an enormously interesting field operation, because one just tended historically to regard a habitat as a potash deficient one or a nitrogen deficient one, but on your interpretation a habitat will be deficient in something different for each of the species in equilibrium. I wonder whether there is field evidence yet that might suggest this sort of effect?

<u>VAN DEN BERGH</u>: This is a very important aspect of our research. Deficiency symptoms in fast-growing crops are well known, but under marginal condition with slowly growing vegetation the same deficiency might have an entirely different appearance. This means that we do not even know what symptoms to look for. Besides discolourings, deviating growth forms might be of greater importance for the detection of the kind of deficiency that occurs in natural communities.

<u>GRIME</u> (Sheffield): I think there is a danger in work of this kind when one moves on to a question of mechanism. The idea that different plants require different quantities of particular elements may be true, but it may surely also be a question of the spending of the nutrient when it is got actually into the plant. Don't you think that in one case the plant is almost immediately investing the nutrient in growth and in the other case we have what I would like to call a stress tolerator, which is accumulating the nutrient perhaps for future occasions?

<u>VAN DEN BERGH</u>: It is hard to translate your question into our observations. You are talking about total quantities taken up, whereas our determinations concern the minimum content, that is the content in the plant when growth ceases due to lack of a particular mineral. On the other hand, and that is a real problem for us, the various ions differ in their mobility within the

plant. For example, in case of shortage, potash can be redistributed from the old leaves to the young ones, whereas calcium remains in the old tissues. Hence, the Ca content of the different parts of the plant may vary considerably.

CAVERS (Ontario): I am curious to know whether there is a relationship between plant form and the nutrient availability. Is it possible that a plant might grow in a certain way because it has a certain supply of nutrients?

VAN DEN BERGH: As I mentioned in my answer to Professor Harper, wo don't know yet. We have to start to study the symptomatology of nutrient deficiencies in natural communities.

The experimental approach to ecological problems

1. INTRODUCTION

The beginning of all science of course lies in the description and classification of facts and phenomena. Much observation and description remain to be done in ecology, but we also want to understand more about the interrelationships between plants and their surroundings. There are several ways in which we can approach such relationships. To arrive at a predictive ecology, the ecologist has traditionally used the correlational method. Unfortunately, both the plant and its environment are of almost infinite complexity and consequently their interrelationships are equally complex and difficult to unravel. This is where the physiologist steps in. His experimental method enables him to establish causal relationships, despite the complexity of a situation in which a multi-dimensional unit, the plant, interacts with an ever-changing complex of environmental factors.

In the first half of this century there was very little contact between ecologists and plantphysiologists. Their domains were very different: the field and the laboratory, the whole vegetation in contrast with the individual plant or even an individual process, the cosmic problems of forest or tundra as against the esoteric problems of tropisms and metabolism, on the one hand the macro-world and on the other the molecular world. The ecologist was equipped with boots, raincoat, and vasculum; the physiologist was strictly an indoor man wedded to his bunsen burner, coleoptile, and microscope. But some ecologists took a pH meter into the field and the physiologist brought sun and shade leaves into the laboratory, and today fully-equipped field laboratories move where formerly ecologists only set a hesitant foot. Now a new breed of scientists has emerged, the ecophysiologist, equally at home in the field and in the laboratory. But his laboratory is no longer in a multistoried building but in a chamber on wheels, and the plant is no longer transported to the laboratory, the laboratory is brought to the plant. This makes it possible not only to study well-rooted plants or even trees with respect, for example, to their water and gas exchange, but also to bring the microscope to the plant. All this has opened completely new horizons for me personally, for instance the world of the root and the surrounding soil.

Since I have only become aware of this soil-root environment so recently, I do not want to go into this subject in any depth, because I feel myself still too much a "babe in the roots". Yet, I have become aware of completely new relationships in a world largely dominated by fungi, not just as decomposers or parasites but also as builders and symbionts. Under the harsh conditions of the desert and the dunes, where the fragile root hairs usually do not survive, hyphae and rhizomorphs seem to have taken over the function of root hairs, and supply the desert plants with water and nutrients, in the same way as the foresters have found in the case of trees. I found a completely unexpected function of the soil fungi in the desert, where they play a dominant role in soil binding and dune fixation. It is not the rather poor or deep root systems of the desert plants which hold the sand together; this is done by the mycelial mats in the surface layers of the desert soil in which each particle is woven into a mycelial network strong enough to resist wind erosion (WENT & STARK 1968).

There is another aspect of ecophysiology which is very important. It is the *trait d'union* between the field problem and its understanding in terms of causal relationships. It is in the field, in nature, that I find the problem I want to understand. For this understanding I need analysis in the laboratory. But then I have to go back to nature, to see whether I have come up with the real explanation, and whether I have gained real understanding. Without the intermediate laboratory phase of the analysis I would feel insecure, floundering in a sea of uncertainties and a multitude of possibilities.

The place where ecologists and physiologists meet at present is in the phytotron, a series of controlled-environment rooms where plants in a reproducible condition can be exposed to controlled environments. In a phytotron not only air temperature but also radiation, humidity, wind, and several other factors governing or influencing organismal behavior can be maintained at the desired levels. The greater the number and the wider the range of the environmental factors, the more useful a phytotron is.

I do not, of course, want to discuss the phytotron here, but I do want to say a few things about its significance for the understanding of the role of environment in the life of a plant. The phytotron does not attempt to reproduce the natural environment in all its details: for those conditions we use the natural environment itself. But if we want to know the effect of a particular temperature or temperature sequence or humidity or light condition, we are able to maintain this condition for any desired length of time in a phytotron.

2. GROWTH CONTROLLED BY PHOTOPERIOD

Before air conditioning for plants was available, only one factor in the environment of plants was controllable, and this was the photoperiod. Therefore, a fairly large body of information on long- and short-day responses of plants is available. But as yet little is known about interrelationships between temperature and photoperiod, except for the effects of chilling on subsequent photoperiodic responses. For many plants it is known that the preliminary treatment with temperatures below 10°C will make them responsive to long days.

Much more subtle temperature relationships with flowering and vegetative response are found in biennial and perennial plants. For instance, the older commercial varieties of strawberries were strictly short-day plants at high summer temperatures. Yet in for example northern Sweden, where they are never exposed to short days during their growing season, they flower abundantly, because at low temperatures they initiate flowers even under long days. A similar temperature-photoperiod interrelationship is found in the tuber formation of potatoes. In greenhouse experiments (all carried out at high temperatures) potatoes were found to be strictly short-day plants. Yet in most of the main potato-growing areas (The Netherlands, Ireland) potatoes are harvested toward the end of the summer, after experiencing only long days during their growing period. Thus, as short-day plants one would not have expected them to produce tubers until late in the autumn. This enigma was solved in a phytotron, where it was established that at low night temperatures tuber formation occurs optimally during long days (WENT 1957).

I will not give other examples of temperature control of photoperiodic responses. We know much less about relationship between the photoperiodic response and light intensity. What is the critical light intensity at which the plant switches from its light-metabolism to its dark-response? This could be expected to occur sooner in cloudy than on sunny days, and to be different for plants growing in full sunlight than for those growing in deep shade. I demonstrated this in the case of the coffee tree.

Coffee (*Coffea arabica*) is a strictly short-day plant in its flower initiation. After 75 days under an 8-hour photoperiod, coffee plants had all developed flowerbuds on the nodes of their plagiotropic lateral branches. Three other groups of coffee plants were also exposed to 8 hours of full daylight, but received daily in addition 4 hours of dawn and 4 hours of dusk illumination at 10, 100 or 1,000 lux. The plants with 1,000 lux dawn-and-dusk did not initiate any flower buds and behaved as if they were kept under long days. At 10 lux they differentiated flowerbuds as if they had been under short days, and at 100 lux they just barely started flower initiation. This means that coffee plants grown under shade trees receive effectively a shorter daylight treatment than when grown in full sun. This makes it possible to grow coffee near the equator where it is never exposed to short days except under shade trees. But in Sao Paulo, Guatemala, or Hawaii, where near the tropics of Cancer and Capricorn they are exposed to natural short days in winter, coffee plants produce well without shade trees (WENT 1957).

This shade effect also explains another flowering anomaly. *Cestrum nocturnum* is a flowering shrub belonging to the long-day/short-day type, that is to say,

it only initiates flowers under short days if it has previously been exposed
to long days. Therefore, it flowers abundantly in the autumn, but occasionally
a few flowers are produced in the spring, presumably when a period of bright
days in April is followed by a cloudy period producing effective short days
after the long days in April.

It is not just growth or flowering which can be studied in a phytotron. As
an example of the other possibilities, let me discuss taste. In the past, most
investigations were limited to the effects of soil composition and fertilizer
treatment on taste development, and the results were disappointing. But when
strawberries were grown under different temperature conditions in the Pasadena
phytotron, the typical strawberry aroma developed only in the coolest growth
chambers, and no trace of strawberry smell was observed in any of the warmer
greenhouses. It turned out that the aromatic taste of strawberries was
completely linked with aromatic smell and did not develop at all in berries
grown at temperatures of $17^{o}C$ or higher. When strawberries were scored for
taste, solely on the basis of the aroma and disregarding the sugar or acid
content, only fruits that had ripened for some time at $10^{o}C$ in the presence of
light were rated tasty. Full flavor developed in any berries exposed to 10,000
lux at $10^{o}C$ for one or more hours, no matter how high the photo- or
nyctotemperatures had been during the remainder of the 24-hour day. At 5,000
lux an exposure of at least 2 hours was required at the low temperature, but
the berries never developed full flavor. This explains why the first
strawberries ripening in spring are the best-tasting, that is, when the
temperatures during the first 1-2 daylight hours are only $10^{o}C$, whereas during
the summer, with much warmer morning hours, strawberries have an insipid taste
except in northern Sweden or Alaska. If the same rule of cool mornings for
taste development holds for other fruits, superiority of northern-grown apples
would be expected (WENT 1957).

3. GROWTH CONTROLLED BY TEMPERATURE

To come to a real understanding of the behavior of plants, both the ecologist
and the physiologist have to make concessions. If, for instance, we want to
determine the role of temperature in the fruit production of the tomato, the
physiologist has to simplify the laboratory analysis of the temperature
response, by breaking it up into discrete ranges, as A.H. Blaauw did in his
analysis of the response of bulb crops to temperature. In tomato analysis the
first simplification came from the realization that during the day and the
night, different processes with different temperature requirements were in
control. Thus, the optimal phototemperature had to be determined independent of
the night temperature. Since it also was found that the response to the
nyctotemperature was about the same for a constant temperature as for two
different temperatures averaging the constant temperatures, provided their

difference did not exceed about $5^{\circ}C$, experimentation became rather simple. It was possible to grow a tomato plant that produced like a field-grown plant by subjecting it to a day temperature similar to the average outside day temperature and during night to a temperature equal to the average temperature in the field during night. How could the ecologist cope with the problem of averaging the continuously changing temperature occurring in the field? This proved to be relatively simple. Instead of integrating thermograph records, which is laborious, we can calculate the effective day temperature by subtracting one-fourth of the difference between the maximum and minimum temperatures from the maximum temperature. In a similar way the effective night temperature can be calculated by adding one-fourth of the difference between the maximum and minimum temperatures to the minimum temperature. Since most meteorological stations record both minimum and maximum temperatures, the effective day and night temperatures can be calculated for almost any location near a meteorological station.

To return to the tomato, it was found that, in the laboratory the highest tomato production occurred at a nyctotemperature of $17^{\circ}C$, and that field production was highest when the effective night temperature was $17^{\circ}C$. Some tomato varieties would set in both the laboratory and in the field at a higher temperature, and in a breeding program in the laboratory it proved possible to introduce this higher optimal temperature response into a commercial variety which now is used in tomato production in warmer climates.

4. GROWTH CONTROLLED BY CIRCADIAN RHYTHM

For the solution of these and many other ecophysiological problems a phytotron is essential, especially when the interaction of factors becomes more complex. I would like to discuss one problem in particular, the existence of which was not even suspected before the advent of the phytotron. This problem concerns the role played by circadian rhythms in the temperature response of plants. Only when plants can be grown entirely in artificial light can we escape the normally unbreakable straight-jacket of the 24-hour day-night cycle imposed upon us by the rotation of the earth. Quite unexpectedly, it was found that without a circadian rhythm of light-darkness tomato plants in continuous light soon showed declining growth, became chlorotic, and ultimately died. These abnormalities could be prevented by interrupting the continuous light with periods of darkness, but only if the interruptions came at intervals of 24 hours. Plants could not be "trained" to a regime of say 6 hours light and 6 hours of darkness or 18 hours light and 18 hours dark. At first it was suspected that these dark interruptions were needed to improve whatever light reaction was critical, but this was disproven by another observation. Tomato plants grew normally in continuous light when they were subjected to a temperature cycle of 24 hours, being kept at an optimal temperature of $26^{\circ}C$

which was interrupted daily by periods of 4-8 hours at 10°C. A 6-hour dark
interruption turned out to be just as effective as a 6-hour low temperature
interruption. It was concluded that a tomato plant needs a rhythmicity of
either temperature or light on a 24-hour basis to develop normally. This means
that the tomato has an internal circadian rhythm which is so deeply ingrained
that the plant cannot develop normally unless this internal rhythm is
synchronized with an external 24-hour rhythm. Seemingly to complicate the
picture, it was then found that this internal circadian rhythm had a
temperature coefficient of 1.2, so that at higher temperatures the cycle of
internal rhythm was less than 24 hours and at lower temperatures more than 24
hours. This suddenly explained a number of very puzzling observations (WENT
1962).

It had been known for a long time that many tropical plants could not be
grown in cool climates, but gradually died even though they were never exposed
to freezing temperatures. They just could not tolerate prolonged exposure to
temperatures of 5° or 10°C. Similarly, it was known that many plants of cool
climates would die in the tropics even though they could stand occasional
temperatures of 30°C. A biochemical explanation of this behavior could not be
given, because no enzyme systems that failed or became toxic at these otherwise
physiological temperatures were found in these plants.

However, phytotron experiments showed that there was nothing wrong with an
African Violet (*Saintpaulia ionantha*) which died when kept for many months at
10°C, except the circadian rhythm. These plants develop normally at 10°C if
they receive 16 hours of light followed by 16 hours of darkness. Conversely, a
cool-climate plant like *Baeria chrysostoma* dies on a 24-hour light-dark rhythm
at 26°C but survives at that temperature on an 18-hour rhythm. This means that
the general response of plants to temperature is much more complex than was
originally supposed. In the case of a tomato plant, for instance, half of its
growth response to temperature can be accounted for by the temperature
coefficient of physiological and biochemical processes, and half by the effect
of temperature on its circadian rhythm. Therefore, it is not a low or high
temperature which these plants cannot tolerate, but the wrong circadian rhythm
(WENT 1962). During the long geological eras in which these plants grew in
tropical or temperate climates, this circadian rhythm was so strongly imprinted
on them at the prevailing temperatures that they cannot easily adapt to a
different rhythm.

5. GROWTH CONTROLLED BY PHOTOSYNTHESIS

In the preceding sections I have tried to analyze the physiological basis of
certain temperature adaptations of plants on the basis of phytotron studies.
Let me now use the same approach to analyze other basic processes in plant
development. I shall begin with photosynthesis. The information on the extent

to which the photosynthetic process limits plant development is very conflicting. Ever since Wiesner, three-quarters of a century ago, we have been talking about shade plants which are presumably saturated at low light intensities, and sun plants able to utilize full sunlight intensity. However, in studies such as those in which BLACKMAN & WILSON (1951) shaded plants in field experiments, any amount of shading reduced productivity. It seemed as though both sun and shade plants were saturated only at full sunlight intensity. My phytotron experiments were in apparent contradiction with their findings, in that under long-term exposure all leaves, whether of strawberries, tomatoes, or beets, were saturated with light at or slightly above 10,000 lux (about one-tenth of full sunlight intensity). Still different results have been obtained in the most recent work on photosynthesis, based on measurement of the rate of gas exchange of individual leaves, which indicated that in many plants light saturation did not even occur in full sunlight. This disagreement is probably to be ascribed to the entirely different techniques applied in these studies.

Forty-five years ago, EMERSON (1932) had already demonstrated in his flashing light experiments that the photosynthetic process consists of at least two steps: a light reaction followed by a dark reaction, the former being completed in a millisecond but only at exceedingly high intensities. Since Emerson's basic work, other successive light-dark reactions have been found, such as Bunning's photophil and skotophil phases of growth, the occurrence of a dark reaction usually delaying the light reaction. One of these delaying dark reactions in photosynthesis was already recognized fifty years ago by KURSSANOV (1933), when he observed that photosynthesis was slowed down by the accumulation of photosynthetic products, which means that the rate of CO_2 reduction can be higher in the morning than in the afternoon, when leaves are almost saturated with carbohydrates. These carbohydrates have to be translocated out of the leaves (usually during night) before high rates of photosynthesis can be reached again the next day. In addition, two processes are known to counteract high photosynthetic rates very early in the day. The first of these is the circadian rhythm, which slows down light reactions in the skotophil phase. The second process can be observed in the rate of sucrose production in tomato leaves in the early morning light. After a two-hour lag period, sucrose is produced at a maximal rate for 2-4 hours (WENT & ENGELSBERG 1946), after which this production decreases again in the later hours. This suggested that dry-matter production in the tomato proceeds at different rates at different times of the day. This was tested by WENT (1946) in a field experiment as follows.

Growth and total dry matter were measured in five groups of tomato plants, four of which were exposed to light for 5, 6, 7, 8 or 14 hours daily and covered with black cloth the rest of the day and during the night; the sixth group was not covered at all. As usual in such practical experiments the problem was complex, because although the outside temperature and that in the enclosures were the same during the night, during the afternoon the cloth-covered plants received a higher nyctotemperature than the uncovered ones. Since the experiment was performed in Pasadena during the spring, when the night temperatures are too low for optimal tomato growth and fruit set, the controls did not grow at an optimal rate and barely set fruit. Nevertheless, the results clearly showed that adequate photosynthesis had occurred during the first 7 hours of daylight. The plants given 5, 6, 7, 8, and 14 hours of daylight had mean weights of 65, 325, 680, 500, and 310 grams and mean fruit weights of 0, 3, 260, 140, and 25 grams, dry weight and fruit production being optimal for the plants given 7 hours of light each day (WENT 1946).

This and many other experiments have shown that measurement of the rate of CO_2 reduction in one particular period during the daily photosynthetic cycle cannot possibly establish whether a plant can utilize more or less light; this utilization depends on the physiological condition of the leaf, which varies throughout the day. Since most of the experiments done on rates of photosynthesis with gas-analytical methods have been carried out in the morning, when rates of CO_2 reduction are highest, the rates obtained by such instantaneous determinations of photosynthesis are undoubtedly too high.

6. PHOTOSYNTHESIS CONTROLLED BY GROWTH

To avoid this possible source of error, I suggest that photosynthesis should be measured over a long experimental period, for instance 1-2 weeks. To avoid tying up CO_2 analysers for weeks at a time, some other method should be used. Theoretically, the very best method is unquestionably the measurement of fixed chemical energy, which in most plants is equivalent to dry matter production. Unfortunately, this method is hardly possible with unstandardized plants in the field (except certain agricultural crops); it requires laboratory tests with properly matched groups of uniform plants.

In the Earhart phytotron in Pasadena a standard method for growing tomato plants for dry weight increase determinations was worked out. The plants were grown under optimal temperature and nutritional conditions in such numbers and groupings that statistical uniformity was attained. Under the most favorable light conditions the plants reached a photosynthetic efficiency of 9.2 per cent, that is to say, 9.2 per cent of the light energy falling on these plants over a period of 1-2 weeks could be harvested as chemical energy. This is not the highest efficiency for the photosynthetic process but it is for plant dry matter production, and indicates that the plants were grown under the most favorable conditions. The question was then raised as to whether, if it was not due to photosynthetic CO_2 reduction, this limitation of dry matter production could be attributed to some other process or processes occurring in the growth of tomato plants.

Fig. 1 probably gives the clearest indication of where the control of dry matter production lies. Groups of 24 containers with standardized tomato plants were grown in different colors of light. To achieve light saturation, use was made of fluorescent tubes with different "fosfors" which produced light over a wide range of wavelengths, in general 100 mμ but hardly overlapping. For all colors except green, light saturation was achieved. Before light saturation was reached, the efficiency of light utilization (the slope of the curve) for blue, red, and a mixture of blue and red light was the same, as had been found in

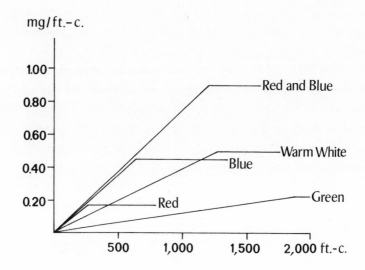

FIG. 1. *Dry matter production (produced in 6 days per square of tomato plants) as a function of the incident light intensity. Colors indicate the type of fluorescent lamp used; for "Red and Blue" two of each color were used. Data of* S. Dunn (WENT 1957).

many other experiments done with different techniques. The efficiency of the green light was much lower, because of its lower absorption. The white fluorescent light was slightly less efficient, because of the mixture in the white tubes of red and blue light with green. The main difference between blue and red lies in the saturation intensity, which is much higher for the blue than for the red, but is far outstripped by the combination of red and blue light. This is due not to the efficiency of the light in energy transformations but to the fact that the plants growing in red light become very spindly, in blue sturdy but short, and largest in the combination of red and blue light. Thus, the efficiency of the light is due to the utilization of the photosynthates in growth. This was confirmed in other experiments in which growth was not controlled by light color but by restricting it in other ways, e.g. by cutting off roots, which reduces both growth and dry matter production. It is not a question how much photosynthate is produced, but how it is used.

This suggests that we should seek ways to improve growth processes rather than to increase photosynthetic CO_2 reduction if we want to increase plant production (WENT 1957).

I had reached a somewhat similar conclusion after my first comprehensive investigation in airconditioned greenhouses. The first phytotron-like structure I built forty years ago -the Clark greenhouse in Pasadena- was used to obtain maximum growth and production, in this case in tomato. After years of work, higher productivity could not be achieved by further changes in temperature, light, water, nutrition, chemicals, root medium, breeding, etc. I then tried to identify the processes which limited growth under these optimal growing conditions. At that point I was no longer led by my own preconceived ideas, but entirely by the plant (WENT 1945).

Up till that time I had always worked with plants under conditions in which auxin was limiting for growth, conditions which I had deliberately chosen. Therefore I had become too convinced of the importance of auxin for growth, it is hardly surprising that I received one of the greatest shocks in my scientific life when it turned out that auxin was not an important factor limiting growth of the tomato plant. The decisive factor proved to be the rate at which photosynthates were translocated from the leaves to the growing areas: roots, stems, and fruits. This means that growth limitation is basically controlled by the utilization of photosynthates and not by photosynthesis itself. The lesson to be drawn from this experience is that we should not have preconceived ideas about the factor or factors involved in controlling a plant or a process: we should be led by the plant and not by our own (clever) ideas.

7. SATURATING LIGHT INTENSITY FOR PHOTOSYNTHESIS

I have already referred to the perhaps unexpected fact that for most plants investigated in the Pasadena phytotron the saturating light intensity was around 10,000 lux, which is only one-tenth of full sunlight. This was determined for tomatoes, sugar beets, strawberries, *Bryophyllum,* and a few other plants; all on the basis of the dry matter production over a 1-2 week period. The leaves were exposed as much as possible in a single layer perpendicular to the incident fluorescent light. If in nature all leaves were exposed in this position to sunlight at 100,000 lux, there would be an enormous waste of solar energy, since only 10,000 of the 90,000 lux absorbed light would be used in photosynthesis. If, however, the leaf was exposed at a 30° angle, the average light intensity in full sunlight would be only 50,000 lux with a wastage of 40,000 lux, and at an angle of 8° theoretically no light would be wasted.

In nature only shade plants place their leaves perpendicular to the incident light, which intensity is so much lower that most of it can be used. Also, they usually have all their leaves in a single layer, because there would not be enough light for effective photosynthesis in a second layer underneath their leaf mosaic. The compensation point for light utilization is generally 1,000-2,000 lux. In any plants which normally grow in full sunlight, such as trees and most crop plants, the leaves are never placed perpendicular to the incident sun rays. Many plants, specifically *Leguminosae,* even have a mechanism by which at high incident light intensity the leaves assume a position parallel with the light. In sugar cane and corn, the leaves or at least their upper parts usually hang down making an acute angle with the noon sunlight. In sugar beets the young leaves also stand up vertically, and the older leaves, which are spread out almost horizontally, are shaded by the younger leaves. Needles of conifers and the hanging *Eucalyptus* leaves are never placed perpendicular to sunlight. The leaves of quaking aspen not only hang down, but are exposed to flashing light, which would increase their photosynthetic activity (WENT 1958).

At low light intensities, below the compensation point, there is an excessive weight loss in tomato plants, even exceeding the loss suffered in complete darkness. This may be due to photorespiration that is not balanced by photosynthesis, and is probably the herald of premature death. This would mean premature death of the lower leaves, which do not receive enough light to continue to be productive. It would also mean that the plants prune off their photosynthetically ineffective leaves and merely keep the optimal number of leaves to utilize all the light to which they are normally exposed. This is illustrated by the contrast between a plant growing in a greenhouse in full daylight and a plant growing in artificial light of 15,000 lux: the former has many more leaves, even though the upper leaves of both plants are light-saturated and look comparable. The same mechanism probably causes self-pruning of trees where the lower leaves and those nearest the trunk are shed instead of living a parasitic existence at the cost of the rest of the tree.

This behavior is of considerable importance to ecologists. Plants are not adapted to a particular light intensity, they adjust themselves to an existing light intensity; thus, the same plant species, even the same ecotype, can also live just outside a forest or at different distances from its edge, and in general plants can tolerate a wide range of light intensities.

8. GROWTH CONTROLLED BY RAIN

Thus far, I have dealt mainly with an experimental approach to ecology for which a phytotron is essential. Fortunately for those who do not have access to a phytotron, it is possible to approach many ecological problems without such facilities but still using physiological thinking and employing the experimental approach. This has been demonstrated in this monograph by many investigators, among them Woldendorp, Ernst, and Brouwer. I would like to add

just one more example, in which the main instrument is a lawn-sprinkler.
Anyone who has made observations in a desert has been struck by the fact that
in some years the surface of the sandy desert plains are covered by an
extensive vegetation of annuals, all of which burst into bloom simultaneously,
whereas in other years there may not be a single annual in evidence. The drier
the desert, the more infrequent these blooming years. In Death Valley, on the
border between California and Nevada, the intervals between good flowering
years perhaps range from 5 to 20 years. The annual rainfall fluctuates widely
from year to year in Death Valley, ranging from 0 to over 100 mm/year, with an
average of 43 mm. What factor or factors cause the seeds of these desert
annuals, which have been lying dormant throughout these 5-20 years, to abandon
their inactivity suddenly and all at the same time? One would immediately
suspect moisture or rain, but with an average of five or more rains a year,
how did all seeds select the same rain? A correlational analysis soon
suggested that only a rain amounting to more than 30 mm was responsible,
provided it fell in the month of November. When this problem was taken to the
laboratory, a few facts immediately became clear (JUHREN *et al*. 1956). For one
thing, none of the annuals growing in a rainy year in Death Valley germinated
when watered by soaking alone. But when the seeds were rained upon by a
lawn-sprinkler for a sufficient time, they all germinated provided they
received the equivalent of 30 mm rain over a sufficiently long period, such
as 10 hours. It was then found that these seeds contained water-soluble
inhibitors which could be leached out by a prolonged period of artificial rain.
The response to a "November" rain was controlled by the temperature to which
the seeds were exposed afterward. Summer annuals germinated exclusively at
high temperatures, e.g. 26°C (*Pectis papposa, Bouteloua* spp.). December
germinators, such as *Gilia aurea* and *Eriophyllum wallacei*, needed low
greenhouse temperatures. But the most spectacular desert annuals, germinating
late in the fall, such as *Garaea canescens,* developed in the greenhouse at
intermediate temperatures. Thus, depending upon the temperatures prevailing
after the artificial rain, quite different plants developed. This was not a
question of survival of the cold- or warm-adapted plants, but rather of
differential germination. Therefore, in each year when germination occurs
after a sufficiently heavy rain, the composition of the annual vegetation in
our deserts is different. There are years when *Plantago insularis* is dominant
in a certain location; in other years it is *Oenothera clavaeformis* or
Chaenactis carphoclinia.

There are many other mechanisms besides germination inhibitors by which
desert plants delay their germination until the arrival of a rain of
sufficient intensity to insure their survival in a normally dry climate. Many
of the shrubs and trees that grow in dry washes are hard-seeded, i.e., their
seedcoat has to be abraded before they can germinate. This abrasion can only
occur after a heavy rain, when a slurry of water, sand, and stones runs down

an otherwise dry wash and soaks the soil of the wash to a considerable depth.

To me, the most amazing phenomenon I observed in the desert was the lack of competition between desert annual plants. Sometimes, after a particularly heavy rain, excessive numbers of seedlings develop. Under these conditions, essentially the whole seed pool of the desert develops, and thousands, even up to 50,000, seedlings per square meter may come up, resulting in a green cover of the soil composed solely of cotyledons. Under such conditions one would expect a fierce competition for light, food, water, and space between the crowded seedlings, but this is not the case. On average, 45 per cent of all seedlings not only survive but manage to flower and fruit. Whereas normally each germination results in 10-20 new seeds, under these crowded conditions perhaps only one or two viable seeds per germination are formed, and all plantlets share all resources equally. Therefore, there is no selection during the vegetative and reproductive stages in these desert plants, and since most of their seeds have an extraordinarily long life span, one would expect an ever-increasing seed mass in deserts. That is obviously not the case; we found that after a good seed year the seed population is reduced to normal levels by the activities of seed-eating rodents and harvester ants. If one still wanted to use the term selection pressure, one would have to state that for desert annuals, selection pressure is exerted by seed-eating animals. In general, it can be stated that ideas on Darwinian evolution must be drastically revised, certainly for deserts.

9.REFERENCES

BLACKMAN, G.E. & G.L. WILSON, 1951 - Physiological and ecological studies in the analysis of plant environment. VI. The constancy for different species of a logarithmic relationship between net assimilation rate and light intensity and its ecological significance. *Ann. Bot. N.S.*, 15, 64-94.

EMERSON, R. & W. ARNOLD, 1932 - A separation of the reactions in photosynthesis by means of intermittant light. *J. Gen. Physiol.*, 15, 391-420.

JUHREN, M., WENT, F.W. & W. PHILLIPS, 1956 - Ecology of desert plants. IV. Combined field and laboratory work on germination of annuals in the Joshua Tree National Monument, California. *Ecology*, 37, 318-330.

KURSSANOW, A.L., 1933 - Ueber den Einfluss der Kohlenhydrate auf den Tagesverlauf der Photosynthese. *Planta*, 20, 535-548.

WENT, F.W., 1945 - Plant growth under controlled conditions. V. The relation between age, light, variety and thermoperiodicity of tomatoes. *Am. J. Bot.*, 32, 469-479.

WENT, F.W., 1946 - Effects of temporary shading on vegetables. *Proc. Am. Soc. hort. Sci.*, 48, 374-380.

WENT, F.W., 1957 - *The experimental control of plant growth*. Ronald Press, New York.

WENT, F.W., 1958 - The physiology of photosynthesis in higher plants. *Preslia*, 30, 225-240.

WENT, F.W., 1962 - Ecological implications of the autonomous 24-hour rhythm in plants. *Ann. N.Y. Acad. Sci.*, 98, 866-875.

WENT, F.W. & R. ENGELSBERG, 1946 - Plant growth under controlled conditions. VII. Sucrose content of the tomato plant. *Archs. Biochem.*, 9, 187-200.

WENT, F.W. & N. STARK, 1968 - The biological and mechanical role of soil fungi. *Proc. natn. Acad. Sci. U.S.A.*, 60, 497-504.

QUESTION: You mentioned that growth of the tomato plant depended entirely on the night temperature. This observation does not necessarily apply to all plants or every region involved, since you will have noticed that day temperature often is important as well.

ANSWER: Each plant has different factors to which it responds. The pea responds to day temperature and not to night temperature. Several varieties of tomato respond to a high temperature as well, a factor which was bred into a commercial variety to make it possible to grow tomatoes in hot climates, for instance in Texas. In general, however, the tomato has a very limited range of night temperatures in which it will set fruit.

QUESTION: Do you want to say that these desert plants suffer from stress? They are so well adapted to this environment! Do they live in a stressed environment?

ANSWER: Stress is one of those words I do not like. What do you mean by stress? Actually, the desert provides the optimal conditions for the growth of desert plants. They are typically adapted as regards temperature, water economy, etc. Imitation of desert conditions in a climate-room gives exactly the same plants. Therefore, you have to define what you mean by stress. For man, the desert is a stressful environment. The desert animals who live there are very happy; they are not under stress. Only under extreme conditions may one speak of "stress".

QUESTION: We both attended the lectures of Professor NIERSTRASZ in Utrecht. He emphasized the point that survival is not a question of fitness but of chance. Do you think the same applies to annual plants in the desert?

ANSWER: I do not think the factor "chance" is very much involved when the organism is completely adapted. I will give you an example of the misunderstanding that can arise. A plant is not particularly adapted to a certain light intensity, but adapts to the intensity under which it normally lives. In the tomato, photosynthesis in the leaf is saturated at 10 per cent of full sunlight. It does not waste the extra light, however, since the leaves are almost never perpendicular to the direction of the sunlight. Especially older leaves do not stand upright but hang to some degree (corn, tomato, sugar beet). The plants adapt themselves to the prevailing conditions.

QUESTION: Stress conditions must occur in the field, or how would you otherwise explain the spacial problems observed there?

ANSWER: Spacial problems do not exist for annual plants; we notice them in perennials and in shrubs. Under favourable rain conditions in the desert we can see millions of seedlings of the shrub *Larrea divaricata,* the creosote bush, but these seedlings are not found in the immediate vicinity of the shrubs themselves, because of a germination-inhibiting excretion product that we have been unable to identify. There is no effect of the surrounding seeds on germination, just as is the case with agricultural plants.

QUESTION: Do you think use of the word "stress" may still be meaningful to explain why certain areas have a very low productivity or why so many plants are absent in a certain environment?

ANSWER: In most cases the problem is not that only a few species survive but rather that relatively few germinate. Once a tomato plant has germinated in the wild (this species has a high temperature requirement for germination), it survives and it rarely happens that it is actually killed as a plant. It might be attacked by fungi or later by animals, but I would not call this stress. Stress should only be referred to if you can express the phenomenon in quantitative terms, e.g. growth.

QUESTION: What controls the density of seedlings of annual desert plants?

ANSWER: This is entirely determined by the number of seeds present. Densities of 50,000 seedlings per m^2 have been reported, but a density of 5,000 is more usual. Of the 50,000 seedlings, at least 20,000 survived and in such cases the production of seeds per plant will be reduced considerably by the crowded conditions. In these plants "competition" is not between species but within the species.

I would like to express my admiration for Professor HARPER's work as regards the quantitative aspects and the numerous measurements on plants included in his research. In my opinion, ecologists should pursue this line much more than is the case at present.

Chemical soil factors determining plant growth*

1. INTRODUCTION

Much has been written about the plant-soil relationship in the context of ion uptake and the transport and utilization of chemicals by plants. These papers emphasize the diversity of soils and the diversity of plants. There are also many other problems, such as the mineral nutrient supply in space and time and the competition, both between and within species, for the elements in the soil. In contrast to other abiotic factors such as light or temperature the chemical situation in soils is complex, if only because of the presence of more than hundred chemical basic elements. This situation is made essentially more complicated by combination of these chemicals and by the organic compounds formed, for instance the humic acids of soils.

If we restrict the problem to the basic elements, we can arrange these chemical elements according to their frequency and their function in plant metabolism (Table 1). The current classification of these elements according to their function into (a) major or macronutrients, (b) minor or micronutrients, and (c) oligo-(trace) elements, reflects a large and small requirement, and no physiological need for these elements by plants, respectively. Whereas all of the classical macronutrients (C, H, O, N, P, S, K, Ca, Mg) are universally required by plants, not all of the micronutrients (Fe, Mn, Cu, Zn, Mo, B, Cl, Na, Si, Co), and especially silica and cobalt, have been shown to be necessary for all plants. For the functioning of ecosystems, the plant also has to supply animals with certain elements which are -at least at times- not required for plant metabolism such as J, F, Se, V, and perhaps Cr (LEE 1975). Another point here is that oligo-elements such as aluminium are sometimes present in the soil in larger amounts than micro- or macronutrients and may influence plant growth and thus plant distribution. The individual plant has only a very small possibility to escape from an environment, in which it has been established by a propagule which means that it must have evolved special adaptations enabling it to endure the quantity and quality of these elements both in space and in time.

With respect to soils, little is known concerning the geochemistry of the elements and the soil factors governing their distribution. In terms of nutrient

---

* Dedicated to Prof. W. Baumeister

TABLE 1. *Frequency of occurrence of the chemical elements in the rock material of the earth's crust (mM), with indication of the macro- (*) and micronutrients (**) for plants and the additional micronutrients (***) for animals*

10^3		10^2		10^1		10^0		10^{-1}		10^{-2}	
O	29.5*	K	7.2*	Ti	9.8	S	9.7*	Zn	9.2**	Pb	7.2
Si	10.9**	Ca	7.1*	F	3.8***	Cl	9.3**	B	8.3**	Sm	5.7
Al	2.9	H	6.9*	C	2.7*	Li	4.3	Ni	7.5	Gd	5.6
Na	1.1**	Fe	6.3**	P	2.6*	Ba	4.3	Cu	5.4**	Pr	5.4
		Mg	5.7*	Mn	1.3**	Sr	3.3	Y	3.8	Th	4.7
						V	1.9***	La	3.2	Dy	3.8
						Zr	1.8	Sc	3.1	Br	3.6
						N	1.4*	Ga	2.5	As	2.3
						Rb	1.4	Nb	2.2	Sn	2.5
						Cr	1.3	Be	2.2	Er	2.0
								Nd	2.1	Cs	2.0
								Co	2.0**	Yb	2.0
										Ta	1.9
										Ge	1.8
										Hf	1.7
										U	1.5
										Ho	1.1
										Mo	1.0**
										J	0.4***
										Se	0.1***

reactions, clay forms the most important part of the soil, but its components are so small that only the recent development of X-ray diffraction techniques may make it possible to resolve many of the current problems on nutrient availability (GIESEKING 1975). With respect to the plant-soil relationship, the concept of the availability of nutrients has been prominent in the attempts to measure the amount of available chemical elements in soils by chemical extraction procedures. In the present state of our knowledge concerning the plant-soil relationship, we should deliberately realize that despite all of the possibilities offered by chemical extractions, the availability of nutrients ultimately depends on the properties of the plant. Besides, plants vary greatly in their absorption and excretion activities. Next to the available quantity, the chemical quality of the elements has to be considered. However, our knowledge of some compounds, such as humic substances, is restricted to series of chemical similarity, in this case fulvic acids, humic acids, or hymatomelanic acids (FLAIG *et al.* 1975) and nearly nothing is known about the basic substance itself. This is hardly surprising because the chemical constitution of these compounds depends on the interaction between various biological pathways and on environmental factors.

Furthermore, there are the allelopathic substances interfering in mineral nutrition. These compounds, which are excreted by plants and are accumulated in the soil, are effective inhibitors of the growth of other plants (WHITTAKER 1970; RICE 1974; QUINN 1974).

These few remarks will suffice to show the impossibility of covering the broad field of the chemical aspects of the plant-soil relationship here. I shall therefore confine the discussion to some of these aspects.

2. HALOPHYTES AND MARITIME SALINE SOILS

The first example of the effect of mineral nutrients on plant growth refers to an environment where the relevant ions are more or less completely available to the plants in large quantities. These ions are chloride and, to a certain degree, sodium and potassium. Many ecological and physiological experiments have been carried out with plants of saline environments (for reviews see: WAISEL 1972; RANWELL 1972; CHAPMAN 1974; REIMOLD & QUEEN 1974; FLOWERS 1975; POLJAKOFF-MAYBER & GALE 1975). But are the interactions of the halophyte with its saline habitat understood? A synthesis of the research done on different biological levels may supply the answer to this question.

I shall focus on one of the halophytes, *Suaeda maritima,* which occurs on salt marshes in Europe. In the field, optimal growth is observed in those places where there is a strong daily fluctuation of the tidal submergence, but the seasonal variations in sodium and chloride content are low (ESSING 1972). In this environment *Suaeda maritima* grows in monoculture, sometimes mingled with *Salicornia europaea, Spartina townsendii* and *Puccinellia maritima*.On the middle salt marsh with the plant community *Puccinellietum maritimae* characterized by a great seasonal variation of the salt level, the annual biomass production of *Suaeda maritima* is low, sometimes amounting to no more than 0.01 g dry matter m^{-2} (KETNER 1972). The germination capacity to *Suaeda maritima* is the same in both saline and non-saline environments (BINET 1968; BOUCAUD & UNGAR 1976). In physiological experiments, the growth of *Suaeda maritima* is promoted by sodium levels well in excess of amounts needed by non-halophytes (PIGOTT 1969). Comparison of the biomass production of *Suaeda maritima* in the natural habitats with that in sand culture experiments (FLOWERS 1972; YEO 1974) shows a discrepancy between the physiological and the ecological optimum (ELLENBERG 1958). Optimal growth in physiological experiments proves to be confined to lower salt concentrations (Fig. 1). The results of sophisticated experiments such as in-vitro culture of isolated tissues (HEDENSTRÖM & BRECKLE 1974) also indicate a lower tolerance for sodium chloride than is found in the total plant. Last, but not least, at the subcellular level, cytoplasmic enzymes such as nitrate

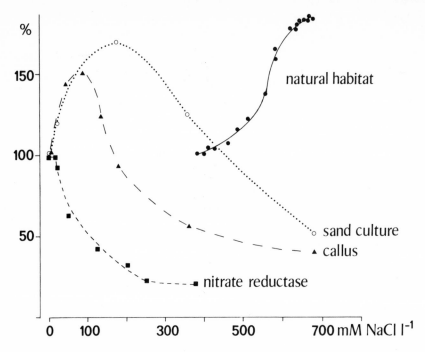

FIG. 1.   Suaeda maritima. *Effect of increasing concentrations of NaCl on the growth of plants in the field* (ERNST *unpublished*) *and in sand culture* (YEO 1974), *on the growth of callus* (HEDENSTRÖM & BRECKLE 1974), *and on the in-vitro activity of nitrate reductase in leaves* (ERNST *unpublished*). *Control values at 0 mM NaCl l^{-1} are defined as 100%*

reductase, which is a key enzyme in nitrogen metabolism, are the least able to tolerate sodium chloride, which is in agreement with the in-vitro behaviour of the enzymes of other halophytes (AUSTENFELD 1974; FLOWERS 1975). The implication of these findings is the assumption that metabolic regulation is necessary for the maintainance of a low level of salt in the cytoplasm; otherwise, salts could disrupt the structure and/or function of enzymes or other macromolecules and thus damage cellular organelles and hamper metabolism. One of the pre-conditions for this regulation is a compartmentation of ions within the cell and possibility for the storage of ions. A form of compartmentation is the selective accumulation of salts in the vacuoles of halophytes. This requires a special mechanism for the arrangement of an osmotic equilibrium between cytoplasm and the vacuoles. The osmotic pressure of the cytoplasm in the cell is found to be generated by an accumulation of proline (STEWART & LEE 1974), glycerol (BEN-AMOTZ & AVRON 1973), or betaine (STOREY & WYN JONES 1975; WYN JONES *et al*. 1977); some of these compounds also play an effective role under various stress conditions such as drought (proline: BARNETT & NAYLOR 1966) or flooding (glycerol: VESTER 1972). What distinguishes halophytes from plants of non-saline habitats is not the ability to withstand physiological drought, but their ability to regulate selectively ion concentrations. Halophytes, in particular, have to cope with an excess of sodium chloride in their environment that impedes the acquisition of other essential nutrients, for instance potassium.

For this reason, we shall focus here on two of the alkali metals in this environment, i.e. sodium, which seems to govern growth in halophytes (e.g. *Salicornia europaea,* BAUMEISTER & SCHMIDT 1962; and *Aster tripolium,* WEISSENBÖCK 1969) and is abundantly available in the salt marsh; and potassium, of which large amounts are required for optimal plant development, but which occurs in small amounts in saline environments. Despite this imbalance, the mechanisms of ion absorption in roots are sufficiently specific to permit the absorption of an ion (potassium) occurring in a low concentration in the presence of an ion (sodium) occurring in a high concentration. This means that in *Suaeda maritima* there is an enrichment of potassium relative to the concentration in the soil solution. In the root of this halophyte the Na/K-ratio is 3.7 as against 22.5 in the soil solution, which suggests a competitive advantage of the potassium uptake mechanism (Table 2). Even in the leaves with sodium

TABLE 2. *Concentration of sodium and potassium in the soil solution of a lower salt marsh site (Wadden Sea) in relation to the levels in the cell sap of leaves and in plant organs of* Suaeda maritima

Factor	Soil solution	Cell sap	Enrichment	Leaves	Shoots	Roots
	(mM l^{-1})		factor	(mM kg^{-1} dry weight)		
Na	607.8	564.0	0.93	4350	1925	2001
K	27.0	58.1	2.15	450	393	537
Na/K-ratio	22.5	9.71		9.67	4.90	3.72

storage in the vacuoles, the Na/K-ratio is lower than in the soil solution, as has been confirmed by results obtained in various European salt marshes (CAPPELLETTI PAGANELLI 1967; ESSING 1972; ALBERT & KINZEL 1973; ALBERT & POPP 1977). Within the plant, the Na/K-ratio is generally higher in the leaves than in the roots indicating a more effective translocation of sodium than of potassium (*cf.* HALL *et al.* 1974 for rubidium). The rate of the potassium uptake (V_{max} 50.1 $\pm$ 7.9 µM K h^{-1} g^{-1} fresh weight from a solution with 8-20 mM K l^{-1}) is higher than the rate of sodium uptake (V_{max} 28.3 $\pm$ 4.7 µM Na h^{-1} g^{-1} fresh weight from a solution with 100-500 mM Na l^{-1}), but the rate of sodium uptake in halophytes (see FLOWERS 1975) is of the same order as that of glycophytic barley roots (RAINS & EPSTEIN 1967).

To understand halophytes, the function of sodium chloride in their metabolism must be known. Despite the extensive research on these plants, uncertainty prevails. A supply of chloride ions, however not of sodium ions may change the balance between the C_3 and C_4 carbon fixation pathways by regulation of the activity of phosphoenolpyruvate carboxylase, thus stimulating CO_2-fixation (BEER et al. 1975; TIKU 1976) and reducing transpiration. (Na-K)-activated ATPases may also play a role, but they do not explain a special physiological function of sodium in the metabolism (KYLIN & QUATRANO 1975). This problem can only be solved by applying the ecotype concept developed by TURESSON (1922), according to which the smallest overall difference in basic metabolism is ensured; it cannot be approached comparing peas or beans with halophytes (FLOWERS 1972; AUSTENFELD 1976).

Another unsolved ecological problem is the difference between optimal biomass production of Suaeda maritima in the field and in the greenhouse. Extrapolation from fertilization experiments (STEWART et al. 1973) suggests that Suaeda maritima demands high amounts of nitrogen which do not occur in the upper zone of salt marshes. The in vivo nitrate reductase activity in Suaeda maritima was found to be as much as 50 times higher in plants from the lower salt marsh (5.55 μM NO_2 h^{-1} g^{-1} fresh weight) than in plants from the upper marsh (STEWART et al. 1972). The sparseness of this species in environments with lower salinity is due to low nitrogen levels, not to the low salinity. Nevertheless, despite strong growth of Suaeda maritima at low salinity in the laboratory, the equivalent biomass of the production in the field is not reached. This is due to the omission from the physiological experiment of one of the main environmental factors, i.e. temporary submersion. As shown for Salicornia europaea (LANGLOIS 1971), the temporary submersion of shoots induces changes in protein and carbohydrate metabolism and causes perhaps also leaching of salts from the plants. Therefore, physiological and ecological optima may not be compared unless the experimental design is adequate, in this special case also with respect to the supply of boron (BRECKLE 1976), bromine (PEDERSEN et al. 1974), and iodine (FOWDEN 1959). Until more is known, every shift of the ecological optimum above the physiological optimum must be taken to indicate unsufficient knowledge of the ecology of the plant and its environment even when it seems to offer a quite simple example of an "extreme" environment (see also the section on the qualitative aspects of mineral elements).

3. OXIDATION STATUS AND THE AVAILABILITY OF NUTRIENTS

For most of the mineral elements under consideration, there is a difference between the total amount in the soil and the amount available to plants. This is the most important complication encountered in the study of the mineral nutrition of plants in the field. One of the factors determining availability is the oxygen supply in the soil, which is unfavourably affected by an excess of water. Flooding -dependent on its frequency, seasonality, or permanency- displaces air in the soil and permits the development of a reducing system rather than oxidizing conditions. The sequence of events after flooding of a mineral soil is shown in Fig. 2. The first consequence of waterlogging is the

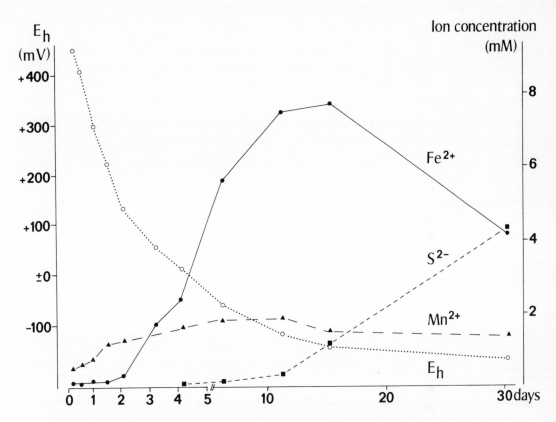

FIG. 2.   *Redox potential and contents of sulphide and of water soluble and exchangeable ions of Mn^{2+} and Fe^{2+} of a soil, in relation to the saturation time with seawater. (Reproduced in a modified form, with permission, from BRÜMMER 1974)*

limited supply of oxygen, which leads certain soil microorganisms to make use of alternative electron acceptors for their respiratory oxidation. Nitrate is the first nutrient to be used as an oxygen source by anaerobic microorganisms (PATRICK & MAHAPATRA 1968). Further lowering of the redox potential causes a reduction of manganese-(IV, III)-oxides to the water-soluble and plant-available manganese-(II)-ions, which occur already within two days of waterlogging. Therefore, frequent fluctuations in the oxygen status of the soil increase the availability of this heavy metal. When the redox potential of the soil decreases to about 150 mV, apart from amorphous iron, iron-(III)-oxides are reduced to the water-soluble Fe-(II)-ions, which usually does not start until the third of the fifth day of flooding. Persistence of waterlogging conditions causes the reduction of sulphate to soluble sulphides such as S^{2-}, HS^{-}, or H_2S. These sulphides not only react with the plant-available iron and thus create an iron deficiency, but are considered to be highly toxic to plants and may be responsible for the vegetation differentiation in wet grassland communities (YERLY 1970).

Iron in the ferrous form gives waterlogged soils their black, grey or blueish-green colour, whereas the ferric state is characterized by a brown, red or yellow colour. This colour difference makes it possible for the ecologist to estimate the efficiency of oxygen transport from the shoot via the aerenchyma to the root. If the oxygen loss by diffusion from the roots is sufficiently rapid and the demand for oxygen by the microorganisms in the vicinity of the roots is lower than the diffusion rate, an oxygenated zone will build up around the roots. Waterlogged soils offer an example of how microorganisms create a stress environment and the way in which higher plants can restore the situation, in this case via an efficient aerenchyma. But not all plant species are capable of evolving an efficient oxygen transport system. For an analysis of the impact of waterlogging on plant growth, we will consider a gradient from a dry to a wet environment and limit ourselves to the effects of mineral nutrition. (For the effects of oxygen deficiency, see the paper by BROUWER in this volume).

The effect of waterlogging on plants in a mineral soil over a period of 90 days was investigated by comparing plants with a different tolerance for flooding. Inundation has a tremendous effect on the biomass production of the common weed *Stellaria media* (Table 3); it decreases the growth of *Holcus lanatus* slightly

TABLE 3. *Effect of waterlogging on biomass production (mg d.w.) and the iron and manganese content (μM g^{-1} dry weight) of* Stellaria media *(non-tolerent),* Holcus lanatus *(moderately tolerant), and* Juncus articulatus *(very tolerant)*

		Stellaria		*Holcus*		*Juncus*	
Biomass	dry*	3,508 + 448		2,271 + 105		2,143 + 251	
	wet	136 + 59		1,573 + 97		2,575 + 526	
		root	shoot	root	shoot	root	shoot
Fe content	dry	12.39	1.46	18.44	1.97	9.67	2.54
	wet	19.46	1.43	27.57	2.86	17.01	2.65
Mn content	dry	1.47	0.86	1.49	0.75	0.38	1.07
	wet	1.49	1.13	1.89	0.69	2.17	3.49

* During the 90 days of the experiment the wet soil conditions were maintained
  by keeping the water-table at the soil surface and the dry soil was kept at
  70% of its field capacity

and stimulates the growth of *Juncus articulatus*. Waterlogged plants of all three species take up more iron from the soil than do plants of the same species in drier conditions. The increase was particularly marked in the roots, as had already been found for other graminoids and herbs (JONES & ETHERINGTON 1970; JONES 1972; ROZEMA & BLOM 1977). The degree to which internal concentrations of these heavy metals affect the metabolism of the plant seems to be dependent on the internal compartmentation, as reported for *Ericaceae* (HENRICHFREISE 1973; BAUMEISTER & ERNST 1978), and remains to be determined for all other species. In our experiment the same total amount of iron and manganese caused toxicity in *Stellaria media* and promoted growth in *Juncus articulatus*.

The liberation of soluble divalent manganese and iron not only determines vegetation differentiation on dunes and dune slacks (JONES 1972) and wetlands, but also regulates the distribution of woodland plants of well-drained micro-habitats, for instance *Mercurialis perennis* (iron-sensitive), and those of less well-drained situations, for instance *Primula elatior* (iron-tolerant), as demonstrated by MARTIN (1968).

4. SOIL REACTION AND THE AVAILABILITY OF ALUMINIUM

The emphasis put on pH values in ecological studies is due more to the ease of determination than to the complex correlation of pH values with many interacting mineral elements. Hydrated ions of aluminium, the third most common element in the earth's crust, play an important role in the buffering of soils against excess acidity. In all primary and secondary clay minerals, aluminium is essential for the chemical structure and the exchange capacity of the soil more or less saturated with H^+, Al^{3+} or hydroxy-aluminium polymers.

As demonstrated by the analysis of twenty-five woodland soils in Germany, exchangeable aluminium tends to increase with increasing hydrogen ion concentrations, i.e. pH values lower than 5, which is in good agreement with the results of experimental studies on this problem (SCHEFFER & SCHACHTSCHABEL 1976). The exchangeable amounts increase from 0.4 mM at pH 6.0 to more than 90 mM Al kg^{-1} dry weight at pH 3.0, independent of the total amount of 437 $\pm$ 100 mM kg^{-1} (Fig. 3). The same tendency is found for the water soluble fraction. Despite these known chemical aspects of acid soils, few attempts have been made to analyse the effect of aluminium ions on plant growth. Most plants are very susceptible to aluminium, especially those growing on soils with a pH of 5.0 and higher (CLYMO 1962; HACKETT 1965; CLARKSON 1966, 1969; HENRICHFREISE 1976). The growth of roots particularly is affected, since aluminium interferes with DNA synthesis (SAMPSON *et al*. 1965). These results have been confirmed by the comparison of two populations of *Agrostis tenuis,* one collected from an acid (pH 3.2) and the other from an alkaline site (pH 7.2). Plant growth was strongly reduced by aluminium in the alkaline population, but was only weakly decreased in the acid population. These results suggest that in soils with lower pH

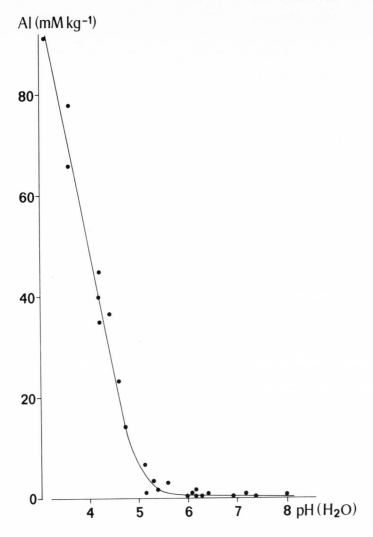

FIG. 3.   Relationship between pH (H₂O) and the (NH₄-oxalic acid)-
extractable amount of aluminium ions of woodland soils in
northern Germany. The total amount of aluminium is 437 ±
100 mM kg⁻¹ dry weight

values aluminium will form a barrier to the colonization by plants originating
from neutral or alkaline habitats. But even within species of acid sites there
seems to be a dramatic differentiation in the sensitivity to aluminium
(HENRICHFREISE 1976; GRIME & HODGSON 1969).

The ecological importance of pH values is often exaggerated. Other basic soil
factors are of greater importance than the hydrogen ions. This may be illustrated
by data on *Deschampsia flexuosa*. According to ELLENBERG (1958) this plant has
its ecological optimum in the vicinity of its physiological minimum. However,
GIGON & RORISON (1972) found good indications that a physiological optimum of

pH 7.2 can only be achieved by supplying nitrogen as ammonia, not as nitrate. Under natural conditions, ammonium ions are not present at this pH (DIERSCHKE 1974), which means that the growth optimum of this species depends not on pH values but on the oxidation status of nitrogen. Furthermore, *Deschampsia flexuosa* has a high demand for iron, but at high pH values there is a decrease in availability of this metal (HENRICHFREISE 1976).

5. COMPLEXATION OF METALS AND PLANT GROWTH

Susceptibility of plants to high concentrations of heavy metals in the soil is a question not only of quantity, but also of quality. The availability of heavy metal cations is influenced in characteristic ways by the soil environment. Under acid conditions these cations have the highest solubility and are most available to plants. If they are tightly bound or fixed to silicate clays they are unavailable. On soils containing organic matter, organic ligands may form stable complexes (chelates) with metal ions. Plants growing on heavy metal rich soils are more or less specifically tolerant to the particular heavy metal predominating in the habitat of the plant to which they belong (for reviews see ANTONOVICS *et al*. 1971; ERNST 1974, 1976; and PROCTOR & WOODELL 1975). But as in halophytes, there seems to be a discrepancy between the ecological and the physiological growth optima with respect to the metal concentration (Fig. 4). *Silene cucubalus,* one of the prominent species in such habitats, appears to tolerate up to 21.1 mM water-soluble zinc in its habitat, whereas in water culture the plant can only tolerate up to 2 mM $ZnSO_4$ with a growth optimum at 0.8 mM Zn (BAUMEISTER 1954; ERNST 1974). In contrast, cell tissues of shoots and leaves are zinc resistant up to 200 mM Zn (GRIES 1966), whereas the enzyme nitrate reductase only tolerates 0.0015 mM Zn *in-vitro* (ERNST *et al*. 1974; MATHYS 1975), which indicates a strong compartmentation within the cells. This situation is complicated by the different responses of ecotypes from normal and zinc-enriched environments; the non-tolerant populations have optimum growth at concentrations, where the zinc-tolerant populations grow under suboptimal conditions. The low zinc efficiency of zinc-tolerant plants is not associated with a lower uptake of this element -as is the case for differences in iron efficiency (BROWN *et al*. 1971)- but it is related to internal competition between zinc complexation (followed by transport to the vacuole system) and the zinc-stimulated enzymes (ERNST 1976). Non-tolerant plants lacking this tolerance mechanism have a higher zinc efficiency and show better growth at low zinc concentrations, but have a lower zinc resistance than tolerant plants.

Let us restrict the problem to the occurrence of tolerant populations of *Silene cucubalus* in the field and their growth in the laboratory. A careful analysis of the water-soluble zinc in the soil revealed the impact of complexation. Dependent on the organic status of the soil from 27.6 up to 100 per cent of the zinc was organically complexed (ERNST 1974). In the water culture experiments

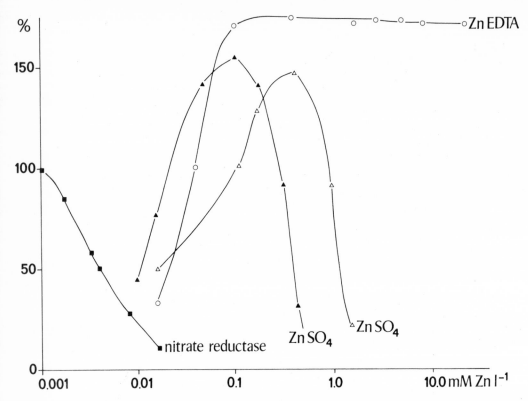

FIG. 4.　Silene cucubalus: *effect of increasing concentrations of zinc on the in-vitro activity of nitrate reductase* (ERNST *et al*. 1974) *and on the growth of zinc-tolerant (open symbols) and non-tolerant (solid symbols) plants in the presence of ionic and complexed zinc* (ERNST 1968, 1974, *and* unpublished). *Control values of the optimum of activity and growth of a non-tolerant population at 0.03 mM Zn are defined as 100%*

with zinc in the ionic and the complexed form as Zn-EDTA - EDTA is one of the best known chelating agents and has been applied to soils as an extractant of heavy metals (LINDSAY 1972) - no growth reduction was found up to 9.2 mM Zn in the presence of the metal bridge complex (ERNST 1968). This indicates that there is no difference between the physiological and ecological optima.

So far, most experiments with complexing agents have been carried out with artificial complexes such as EDTA (aluminum-EDTA: BARLETT & RIEGO 1972; iron-EDTA: BROWN & AMBLER 1974; manganese-EDTA: HENRICHFREISE 1976), all of which reduce the toxicity of the element. To approximate the natural situation we carried out some experiments with natural complexes, such as fulvic and humic acids for comparison with the artificial compounds (FRANCKE 1976; FIT 1977). The first results obtained with the natural compounds (Fig. 5) indicate that metal complexes with fulvic acid have the same or even more negative effects on plant growth than metals in the ionic state. However, the effect of humic

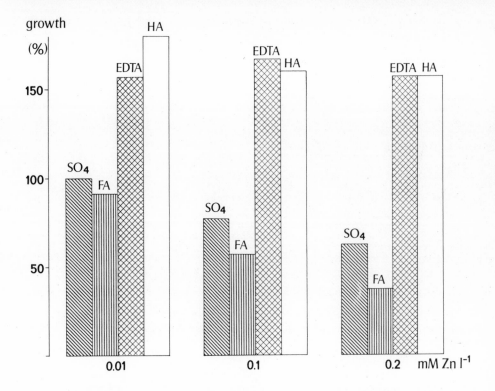

FIG. 5.   *Effect of ionic and complexed zinc on the growth of a non-zinc-
tolerant population of* Agrostis stolonifera. *Growth in a culture
solution with 0.01 mM ZnSO4 was calculated to 100%. Complexing
agents: EDTA = ethylene diaminetetraacetic acid, HA = humic acid,
FA = fulvic acid* (FRANCKE 1976)

acid is generally comparable with that of EDTA, both lowering the toxicity of
high amounts of zinc (ERNST 1968) and copper (SCHILLER 1974) and stimulating
plant growth.

Both, fulvic and humic acids are essential parts of the humic substances
synthesized by the humification process in every soil (FLAIG *et al*. 1975). At
present we have a rough idea on the ratio of humic acid to fulvic acid in soils.
The fulvic acid content of soils is inversely related to the content of humic
acid; the latter predominates in rendzina soils, whereas fulvic acids are
characteristic for podzolic and brown soils. Humic and fulvic acids are occurring
in most of the environments and are able to interact with all metals, so that
they are certain to become important objects in the future plant ecology. The
effect of these substances on the soil should be evident from a podzol profile.
Beneath a 40-60 year old coniferous woodland on the sandy soil of the Veluwe
region of The Netherlands fulvic acids and other organic compounds have caused
loss of iron and manganese from the upper soil horizon and enrichment of these
metals in the lower horizons (Plate 1). The difference in colour between horizons

rich and poor in iron and manganese makes this process visible, but it occurs
in the same way for other transportable nutrients such as potassium, magnesium,
phosphorus, nitrate, and sulphate which have no colour. Therefore, real progress
in the analysis of ecosystems or small parts of them, especially the so called
moderate environments, is only to be expected from investigations of the local
differentiation of the soil profile.

6. NUTRIENT AVAILABILITY IN SPACE

6.1. SOIL PROFILES

Undisturbed soils are characterized by a profile which is the product of the
interaction of the atmosphere and biosphere on the one hand, with the lithosphere
and hydrosphere on the other hand. The soil profile may be quite simple or very
complicated, as already shown for podzolic soils. Anyway, the profile determines
the distribution of minerals in the soil, as can be demonstrated for an acid
(podzolic) and a neutral (rendzina) soil (Fig. 6). Independent of the soil type,
exchangeable calcium, magnesium, and potassium, phosphorus and available nitrogen
in the form of nitrate are not equally distributed over the various soil horizons.
The biological accumulation of these elements in the top layers is evident in
these soils, as has repeatedly been demonstrated for macro- and micronutrients
in forest soils (see e.g. WRIGHT *et al.* 1955; SHAROVA 1957; SCHMIDT 1970;
DIERSCHKE 1974; HINNERI 1974). The soil habitat exhibits small-scale mosaics
varying in nutrient availability which can interact with the rooting behaviour
of plants.

If we consider the distribution of roots in a beech wood on calcareous rocky
soil (rendzina type) we can distinguish between a group of plant species rooting
only in the upper 5 cm of the A_{h1} horizon such as *Anemone nemorosa* and *Galium
odoratum,* a group rooting in the A_{h1}/A_{h2} horizon to a depth of 10-15 cm, for
instance *Mercurialis perennis, Milium effusum, Carex sylvatica*, and *Sanicula
europaea,* and those plants whose roots penetrate to a depth of 30 cm, but not
beyond the lower limit of the C_v horizon (*Allium ursinum, Arum maculatum,*
shrubs, and trees). This distribution of the roots in the soils has the advantage
of decreasing the competition for space, but it creates different nutritional
situations for the species in question. The heterogeneously distributed nitrate
may be suitable for a further analysis of this aspect, because these woodlands
are rich in nitrate, permitting luxury consumption. In fact, analysis of the
cell sap of leaves as a storage pool for nitrate (SCHNURBEIN 1967) showed that
nitrate was more abundant in the leaves of the flat-rooted species *Anemone
nemorosa* and *Galium odoratum* than in the deep rooted *Allium ursinum* and *Arum
maculatum,* independent of the calcareous woodland sites sampled (NIESKE 1973;
SUTTNER 1974; PLADEK-STILLE 1974). The nitrate concentrations were in *Anemone
nemorosa* 29.9 µM, *Galium odoratum* 26.5 µM, *Mercurialis perennis* 11.0 µM,
Sanicula europaea 15.7 µM, *Allium ursinum* 12.7 µM, and *Arum maculatum* 15.4 µM.

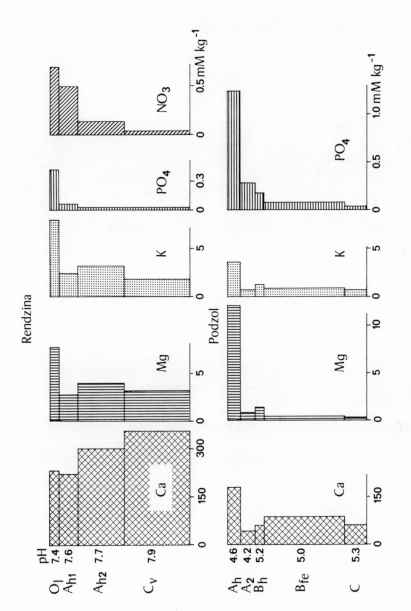

FIG. 6. *Availability of nutrients in a rendzina soil profile on Cenoman chalkstone and of a podzol profile on Gault sandstone in the Teutoburg Forest near Brochterbeck. Availability was measured: for nitrate in an aqueous extract, for potassium and phosphorus on the basis of a lactic acid (pH 3.2) extraction, and for calcium and magnesium from exchange with 1 M ammonium acetic acid (pH 7.0)*

PLATE 1. *Irregular distribution of humic and fulvic acid in the vicinity of roots of* Betula alba *in a podzol profile of the sand in the Veluwe area near Zwolle*

6.2. LOCALLY CONCENTRATED LEACHATES

Nutritional differentiation in plant communities is not only dependent on the soil situation, but can strongly be influenced by the plants themselves. One of the factors involved here is the canalization of water in a woodland system. Everyone knows the difference between the two sides of the trunk of a beech or oak, one exposed to the rain, the other in the rain shadow, both differentiated by their colonization with epiphytes. Besides water there is also a difference in nutrients, due to the input and output of leachates. Leaching is defined as the removal of substances from plants by the action of aqueous solutions such as rain, mist, dew, and fog. Most authors have losses of substances from foliage in mind (for a review see TUKEY 1970). Under the influence of the branching system of the tree, some part of the leachates collects on the trunk representing the stemflow. This water is considerably enriched with nutrients as compared with the normal rain water and throughfall (Table 4), as has been

TABLE 4. *Mineral content of water from rainfall, throughfall, and stemflow (μM l^{-1}) and of bark (μM g^{-1} dry weight) from beech (Fagus sylvatica) and oak (Quercus robur) in the rain shadow (s) and rain channel (c). Site: calcareous woodland near Münster (Germany), 1972 (Data from SCHMITZ 1973 and ERNST unpublished)*

Element	Season	Rainfall	Throughfall	Stemflow		Bark			
				Quercus	Fagus	Quercus		Fagus	
						s	c	s	c
K	winter	14.1	75.7	260.9	69.3	11.4	7.5	22.6	36.7
	summer	20.9	63.9	339.4	94.6				
Na	winter	63.5	166.5	238.0	127.0	8.2	5.6	3.7	3.2
	summer	31.7	79.6	235.4	124.1				
Mg	winter	17.9	49.6	92.8	26.8	3.5	3.3	10.4	10.7
	summer	21.7	75.8	101.3	47.5				
Ca	winter	42.8	90.0	500.0	80.7	599	387	369	372
	summer	40.5	141.3	498.0	96.5				
P		–	–	–	–	4.8	6.7	12.2	13.9

described for mixed oak woodlands in Belgium (DENAEYER-DE SMET 1969) and England (CARLISLE et al. 1966). The stemflow of oak contains distinctly higher amounts of calcium, potassium, sodium, and magnesium than that of beech. Bark analysis of the rain channel and the rain shadow areas of beech and oak (0.5 m above soil surface) supported the suggestion that the leachates derive not only from leaves, but also from the bark itself (SCHMITZ 1973). This stemflow provides an appreciable source of nutrients for plant growth, for instance for mosses beneath forest trees, but may also induce ecological disturbance lowering the pH and adding less favourable elements in the vicinity of the stems. The latter effect can be observed, if stemflow of trees is strongly channelled and a zone bare of herbs and grasses occurs.

6.3. DIFFERENTIATION OF LEAF LITTER IN SPACE AND TIME

One of the factors which builds up the local pattern of nutrients is the leaf litter of the various plant species. The divergent rooting behaviour of plants in a woodland leads to the exploitation of different nutrient pools. Most of the nutrients in the leaves of trees originate from a greater depth than those of the flat-rooted herbs. As a result, the annual supply of leaf litter, especially that of trees, has a strong impact on the herbs of the humus rich horizon (A_h). If we regard leaf litter as a form of fertilization of the herb layer, the seasonal litterfall may be compared with agricultural fertilization practices. In the beech woodland near Münster the dosage of nitrogen is equivalent to a grassland fertilization with 70 kg N/ha. Under favourable natural circumstances potassium too can reach usual agricultural levels. However, the annual addition of phosphorus to the herb layer (0.5 kg P/ha) is only a fraction of the amount given in normal agricultural practice which indicates that phosphorus may be a critical element in such woodlands. This situation is further complicated by the high amount of calcium, which is disadvantageous for the availability of phosphorus and sulphur (Table 5).

TABLE 5.   *Nutrient contents of leaf litter in a calcareous beech woodland (Teutoburg Forest, 1970)*

Species	Dry matter	Nutrient (mg m^{-2})							
	(g m^{-2})	N	P	K	Ca	Mg	Fe	Mn	Zn
Fagus sylvatica	251.1	5,400	221	1,255	5,650	276	50	63	11
Carpinus betulus	29.8	748	81	146	975	70	8	5	2
Allium ursinum (pure stands, June)	45.2	1,440	146	1,012	698	125	20	3	2
Mercurialis perennis (pure stands, Oct.)	32.0	918	68	808	1,280	117	11	1	2

Furthermore, there is a considerable local differentiation of fertilization in woodlands which is dependent on the plant species. The growth pattern of species such as *Allium ursinum* or *Mercurialis perennis* which are clustered in great patches creates a horizontal differentiation in the nutrient concentration, because the litter of *Allium* is rich in phosphorus, but poor in calcium, and *vice versa* for *Mercurialis perennis*. This situation can be further stressed or equalized by the litterfall of the trees, dependent on the species composition and their abundance in these woodlands and on the climatic conditions at the time of shedding, especially on the wind velocity.

In contrast to the normal agricultural situation the plants of the herb layer are predominantly perennials and have mechanisms for the internal re-translocation of nutrients as demonstrated by the efficient translocation of phosphorus from leaves to rhizomes in *Anemone nemorosa* and other herbs as well as in trees (ERNST 1978), unlike potassium in herbs (Table 6). Furthermore, the well developed mycorrhiza of these species also aids in the phosphorus nutrition of plants (see MOSSE's paper in this volume), perhaps with the disadvantage of a too large supply of oligo-elements (ERNST 1978).

TABLE 6. *Translocation of phosphorus from rhizomes to leaves and* vice versa, *in* Anemone nemorosa *in a calcareous beech woodland, as indicated by the concentrations of phosphorus and potassium (mM kg^{-1} d.m.) during the growing season*

		Developmental stage				
		Folded leaves	Unfolded leaves	Flowering	Fruiting	Leaffall
P	leaves	26 + 1	61 + 3	48 + 1	27 + 0.3	19 + 1
	rhizomes	47 + 1.5	34 + 0.2	36 + 1	51 + 2	69 + 2
K	leaves	132 + 3	130 + 3	136 + 1	147 + 9	137 + 7
	rhizomes	110 + 9	98 + 2	96 + 2	110 + 3	110 + 3

7. NUTRIENT AVAILABILITY IN TIME

7.1. MINERALIZATION

The input of nutrients by leaffall both at our research site and in other eco-systems (DUVIGNEAUD 1968; NIHLGÅRD 1972; GLOAGUEN & TOUFFET 1974) demonstrates not only the differentiation of nutrient input in space -horizontal as well as vertical- but also in time. In some of the geophytes such as *Ficaria verna* and *Allium ursinum* leaffall occurs as early as May or June, whereas herbs such as *Mercurialis perennis* shed in October together with the shrubs and trees.

Most of the nutrients in leaf litter, however, are not available just after shedding. Microorganisms act beneficially by transforming unavailable minerals and organic compounds of leaf litter into forms available for uptake by higher plants. The process is called (re)mineralization. The special effects of the soil microflora are highlighted in the paper by WOLDENDORP in this volume, so that I can confine myself to some general statements. Moisture and temperature regulate mineralization, which is promoted by warm and moist conditions and retarded at low values of these abiotic factors. Let us turn now to the seasonal pattern of mineralization of nitrogeneous compounds. There is a wide belief that nitrification is suppressed by temperatures below 5°C (ELLENBERG 1964; WILLIAMS 1968, 1969). This hypothesis is contradicted by the results obtained by ANDERSON et al. (1971), who demonstrated the adaptation of nitrifiers to low temperature. But there is also an ecological phenomenon that argues against it. In all of the ecosystems analysed, nitrogen mineralization was measured throughout the year and as can be seen from Fig. 7 the mineralization rate is not higher in warm and moist summers than in moist winter periods (see JANIESCH 1973; DAVY & TAYLOR 1974; DIERSCHKE 1974). In our calcareous beech woodland, mineral nitrogen accumulated in the soil more or less exclusively in the form of nitrate except at sites with much stemflow where ammonium-nitrogen was also present. This continuous mineralization in woodland soils was also

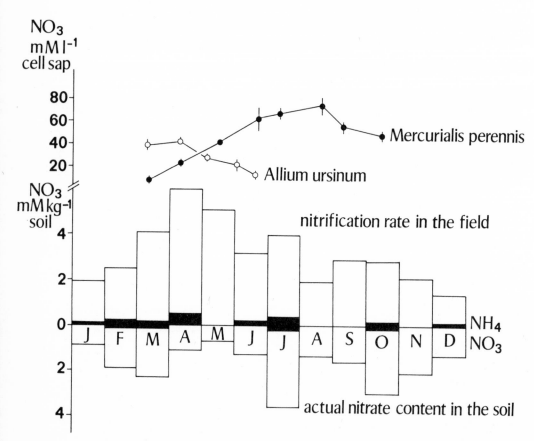

FIG. 7.   *Actual nitrate content (mM kg^{-1}) and nitrification rate in the field (3 weeks) of an A_0-horizon of a rendzina beech forest in relation to the nitrate content of the cell sap (mM NO_3 l^{-1}) from leaves of* Mercurialis perennis *and* Allium ursinum *during one year (January-December)*

demonstrated for phosphate by SCHMIDT (1970). Incubation of soil samples under field conditions -in the absence of higher plants- yielded higher amounts of nitrate nitrogen than the actual soil content at sampling time. Perhaps the most significant feature observed was the pronounced actual concentration of nitrate in the soil, comparable with the values obtained for nitrophytic edge vegetations of Central European forests (JANIESCH 1973; DIERSCHKE 1974). These results seem to contradict the concept of a limited nitrogen supply in these ecosystems. It is conceivable that the species of these soils cannot compensate the nitrate production in the soil by nitrate uptake and that the herbs are not only nitrophytes, i.e. plants growing in a soil rich in nitrogen, but also nitrophilous, which indicates a tendency toward a high nitrogen demand. According to a general interpretation the growth of nitrophilous species is stimulated and optimized by at least 14.7 mM NO_3 (WALTER 1963) which is equivalent to the nitrate content of 1 litre of Knop's culture solution. The supposition on the nitrophilous nature of the above-mentioned species is based on the nitrate concent of cell sap, which is a storage pool for nitrate within the plant. Nitrate uptake was found to be particularly strong so that a considerable amount of nitrate is present in the cell sap of herbs (NIESKE 1973; SUTTNER 1974; PLADEK-STILLE 1974), as shown in Fig. 7. However, the accumulation of nitrate in cell sap, i.e. in the vacuoles of the cell, cannot only be interpreted as a storage pool at times of luxury consumption, but it also suggests an ecological necessity for osmotic adjustment (JANIESCH 1973) or might be a sign of an inefficient activity of the nitrate reducing system, the nitrate reductase.

To elucidate the efficiency of nitrate reduction, nitrate utilization was assessed by the *in vivo* assay for nitrate reductase (STREETER & BOSLER 1972), which reflects the actual reduction of nitrate by the plant and its dependence on the nitrate supply (LEE *et al*. 1974), whereas the *in vitro* technique measures the potential activity of this enzyme and provides an indication for the nitrogen requirement of plants (BAR AKIVA *et al*. 1970). As far as *in vivo* activity of these woodland herbs is concerned, they show an appreciable actual activity which is comparable with that of other species thought to be nitrophilous (LEE *et al*. 1974). Therefore, the accumulation of nitrate in the vacuole might suggest a disturbance in the synthesis of nitrate reductase, which is dependent on the micronutrient molybdenum (NICHOLAS 1961), and might indicate a molybdenum deficiency in these habitats. To check this hypothesis seedlings of *Galium aparine,* a possibly nitrophilous plant of the edge vegetation of these woodlands (DIERSCHKE 1974), were grown in culture solution with increasing amounts of molybdenum (6-6,000 µM Mo). After 4 weeks of growth the activity of nitrate reductase and the nitrate accumulation in the cell sap were markedly affected by molybdenum concentrations higher than 60 µM Mo. Our results (Table 7) support previous reports of molybdenum sensitivity in the nitrophilous *Urtica dioica* and *Chenopodium album* (AUSTENFELD 1969). These results show that the marked accumulation of nitrate in these plants is not associated with and caused by

TABLE 7.   *Effects of increasing amounts of molybdenum on the growth, nitrate content of cell sap (shoots) and nitrate reductase activity of leaves of* Galium aparine *after 4 weeks of growth in a culture solution in a greenhouse at 20° $\pm$ 2°C and 12 hours of light day^{-1}*

Mo concentration in culture solution ($\mu M.l^{-1}$)	Dry matter (mg.plant^{-1})	Nitrate in cell sap ($\mu M.ml^{-1}$)	*In vivo* assay of nitrate reductase ($\mu M\ NO_2\ g^{-1}.h^{-1}$)
6	614 $\pm$ 24	130 $\pm$ 7	1.08 $\pm$ 0.07
60	624 $\pm$ 37	134 $\pm$ 11	1.13 $\pm$ 0.15
600	414 $\pm$ 28	252 $\pm$ 15	0.65 $\pm$ 0.09
6000	180 $\pm$ 18	306 $\pm$ 13	0.35 $\pm$ 0.05

a molybdenum deficiency, but is an index of luxury consumption.

It is also of interest that the mineralization of nitrogen yields not only nitrate, but also a reduced form of nitrogen, i.e. ammonia, depending on various edaphic factors. Although many laboratory experiments have been performed on nitrate and ammonium ions as nitrogen sources (see e.g. EVERS 1964; BOGNER 1968; GIGON & RORISON 1972; VAN ANDEL 1974; LEE et al. 1974) relatively few attempts have been made to compare the performance of plant populations adapted to soils with ammonium or nitrate as nitrogen sources. Populations of *Agrostis stolonifera* from soils with nitrate or ammonia did not differ significantly in the rate of ammonium assimilation, but differed slightly under a nitrate treatment (LEE et al. 1974). The experiments with populations of *Urtica dioica* and *Epilobium angustifolium* from acid (pH 3.8-5.7) and alkaline (pH 7.8-8.1) habitats (Fig. 8) revealed that growth of all populations given an NO_3-N treatment at pH 6.4 was significantly better than in an ammonium-nitrate and a pure ammonium series, both at pH 5.8 (FISCHER 1977). Populations of *Scabiosa columbaria* sampled in different habitats (GIGON & RORISON 1972) also grew better in nitrate than in ammonium treatments. These results suggest the existence of different effects of ammonium and nitrate on plant growth, especially if a surplus of these nitrogen forms is present.

The generally higher susceptibility of plants to ammonium than to nitrate seems to be an ecophysiological paradox in view of the lower energy demand for the assimilation of ammonium. Ammonium ions can be directly incorporated into organic nitrogen compounds, whereas nitrate has to be reduced to ammonium by two enzyme systems (nitrate reductase and nitrite reductase). The nitrate pathway makes the plant dependent not only on the synthesis of these two enzymes, but also on the supply with certain micronutrients. However, in a situation of surplus, nitrate has the advantage of being able to form a storage pool in the vacuoles and of being metabolized on demand. In contrast, ammonium must be metabolized rapidly to prevent metabolic disturbance (see e.g. MATSUMOTO et al.

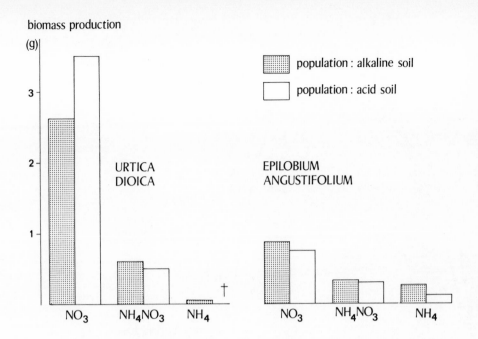

FIG. 8. *Effect of the oxidation state of nitrogen on the biomass production of both an acid and an alkaline population of* Urtica dioica *and* Epilobium angustifolium (FISCHER 1977)

1971). It is therefore of particular interest that the detoxification of ammonia via the formation of acid amides (asparagine, glutamine), ureides, and other compounds (for a review see MOTHES 1958) requires more energy and metabolic reactions than the transport of nitrate to the vacuole system. This aspect of the high susceptibility of many plant species to a surplus of ammonia should be given special attention in further ecological research (KIRKBY 1969).

At this point I wish to make some remarks on those plants which colonize or grow in habitats rich in nitrate. It seems that soils of woodlands and forest borders and skirts are sometimes characterized by such a high nitrification rate that the plants are unable to consume all the soil nitrate. The soil solution around the roots sometimes contains more than 10 mM NO_3, as has been shown for the plant communities *Agropyro-Aegopodietum* (JANIESCH 1973) and *Alliario-Chaerophylletum* (DIERSCHKE 1974). Despite the small number of soil analyses available for these environments, the conclusion of vegetation scientists has been that plants of these soils have a distinctly higher demand for nitrate. These plants can be expected to be nitrophilous. To verify this, I have compiled results on the growth behaviour of these species which are thought to be pronounced nitrogen indicators with an indicator value of 8 or 9 in the scale of ELLENBERG (1974). The behaviour of these N8 plants proved to be highly variable (Fig. 9). Some of the species such as *Rumex obtusifolius*

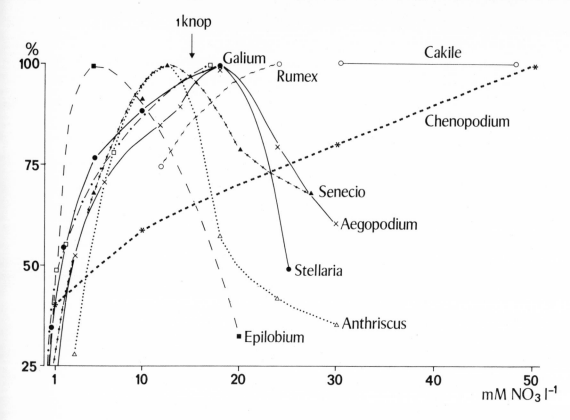

FIG. 9.   *Growth of "nitrophilous" species in water culture with increasing
amounts of nitrate. Optimum growth is defined as 100%. Results for*
Rumex obtusifolius *obtained by* TIJBERG (1974), Epilobium
angustifolium *by* VAN ANDEL (1974), Aegopodium podagraria *and*
Anthriscus sylvestris *by* JANIESCH (1973), Senecio sylvaticus *by*
VERA (1975), Chenopodium album *by* AUSTENFELD (1972), Cakile
maritima *by* ERNST (1969), Stellaria media *and* Galium aparine *by*
ERNST (unpublished)

(TIJBERG 1974) and *Cakile maritima* (ERNST 1969) have a splendid growth up to
at least 25 mM NO₃. Even higher concentrations are demanded by *Chenopodium
album*, a N7 species (AUSTENFELD 1972). Other N8 species such as *Aegopodium
podagraria* (JANIESCH 1973) or *Senecio sylvaticus* (VERA 1975) more or less
tolerate a nitrate concentration of 14.7 mM, but at higher concentrations
those species show a reduced biomass production. However, *Epilobium angustifolium*
regarded as very nitrophilous (N8) is quite sensitive to this amount of nitrate
(VAN ANDEL 1974, 1976). This situation underlines the unreliability of such
speculations on the soil-plant relationship when the factor in question has
not been measured.

If forests are cleared by storms, fire or human activities, the microclimate is changed especially with respect to light and temperature. This generally causes stimulation of nutrient mobilization, in particular nitrification, with a special response of vegetation in these clearfelled areas. The pioneer vegetation establishing in this secondary habitat in western, central and northern Europe is characterized by *Epilobium angustifolium* which is associated with other typical species such as *Digitalis purpurea, Senecio sylvaticus*. Besides differences in the life cycle strategy of *Epilobium* and *Senecio* (VAN ANDEL & VERA 1977), these species also show a quite different response to the nutrient budget of the soil. *Senecio sylvaticus* as an annual species requires large amounts of nitrogen for optimal growth and seed production (12 mM N) and is less nitrogen efficient than *Epilobium angustifolium,* which does not need large amounts of nitrogen, but possesses a certain degree of tolerance against higher nitrogen concentrations. Both plants produce wind-spread propagules, the dissemination further reduces the nutrient budget of the soil (VAN ANDEL 1974), in addition to the removal of the felled trees, and leaching (TAMM *et al.* 1974). However, the loss of nutrients, especially of nitrogen in these clearings is not as dramatic as during the drainage of the Zuiderzee to form the IJsselmeerpolders, when there was an annual loss of nitrogen by fruits of the annual *Senecio congestus* amounting to 60 kg N ha^{-1} (BAKKER 1960; VAN SCHREVEN 1963). In moderate situations as in these clearings, perennial plants such as *Epilobium angustifolium* conserve nutrients in their root stocks and therefore the nutrients of the mineralized leaf litter will be available to the next succession stage. However, our knowledge concerning this aspect of nutrient availability is less than scarce.

8. ACKNOWLEDGEMENTS

I am extremely grateful to Mrs. T.F. Lugtenborg and Mr. H.J.M. Nelissen for their skilful technical assistance, to Mr. G.W.H. van der Berg for preparing the figures, and to Mrs. V.F. Rappange for typing the manuscript.

9. REFERENCES

ALBERT, H. & H. KINZEL, 1973 - Unterscheidung von Physiotypen bei Halophyten des Neusiedlerseegebietes (Österreich). *Z. PflPhysiol.*, 70, 138-157.
ALBERT, R. & M. POPP, 1977 - Chemical composition of halophytes from the Neusiedler Lake region in Austria. *Oecologia,* 27, 157-170.
ANDEL, J. VAN, 1974 - An ecological study on *Chamaenerion angustifolium* L. Scop. *Ph. D. Thesis, Free University, Amsterdam,* 85 p.
ANDEL, J. VAN, 1976 - Growth and mineral nutrition of *Chamaenerion angustifolium* L. Scop. (= *Epilobium angustifolium* L.) on culture solution. *Oecol. Plant.,* 11, 25-40.

ANDEL, J. VAN & F. VERA, 1977 - Life cycle and nutritional behaviour of *Senecio sylvaticus* L. and *Chamaenerion angustifolium* L. Scop. (= *Epilobium angustifolium* L.). *J. Ecol.*, 65, 747-758.

ANDERSON, O.E., BOSWELL, F.C. & R.M. HARRISON, 1971 - Variation in low temperature adaptability of nitrifiers in acid soils. *Proc. Soil Sci. Soc. Am.*, 35, 68-71.

ANTONOVICS, J., BRADSHAW, A.D. & R.G. TURNER, 1971 - Heavy metal tolerance in plants. *Adv. Ecol. Res.*, 7, 1-85.

AUSTENFELD, F.A., 1969 - Einfluss unterschiedlicher Molybdän- und Stickstoff-gaben auf die Nitratspeicherung der nitrophilen Pflanzen. *Staatsexamens-arbeit Biologie, Münster.*

AUSTENFELD, F.A., 1972 - Untersuchungen zur Physiologie der Nitratspeicherung und Nitratassimilation von *Chenopodium album* L. *Z. PflPhysiol.*, 67, 271-281.

AUSTENFELD, F.A., 1974 - Der Einfluss der NaCl und anderer Alkalisalze auf die Nitratreduktaseaktivität von *Salicornia europaea* L. *Z. PflPhysiol.*, 71, 288-296.

AUSTENFELD, F.A., 1976 - The effect of various alkaline salts on the glycolate oxydase of *Salicornia europaea* and *Pisum sativum* in vitro. *Physiologia Pl.*, 36, 82-87.

BAKKER, D., 1960 - *Senecio congestus* (R.Br.) DC. in the Lake IJssel polders. *Acta bot. neerl.*, 9, 235-259.

BAR-AKIVA, A., SAGIV, J. & J. LESHEM, 1970 - Nitrate reductase activity as an indicator for assessing the nitrogen requirements of grass crops. *J. Sci. Fd. Agric.*, 21, 405-407.

BARNETT, N.M. & A.W. NAYLOR, 1966 - Amino acid and protein metabolism in Bermuda grass during water stress. *Pl. Physiol.*, 41, 1222-1230.

BARLETT, R.J. & D.C. RIEGO, 1972 - Effect of chelates on the toxicity of aluminum. *Pl. Soil*, 37, 419-423.

BAUMEISTER, W., 1954 - Über den Einfluss des Zinks bei *Silene inflata* Sm. *Ber. dt. Bot. Ges.*, 67, 205-213.

BAUMEISTER, W. & H. BURGHARDT, 1956 - Über den Einfluss des Zinks bei *Silene inflata* Sm. II. CO_2-Assimilation und Pigmentgehalt. *Ber. dt. bot. Ges.*, 69, 161-168.

BAUMEISTER, W. & W. ERNST - *Mineralstoffe und Pflanzenwachstum.* G. Fischer Verlag, Stuttgart (in press).

BAUMEISTER, W. & L. SCHMIDT, 1962 - Die physiologische Bedeutung des Natriums für die Pflanze. I. Versuche mit höheren Pflanzen. *ForschBer. Landes NRhein-Westf. Nr. 1086, Westdeutscher Verlag, Opladen*, 42 p.

BEER, S., SHOMER-ILAN, A. & Y. WAISEL, 1975 - Salt-stimulated phosphoenolpyruvate carboxylase in *Cakile maritima. Physiologia Pl.*, 34, 293-295.

BEN-AMOTZ, A. & M. AVRON, 1973 - The role of glycerol in the osmotic regulation of the halophilic alga *Dunaliella parva. Pl. Physiol.*, 51, 875-878.

BINET, P., 1968 - Dormances et aptitude à germer en milieu sale chez les halophytes. *Bull. Soc. fr. Physiol. vég.*, 14, 115-124.

BOGNER, W., 1968 - Experimentelle Prüfung von Waldbodenpflanzen auf ihre Ansprüche an die Form der Stickstoff-Ernährung. *Mitt. Ver. forstl. Standortsk. Forstpflanzenzüchtung,* 18, 115-124.

BOUCAUD, J. & I.A. UNGAR, 1976 - Hormonal control of germination under saline conditions of three halophytic taxa in the genus *Suaeda. Physiologia Pl.*, 37, 143-148.

BRECKLE, S.W., 1976 - *Zur Ökologie und zu den Mineralstoffverhältnissen absalzender und nichtabsalzender Halophyten.* J. Cramer Verlag, Vaduz, 169 p.

BROWN, J.C. & J.A. AMBLER, 1974 - Iron stress response in tomato (*Lycopersicon esculentum*). I. Sites of Fe-reduction, absorption and transport. *Physiologia Pl.*, 31, 221-224.

BROWN, J.C., CHANEY, R.L. & J.E. AMBLER, 1971 - A new tomato mutant inefficient in the transport of iron. *Physiologia Pl.*, 25, 48-53.

BRÜMMER, G., 1974 - Redoxpotentiale und Redoxprozesse von Mangan-, Eisen- und Schwefelverbindungen in hydromorphen Böden und Sedimenten. *Geoderma,* 12, 207-222.

CAPPELLETTI PAGANELLI, E.M., 1967 - Il chimismo delle piante di alcune associazioni litoranee durante un ciclo annuale. *Atti Ist. veneto Sci.*

Lett. Art. Scienze mat. e nat., 125, 433-498.

CARLISLE, A., BROWN, A.H.F. & E.J. WHITE, 1966 - The organic matter and
nutrient elements in the precipitation beneath a sessile oak (*Quercus
petraea*) canopy. *J. Ecol.*, 54, 87-98.

CHAPMANN, V.J., 1974 - *Salt marshes and salt deserts of the world*. J. Cramer,
Lehre, 392 p.

CLARKSON, D.T., 1966 - Aluminium tolerance in species within the genus
Agrostis. *J. Ecol.*, 54, 167-178.

CLARKSON, D.T., 1969 - Metabolic aspects of aluminium toxicity and some
possible mechanisms for resistance. *In:* I.H. RORISON (Editor),
Ecological aspects of the mineral nutrition of plants, Blackwell
Scientific, Oxford, p. 381-397.

CLYMO, R.S., 1962 - An experimental approach to part of the calcicole problem.
J. Ecol., 50, 701-731.

DAVY, A.J. & R. TAYLOR, 1974 - Seasonal pattern of nitrogen availability in
contrasting soils in the Chiltern Hills. *J. Ecol.*, 62, 793-807.

DENAEYER-DE SMET, S., 1969 - Apports d'éléments minéraux par les eaux de pré-
cipitations, d'egouttement sous couvert forestier et d'écoulement le
long des troncs (1965,1966,1967). *Bull. Soc. r. Bot. Belg.*, 102, 355-372.

DIERSCHKE, H., 1974 - Saumgesellschaften in Vegetationsgefälle und Standorts-
gefälle an Waldrändern. *Scripta Geobt., Göttingen*, 6, 1-246.

DUVIGNEAUD, P., 1968 - Bisect biogéochimique et composition des nappes foli-
aires en polyéléments biogènes. *Bull. Soc. r. Bot. Belg.*, 101, 129-139.

ELLENBERG, H., 1958 - Bodenreaktion (einschliesslich Kalkfrage). *In:* W.
RUHLAND (Editor), *Encyclopedia of plant physiology,   4. Mineral
nutrition of plants*. Springer Verlag, Berlin, p. 638-708.

ELLENBERG, H., 1964 - Stickstoff als Standortsfaktor. *Ber. dt. bot. Ges.*,
77, 82-92.

ELLENBERG, H., 1974 - Zeigerwerte des Gefässpflanzen Mitteleuropas. *Scripta
Geobot., Göttingen*, 9, 1-97.

ERNST, W., 1968 - Der Einfluss der Phosphatversorgung sowie die Wirkung von
ionogenem und chelatisiertem Zink auf die Zink- und Phosphataufnahme
einiger Schwermetallpflanzen. *Physiologia Pl.*, 21, 323-333.

ERNST, W., 1969a- Zur Physiologie der Schwermetallpflanzen. -Subzelluläre
Speicherungsorte des Zinks. *Ber. dt. bot. Ges.*, 82, 161-164.

ERNST, W., 1969b- Beitrag zur Kenntnis der Ökologie europäischer Spülsaum-
gesellschaften. I. Sand- und Kiesstrände. *Mitt. flor.- soz. ArbGemein.*,
14, 86-94.

ERNST, W., 1974 - *Schwermetallvegetation der Erde*. G. Fischer-Verlag,
Stuttgart, 194 p.

ERNST, W., 1976 - Physiology of heavy metal resistance in plants. *Proc.
Internat. Conf. Heavy Metals in the Environment, Toronto*, 2, 121-136.

ERNST, W. - Ökophysiologische Bedeutung der Mineralstoffe. *In:* W. BAUMEISTER
& W. ERNST (Editors), *Mineralstoffe und Pflanzenwachstum*. G. Fischer-
Verlag, Stuttgart (in press).

ERNST, W., MATHYS, W., SALASKE, J. & P. JANIESCH, 1974 - Aspekte von Schwer-
metallbelastungen in Westfalen. *Abh. Landesmus. Naturk. Münster*, 36, 1-30.

ERNST, W. & H. WEINERT, 1972 - Lokalisation von Zink in den Blättern von
Silene cucubalus Wib. *Z. PflPhysiol.*, 65, 258-264.

ESSING, B., 1972 - Potentieller osmotischer Druck und Ionengehalte der
Presssäfte bei einigen Halophyten des Nordseewattstrandes. *Ph. D. Thesis,
University of Münster*.

EVERS, F.H., 1964 - Die Bedeutung der Stickstoff-Form für Wachstum und
Ernährung der Pflanzen, insbesondere der Laubbäume. *Mitt. Ver. forstl.
Standortsk. Forstpflanzenzüchtung*, 14, 19-37.

FISCHER, C.V., 1977 - Een oecologisch onderzoek naar het stikstofgebruik van
twee populaties van *Chamaenerion angustifolium* Scop. en *Urtica dioica* L.
bij groei op verschillende stikstofbronnen. *Doct. Verslag Plantenoecologie,
Vrije Universiteit*, Amsterdam.

FIT, B., 1977 - Effect van ionogen Cu en Pb en van kunstmatige en natuurlijke
complexoren op een niet metaalresistente populatie van *Agrostis
stolonifera*. *Doct. Verslag Plantenoecologie, Vrije Universiteit*,
Amsterdam.

FLAIG, W., BEUTELSPACHER, H. & E. RIETZ, 1975 - Chemical composition and
    physical properties of humic acid substances. *In:* J.E. GIESEKING (Editor),
    Soil components, Vol. 1. Organic components, Springer-Verlag, Berlin,
    p. 1-221.
FLOWERS, T.J., 1972 - Salt tolerance of *Suaeda maritima* L. Dum. *J. exp. Bot.,*
    23, 310-321.
FLOWERS, T.J., 1975 - Halophytes. *In:* D.A. BAKER & J.L. HALL (Editors), *Ion
    transport in plant cells and tissues.* North Holland Publ.,Amsterdam,
    p. 309-334.
FOWDEN, L., 1959 - Radioactive iodine incorporation into organic compounds of
    various angiosperms. *Physiologia Pl.,* 12, 657-664.
FRANCKE, J., 1976 - Het effect van ionogen Zn, ZnEDTA, ZnFA en ZnHA op een
    niet Zn resistente populatie van *Agrostis stolonifera. Doct. Verslag
    Plantenoecologie, Vrije Universiteit,* Amsterdam.
GIESEKING, J.E., 1975 - *Soil components. Vol. 2. Inorganic compounds.*
    Springer-Verlag, New York.
GIGON, A. & I.H. RORISON, 1972 - The response of some ecologically distinct
    plant species to nitrate and ammonium nitrogen. *J. Ecol.,* 60, 93-102.
GLOAGUEN, J.C. & J. TOUFFET, 1974 - Production de litière et apport au sol
    d'éléments minéraux dans une hêtraie atlantique. *Oecol. Plant.,* 9, 11-28.
GRIES, B., 1966 - Zellphysiologische Untersuchungen über die Zinkresistenz
    bei Galmeiökotypen und Normalformen von *Silene cucubalus* Wib. *Flora,
    Abt. 8,* 156, 271-290.
GRIME, J.P. & J.G. HODGSON, 1969 - An investigation of the ecological
    significance of lime-chlorosis by means of large-scale comparative
    experiments. *In:* I.H. RORISON (Editor), *Ecological aspects of the
    mineral nutrition of plants,* Blackwell Scientific, Oxford, p. 67-99.
HACKETT, C., 1965 - Ecological aspects of the nutrition of *Deschampsia
    flexuosa* L. Trin. II. The effects of Al, Ca, Fe, K, Mn, P and pH on the
    growth of seedlings and established plants. *J. Ecol.,* 53, 315-333.
HALL, J.L., YEO, A.R. & T.J. FLOWERS, 1974 - Uptake and localisation of
    rubidium in the halophyte *Suaeda maritima. Z. PflPhysiol.,* 71, 200-206.
HEDENSTRÖM, H. VON & S.W. BRECKLE, 1974 - Obligate halophytes? A test with
    tissue culture methods. *Z. Pflphysiol.,* 74, 183-185.
HENRICHFREISE, A., 1973 - Manganhaushalt von *Ericaceen. Staatsexamenarbeit
    Biologie, Münster.*
HENRICHFREISE, A., 1976 - Aluminium- und Mangantoleranz von Pflanzen saurer
    und basischer Böden. *Ph. D. Thesis,University of Münster.*
HINNERI, S., 1974 - Podzolic processes and bioelement pools in subarctic
    soils at the Kevo Station, Finnish Lapland. *Rep. Kevo Subarctic Res.
    Stat.,* 11, 26-34.
JANIESCH, P., 1973a- Beitrag zur Physiologie der Nitrophyten. Nitratspeicherung
    und Nitratassimilation bei *Anthriscus sylvestris* Hoffm. *Flora,* 162, 479-491.
JANIESCH, P., 1973b- Ökophysiologische Untersuchungen an Umbelliferen
    nitrophiler Säume. *Oecol. Plant.,* 8, 335-352.
JONES, H.E. & J.R. ETHERINGTON, 1970 - Comparative studies of plant growth
    and distribution in relation to waterlogging. I. The survival of *Erica
    cinerea* L. and *E. tetralix* L. and its apparent relationship to iron and
    manganese uptake in waterlogged soils. *J. Ecol.,* 58, 487-496.
JONES, R., 1972 - Comparative studies of plant growth and distribution in
    relation to waterlogging. V. The uptake of iron and manganese by dune
    and dune slack plants. *J. Ecol.,* 60, 131-139.
JONES, R., 1972 - Comparative studies of plant growth and distribution in
    relation to waterlogging. VI. The effect of manganese on the growth of
    dune and dune slack plants. *J. Ecol.,* 60, 141-145.
KETNER, P., 1972 - Primary production of salt-marsh communities on the Island
    of Terschelling in The Netherlands. *Verhandl. Res. Inst. Nature Manage.,*
    5, 1-181.
KONONOVA, M.M., 1975 - Humus of virgin and cultivated soils. *In:* J.E. GIESEKING
    (Editor), *Soil components. Vol. 1. Organic compounds.* Springer-Verlag,
    Berlin, p. 475-526.
KYLIN, A. & R.S. QUATRANO, 1975 - Metabolic and biochemical aspects of salt
    tolerance. *Ecol. Stud.,* 15, 147-167.
LANGLOIS, J., 1971a- Influence de l'immersion sur le métabolisme glucidique

de *Salicornia stricta* Dumort. *Oecol. Plant.*, 6, 15-24.

LANGLOIS, J., 1971b- Influence du rythme d'immersion sur la croissance et le métabolisme protéique de *Salicornia stricta* Dumort. *Oecol. Plant.*, 6, 227-245.

LEE, J., STEWART, G.R. & D.C. HAVILL, 1974 - Soil acidity and nitrate utilization. *In: Plant analysis and fertilizer problems*. Proc. 7th Int. Coll. Hannover, FRG, p. 229-239.

LEE, H.J., 1975 - Trace elements in animal production. *In:* D.J.D. NICHOLAS & A.R. EGAN (Editors), *Trace elements in soil-plant-animal-systems*. Academic Press, New York, p. 39-54.

LINDSAY, W.L., 1972 - Zinc in soils and plant nutrition. *Adv. Agron.*, 24, 147-186.

MARTIN, M.H., 1968 - Conditions effecting the distribution of *Mercurialis perennis* L. in certain Cambridgeshire woodlands. *J. Ecol.*, 56, 777-793.

MATHYS, W., 1975 - Enzymes of heavy metal resistant and non-resistant populations of *Silene cucubalus* and their interaction with some heavy metals in vitro and in vivo. *Physiologia Pl.*, 33, 161-165.

MATHYS, W., 1976 - The role of malate in zinc tolerance. *Proc. Internat. Conf. Heavy Metals in the Environment, Toronto* 1975, Vol. 1.

MATSUMOTO, H., WAKIUCHI, N. & L. TAKAHASHI, 1971 - Changes of some mitochondrial enzyme activities of cucumber leaves during ammonium toxicity. *Physiologia Pl.*, 25, 353-357.

MOTHES, K., 1958 - Ammoniakentgiftung und Aminogruppenvorrat. *In:* W. RUHLAND (Editor), *Encyclopedia of plant physiology, Vol. 8, Nitrogen metabolism*, Springer-Verlag, Berlin, p. 716-762.

NICHOLAS, D.J.D., 1961 - Minor mineral elements. *A. Rev. Pl. Physiol.*, 12, 63-90.

NIESKE, H., 1973 - Zellsaftanalysen an Geophyten mitteleurpäischen Laubwälder. *Staatsexamensarbeit Biologie, Münster.*

NIHLGÅRD, B., 1972 - Plant biomass, primary production and distribution of chemical elements in a beech and a planted spruce forest in South Sweden. *Oikos*, 23, 69-81.

PATNAIK, R., BARBER, A.V., D.N. MAYNARD, 1972 - Effects of ammonium and potassium ions on some physiological and biochemical processes of excised cucumber cotyledons.*Physiologia Pl.*, 27, 32-36.

PATRICK, W.H. & I.C. MAHAPATRA, 1968 - Transformation and availability to rice of nitrogen and phosphorus in waterlogged soils. *Adv. Agron.*, 20, 323-359.

PEDERSEN, M., SAENGER, P. & L. FRIES, 1974 - Simple brominated phenols in red algae. *Phytochem.*, 13, 2273-2279.

PIGOTT, C.D., 1969 - Influence of mineral nutrition on the zonation of flowering plants in coastal salt-marshes. *In:* I.H. RORISON (Editor), *Ecological aspects of the mineral nutrition of plants*,Blackwell, Oxford, p. 25-35.

PLADEK-STILLE, J., 1974 - Mineralstoff- und Wasserhaushalt von Pflanzen artenreicher Buchenwälder unter Berücksichtigung der Zellsäfte. *Staatsexamensarbeit Biologie, Münster.*

POLJAKOFF-MAYBER, A. & J. GALE, 1975 - *Plants in saline environments, Ecol. Stud.*, Vol. 15, Springer-Verlag, Berlin, 213 p.

PROCTOR, J. & S.R. WOODELL, 1975 - The ecology of serpentine soils. *Adv. Ecol. Res.*, 9, 255-366.

QUINN, J.A., 1974 - *Convolvulus sepium* in old field succession on the New Jersey Piedmont. *Bull. Torrey Bot. Club*, 101, 89-95.

RAINS, D.W. & E. EPSTEIN, 1967 - Sodium absorption by barley roots. *Pl. Physiol.*, 42, 314-323.

RANWELL, D.S., 1972 - *Ecology of salt marshes and sand dunes*. Chapman & Hall, London, 258 p.

REIMOLD, R.J. & W.H. QUEEN, 1974 - *Ecology of halophytes*. Academic Press, New York, 605 p.

RICE, E.L., 1974 - *Allelopathy*. Academic Press, New York, 353 p.

ROZEMA, J. & B. BLOM, 1977 - Effects of salinity and inundation on the growth of *Agrostis stolonifera* and *Juncus gerardii*. *J. Ecol.*, 65, 213-222.

SAMPSON, M., CLARKSON, D.T. & D.D. DAVIES, 1965 - DNA synthesis in aluminium treated roots of barley. *Science*, 148, 1476-1477.

SCHEFFER, F. & P. SCHACHTSCHABEL, 1976 - *Lehrbuch der Bodenkunde,* F. Enke-Verlag, Stuttgart, 394 p.

SCHILLER, W., 1974 - Versuche zur Kupferresistenz bei Schwermetallökotypen von *Silene cucubalus* Wib. *Flora,* 163, 327-347.

SCHMIDT, W., 1970 - Untersuchungen über die Phosphorversorgung nieder-sächsischer Buchenwaldgesellschaften. *Scripta Geobot.,* 1, 1-120.

SCHMITZ, R., 1973 - Mineralstoffhaushalt der Abschlussgewebe mittel-europäischer Laub- und Nadelbäume. *Staatsexamensarbeit Biologie, Münster.*

SCHNURBEIN, C. VON, 1967 - Über den Anteil von Nitrat- und Chlorid an der Zusammensetzung des Zellsaftes von Blütenpflanzen. *Flora Abt. A.,* 158, 577-593.

SCHREVEN, D.A. VAN, 1963 - Nitrogen transformation in the former subaqueous soils of polders, recently reclaimed from Lake IJssel. Water-extractable, exchangeable and fixed ammonium. *Pl. Soil,* 18, 143-162.

SHAROVA, A.S., 1957 - Content of trace elements Cu, Zn, Co and Mn in some soils in Latvia. *Pochvovedenie,* 19-31.

STEWART, G.R. & J.A. LEE, 1974 - The role of proline accumulation in halophytes. *Planta,* 120, 279-289.

STEWART, G.R., LEE, J.A. & T.O. OREBAMJO, 1972 - Nitrogen metabolism of halophytes. I. Nitrate reductase activity in *Suaeda maritima. New Phytol.,* 71, 263-267.

STEWART, G.R., LEE, J.A. & T.O. OREBAMJO, 1973 - Nitrogen metabolism of halophytes. II. Nitrate availability and utilization. *New Phytol.,* 72, 539-546.

STOREY, R. & R.G. WYN JONES, 1975 - Betaine and choline levels in plants and their relationship to NaCl stress. *Pl. Sci. Lett.,* 4, 161-168.

STREETER, J.G. & M.E. BOSLER, 1972 - Comparison of in vitro and in vivo assays for nitrate reductase in soybean leaves. *Pl. Physiol.,* 49, 448-450.

SUTTNER, C., 1974 - Jahreszeitliche Veränderungen im Stoffwechsel von Kräutern mitteleuropäischer Laubwälder. *Staatsexamensarbeit Biologie, Münster.*

TAMM, C.O., HOLMEN, H., POPOVIC, B. & G. WIKLANDER, 1974 - Leaching of plant nutrients from soils as a consequence of forestry operations. *Ambio 3,* 211-221.

TIJBERG, M.M.J.M., 1974 - Autoecologisch onderzoek naar de N-P-K-behoefte van *Rumex obtusifolius. Doct. Verslag Plantenoecologie, Vrije Universiteit,* Amsterdam.

TIKU, B.J., 1976 - Effect of salinity on the photosynthesis of the halophytes *Salicornia rubra* and *Distichlis stricta. Physiologia Pl.,* 37, 23-28.

TUKEY, H.J. Jr., 1970 - The leaching of substances from plants. *A. Rev. Pl. Physiol.,* 21, 305-324.

VERA, F., 1975 - De oecologische betekenis van mineraalrijkdom en mineraal-armoede voor *Senecio sylvaticus* en *Chamaenerion angustifolium. Doct. Verslag Plantenoecologie, Vrije Universiteit,* Amsterdam.

VESTER, G., 1972 - Eine stoffwechselphysiologische Studie der Überflutungs-toleranz bei Bäumen. *Ph. D. Thesis, University of München.*

WAISEL, Y., 1972 - *Biology of halophytes.* Academic Press, New York, 395 p.

WALTER, H., 1963 - Über die Stickstoffansprüche (die Nitrophilie) der Ruderalpflanzen. *Mitt. flor.-soz. ArbGem.,* 10, 56-69.

WEISSENBÖCK, G., 1969 - Einfluss des Bodensalzgehaltes auf Morphologie und Ionenspeicherung von Halophyten. *Flora Abt. B.,* 158, 369-389.

WHITTAKER, R.H., 1970 - The biochemical ecology of higher plants. *In:* E. SONDHELMER & J.H. SIMEONE (Editors), *Chemical ecology.* Academic Press, New York, p. 43-70.

WILLIAMS, J.T., 1968 - The nitrogen relations and other ecological investigations on wet fertilized meadows. *Veröff. geobot. Inst. Zürich,* 41, 70-193.

WILLIAMS, J.T., 1969 - Mineral nitrogen in British grassland soils. I. Seasonal patterns in simple models. *Oecol. Plant.,* 4, 307-320.

WRIGHT, J.R., LEVICK, R. & H.J. ATKINSON, 1955 - Trace element distribution in virgin profiles representing four great soils groups. *Proc. Soil Sci. Soc. Am.,* 19, 340-344.

WYN JONES, A.G., STOREY, R.G., LEIGH, R.A., AHMAD, N. & A. POLLARD, 1977 -
    A hypothesis on cytoplasmic osmoregulation. *In:* E. MARRE & O. CIFFERI
    (Editors), *Regulation of cell membrane activities in plants*, Elsevier -
    North Holland Biomedical Press, Amsterdam, p. 121-136.
YEO, A.R., 1975 - Halophytes. *In:* D.A. Baker & J.L. HALL (Editors), *Ion
    transport in plant cells and tissues*. North Holland Publ., Amsterdam.
YERLY, M., 1970 - Ecologie comparée des prairies marécageuses dans les
    Préalpes de la Suisse occidentale. *Veröff. geobot. Inst. Zürich,* 44,
    1-119.
ZELLNER, J., 1926 - Zur Chemie der Halophyten. *Sber. Akad. Wiss. Math. Nat.
    Kl. Abt.,* IIb, 135, 585-592.

10. DISCUSSION

In reference to the studies on nitrophilous plants, TROELSTRA (Oostvoorne)
drew attention to the concept of the cation-anion (C-A) value of DE WIT *et al.*
(1963). Some plants thrive better with a high (C-A) value (COIC *et al.* 1962).
Therefore, TROELSTRA found it hard to accept the term ammonium toxicity. ERNST
commented that the (C-A) concept is not very useful, because in most of the
analysed examples (e.g. DIJKSHOORN 1957; MENGEL 1965) the value of C is only
estimated as the sum of the macronutrients, which was not the intention of the
initiatior of this concept (BEAR 1950). This is also done in such plants where
other elements, such as silicium, contribute more than 50 per cent to the ion
content of the plants (LEWIN & REIMAN 1969). On the other hand, there are
plants where the (C-A) value is not disturbed, but increased ammonium uptake
is the cause of a decrease in growth rate (e.g. INGESTAD 1976). All in all,
the concept may perhaps be relevant for plants where K^+ has antagonistic
effects to NH_4^+ (BARKER *et al.* 1967; NELSON & HSIEH 1971; PATNAIK *et al.* 1972).

 GRIME (Sheffield) then discussed the mechanism of the positive growth
response to increasingly low concentrations of micronutrients such as zinc or
even oligo-elements such as aluminium. ERNST pointed out that in low
concentrations most of the elements may have a stimulating effect on growth,
although they are by no means essential. The absence of these elements may
cause metabolic disfunction. CHAPHEKAR (Bangor) emphasized this aspect once
more by mentioning the plant populations which are tolerant for heavy metals
such as lead and copper.

 KINZEL (Vienna) then referred to his idea of the physiotypes of plants
(KINZEL 1971). He thought that at the moment the study of the soil-plant
relationship is characterized by a relatively good amount of knowledge about
well-defined soil types but a sparse knowledge of metabolic types of plants.
Research on the latter should be encouraged because it can help us to
understand the ecology of plants.

 With respect to the mineral cycling in ecosystems, BARKMAN (Wijster) then
drew attention to the current practice in national forestry in The Netherlands
of decorticating trees before their removal and using the peeled bark to cover
sandy paths. ERNST agreed with BARKMAN's suggestion that it would be worthwhile

to scatter the bark throughout the forest and thus keep the minerals in the ecosystem. PERSSON (Uppsala) remarked that the discussion on mineral cycling should put more emphasis on the turnover of the below-ground biomass, especially the fine roots, because they contribute twice as much as the leaf litter to the organic matter of forest ecosystems (PERSSON 1975).

There was some argument between MINDERMAN (Arnhem) and ERNST about whether the quality of chemicals in the soil can be determined. ERNST dit not share MINDERMAN's opinion that there is only certainty about the total amount of minerals within the plant, but he believed that by using KINZEL's concept of physiotypes, knowledge about rood exudates and new techniques for chemical soil analysis sufficient progress in the future will be made.

WHITE (Dublin) shifted the discussion to another ecological subject. He asked whether the distinction between ecological and physiological amplitude, which according to ELLENBERG is almost a tenor of contemporary ecology, was really more a distinction made by investigators than by nature, as suggested by ERNST, or is a general principle. ERNST denied the latter, because in most of the re-analyzed examples the published differences are due to an omission of important abiotic factors or to neglecting of the physiological differentiation of plant species in ecotypes. He considered the great discrepancy between the two amplitudes to be a symptom of our ecological ignorance. WOLDENDORP (Arnhem) pointed out that the comparison is useless, because in the physiological experiment only the dry matter is compared with the factor in question, whereas in the field the total fitness of the population can be studied. ERNST commented that this is only partially true, because in good physiological experiments the fitness of a population, too, can be elucidated with respect to its reproduction, genetic structure, competitive ability, and other fitness parameters.

In answer to a question on the necessity to extend the physiological experiments to mycorrhiza, ERNST replied that this should be done, especially where phosphorus nutrition must be taken into consideration (see the paper by MOSSE in this volume). However, besides the beneficial aspect of phosphorus supply by mycorrhiza, it must be kept in mind that via mycorrhiza the input of other elements, especially oligo-elements, is also stimulated, as has been shown for *Neottia nidus-avis* and *Monotropa hypopithys*, which are completely dependent on mycorrhiza.

References

BARNER, A.V., MAYNARD, D.N. & W.H. LACHMAN, 1967 - Induction of tomato stem and leaf lesions, and potassium deficiency, by excessive ammonium nutrition. *Soil Sci.*, 103, 319-327.
BEAR, F.E., 1950 - Cation and anion relationships in plants and their bearing on crop quality. *Agr. J.*, 42, 176-178.
COIC, Y., LESAINT, Ch. & F. LE ROUX, 1962 - Effects de la nature amoniacale ou nitrique de l'alimentation azotic et du changement de la nature de cette alimentation sur le métabolisme des anions et cations chez la tomate. *Annls. Physiol. vég.*, 4, 117-125.

DIJKSHOORN, W., 1957 - A note on the cation-anion relationship in perennial ryegrass. *Neth. J. agr. Sci.*, 5, 81-85.

INGESTAD, T., 1976 - Nitrogen and cation nutrition of three ecologically different plant species. *Physiologia Pl.*, 38, 29-34.

KINZEL, H., 1971 - Biochemische Ökologie-Ergebnisse und Aufgaben. *Ber. dt. bot. Ges.*, 84, 381-403.

LAZENBY, A., 1955 - Germination and establishment of *Juncus effusus* L. The interaction effect of moisture and competition. *J. Ecol.*, 43, 595-605.

LEWIN, J.C. & B.E.F. REIMAN, 1969 - Silicon and plant growth. *A. Rev. Pl. Physiol.*, 20, 289-304.

MENGEL, K., 1965 - Das Kationen-Anionen-Gleichgewicht in Wurzel, Stengel und Blatt von *Helianthus annuus* bei K-Chlorid und K-Sulfat-Ernährung. *Planta*, 65, 358-368.

NELSON, P. & K.H. HSIEH, 1971 - Ammonium toxicity in chrysanthemum: Critical level and symptoms. *Commum. Soil Sci. Plant Anal.*, 2, 439-448.

PATNAIK, R., BARBER, A.V. & D.N. MAYNARD, 1972 - Effects of ammonium and potassium ions on some physiological and biochemical processes of excised cucumber cotyledons. *Physiologia Pl.*, 27, 32-36.

PERSSON, H., 1975 - Deciduous woodland at Andersby, Eastern Sweden: Field-layer and below-ground production. *Acta phytogeogr. suec.*, 62, 1-71.

WIT, C.T. DE, DIJKSHOORN, W. & J.C. NOGGLE, 1963 - Ionic balance and growth of plants. *Agric. Res. Rep. (Wageningen)*, 69, 1-68.

Soil physical conditions and plant growth

1. INTRODUCTION

The environment is a complex of so many factors, all interacting with each other, that it is impossible to isolate any one factor that does not influence another. For the study of environmental effects, however, this complex is usually sub-divided into clearly defined units. One of these units is the soil, which is vitally important for plant growth and development. Soil in itself represents a complicated physical, chemical, and biological system by which the plant is supplied with the water, nutrients, and oxygen it requires for its development.

Although over the centuries plants have adapted themselves to various kinds of soil, the adaptation capacity of certain species is limited. This can be clearly seen when soil properties alter. The nature of the soil determines whether a species will thrive and influences its natural distribution. Within small areas, slight local variations in the soil may be sufficient to affect a plant's chances of survival.

The physical properties of soil are known to be of fundamental importance for plant growth, but much of the literature on the subject is qualitative or vague. This is not surprising in view of the difficulty one encounters in attempting to divide the edaphic factors unambiguously into physical, chemical, and biological classes. Most physical phenomena have important effects on the chemical and biological soil properties and processes, and these in turn influence plant growth.

Soil is a physical system and can be described in terms of grain size, apparent density, porosity, moisture content, temperature, and friability. Plant growth is affected by the amount of moisture and air in the soil and by the temperature of the soil. The composition of the soil can impede or foster root development and shoot emergence. It should be mentioned, too, that the physical features of the soil have certain indirect effects on other edaphic factors such as nutrient supply and pH.

Unlike mobile organisms, terrestrial plants are bound to the soil where the seed has fallen. Plants generally have to cope with a hostile environment and may not survive, but in course of time every type of soil, however hostile, becomes covered with vegetation. Since the type of vegetation depends on the prevailing soil conditions, in a sense each particular vegetation is adapted. In nature, however, plant species are rarely found on the soils whose physical

conditions are optimal for their growth and performance. Comparative experiments
with various plants show that the general shape of the curve representing the
response to the degree of severity of adverse conditions (such as oxygen deficiency,
soil compaction, low soil temperature, high sodium chloride concentrations) is
very similar to all plants, whether or not they are adapted. Minor differences
in a single soil factor are sufficient to cause minute variations in the
occurrence of plant species in the field. Comparative experiments have also shown
not only that plants have a tremendous plasticity that enables them to survive
under adverse conditions but also that species develop different strategies in
order to survive. We do not fully understand many plant-soil relationships
because we do not have sufficient knowledge about:

    a) the physical conditions of the soil in space and time;

    b) the differences in a plant's response at various developmental stages;

    c) the plant's response to changes in the degree of adversity of conditions;

    d) the extent to which a plant's response is determined by the interaction
       of other factors;

    e) methods to assess the effect of minor differences in response over a
       long period in a plant's life-cycle.

One of the serious drawbacks of ecophysiological experimentation is that we
can hardly discern differences amounting to less than 5-10 per cent but in
nature even smaller differences may determine discrimination in the long run,
especially in interspecific competition.

It is therefore with considerable diffidence that I present this paper,
since it cannot solve the two main problems in plant ecology, namely:

    1) Why are certain types of vegetation (species) restricted to a certain
       habitat, whereas others clearly prefer another set of conditions?

    2) How can diversity of species in a vegetation be maintained for a relatively
       long period (measured by human standards) when so many individuals are
       all dependent on the same resources, i.e., light, water, and minerals?

The second of these problems is the more challenging one, since we know that
in contrast to the diversity in nature, competition experiments almost invariably
result in survival of only one of the competing species (DE WIT 1960). The niche
concept, a separation of interests in time and (or) space, was introduced to
reconcile this discrepancy (DE WIT & VAN DEN BERGH 1965; VAN DEN BERGH & BRAAK-
HEKKE, this volume), but we shall have to learn much more about the ways in
which plants behave before these questions can be adequately answered.

If we want to obtain satisfactory answers we must pay close attention to the
complete life-cycle of the species in question, since niche differentiation may
show up in only one of the life stages (germination, seedling establishment,
vegetative growth, generative growth, and dissemination or seed longevity).
Sometimes adaptation to a certain habitat can be due to relatively small
differences in a number of aspects (PEGTEL 1976; PONS 1976, 1977), none of which
alone would fully explain the species' preference for a particular habitat.

Since the ecophysiological approach is based on experience acquired in the field of crop physiology, possibly essential differences in behaviour between natural vegetations and crops must be taken into account. In both kinds of populations the individual responds to the complex of conditions but agricultural practice has selected for uniformity of response, whereas natural selection has often resulted in the maintenance of a certain degree of diversity and plasticity within a population. Moreover, the external conditions are much less predictable for natural vegetations.

2. SEED POPULATION AND SOIL PHYSICAL CONDITIONS

2.1. INTRODUCTORY REMARKS

The number of seeds in a population on or in the soil depends on the rate of dissemination and on the rate at which seeds are lost through deterioration, germination, and consumption. Seed consumption will not be considered in this paper. The proceedings of a recent Nottingham Symposium on Seed Ecology (HEYDECKER 1973) have provided a considerable amount of information about the behaviour of seeds in general; the information is useful to both agriculturists and ecologists. The seed population is important not only because it determines the timing of germination but also because in a given locality it may represent a high percentage of the total number of individuals present in that locality (see also the paper by RABOTNOV in this volume). Depending on the type of dispersal, seeds pass from the plant and the place where they have been produced to a place on or in the soil where they will lie until conditions are suitable for germination and for growth into new plants. For the ultimate success of the seeds, both their longevity and their germination behaviour are important.

2.2. LONGEVITY OF SEEDS

It is well known that several ambiguous factors such as cool temperatures, low oxygen tension, and a low moisture content, all of which tend to decrease metabolic activity, increase the length of time that seeds can be stored (BARTON 1961; HARRISON 1966; ABDALLA & ROBERTS 1968). This finding is surprising, and makes it difficult to explain why many seeds apparently retain their viability longer when buried in moist soil than when kept in dry storage (VILLIERS 1973) (Fig. 1).

In an attempt to reconcile these different types of seed behaviour, VILLIERS (1973) showed that longevity is high at low water content (depending on the species, 5-8 per cent) and at high water content when the seed has fully imbibed. He assumed that ageing phenomena occurred at all levels of water content. These ageing processes are assumed to result in cross-linkage of macromolecules, which render enzymes and membranes non-functional, and in a gradually increasing number of somatic mutations, many of which may cause the production of defective proteins.

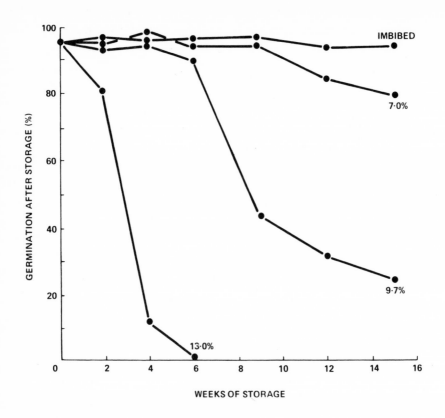

FIG. 1.   *Influence of the water content of seeds during storage on longevity* (VILLIERS 1973)

By comparing the results of germination experiments with those of electron-microscopical studies on the embryos, Villiers concluded that the functioning of macromolecule and organelle repair mechanisms was seriously impaired during air-dry storage at intermediate relative humidities of the air. In practice, the degree of ageing can be related to the amounts and kinds of substances lost by leaching from the seeds during soaking. According to SIMON (1974), when dried seeds were moistened again the initial leakage of electrolytes which normally decreased rapidly when the seeds reached a water content of about 15 per cent persisted due to the globular state of the plasma membrane of these dry seeds. This globular state is ineffective in maintaining gradients of ions and charge across the membrane. After rewetting, the normal non-leaky bilayer structure is restored, a process which would take more time in aged seeds. In imbibed seeds stored in the soil, leakage will not occur while the seed remains viable, provided the membranes remain in good condition. Repeated alternation of drying and wetting, which often occurs in the upper soil layers, could ultimately lead to the complete exhaustion of accumulated solutes and could therefore prove very harmful. In this respect it is interesting to note that the reverse response has been reported more frequently. A number of authors (e.g. KOLLER et al. 1962; HEGARTY 1970; PEGTEL 1976) demonstrated that both germination and establishment improved after seeds of various species were repeatedly wetted and dried. After studying the effects of the duration and number of the drying-wetting cycles, BERRIE & DRENNAN (1971) concluded that germination was more rapid when the seeds were redried after short periods of inbibition. A prolonged wet period before redrying resulted in embryo damage and poor germination, the critical factor being whether or not active cell division had begun in the imbibed embryo. The duration of the periods of imbibition that seeds can withstand depends on the species (CARCELLOR & SORIANO 1972) and the stage of development of the embryo (WOODRUFF 1969). Since these periods cannot be controlled under field conditions, repeated drying will result in a reduction of the seed population.

It may be concluded that seeds lying in or on the soil are subject to deterioration as a consequence of ageing. The rate of deterioration increases with increasing temperature and decreasing water content. Reiterated drying-wetting cycles are especially harmful. In fully imbibed seeds deterioration and ultimate death are postponed by the continuous repair of damaged structures. Although seeds have a fairly efficient repair mechanism, their ability to produce vigorous seedlings gradually declines during the ageing process.

2.3. DORMANCY

The fate of a seed after dissemination depends largely on external conditions and its internal features. In many cases seeds are not able to germinate immediately after dissemination, even when conditions seem to be favourable. These seeds are "dormant" (GORDON 1973). The function of dormancy in determining the timing of growth resumption whenever external conditions become suitable is quite clear, but the phenomenon itself is very complex. For instance, dormancy may depend on a variety of internal features. The seed coat may be highly resistant to the diffusion of oxygen from the environment to the embryo. Furthermore, the seeds coatings may contain substances that inhibit embryo growth and have to be broken down or rinsed out before germination can start. Dormancy may also be governed by an internal hormonal balance between growth-inhibiting and growth-promoting substances in the embryo itself, i.e., a balance which is inadequate for growth initiation (HEMBERG 1949; WAREING 1965; WAREING et al. 1973).

In some cases the degree of dormancy of the seeds depends on the conditions to which the mother plant was exposed during fruit development. Very often, the inability to germinate directly after dissemination disappears for no apparent reason in a few weeks (after ripering). There also a kind of dormancy that can be induced in normally non-dormant seeds by the application of special treatments such as high temperatures (thermo-dormancy) or osmotic stress.

The breaking of dormancy requires a specific sequence of external factors. The literature on dormancy and termination of dormancy is very extensive, and at first gives the impression that each case forms a separate problem. Only recently has some progress been made in formulating more unifying concepts (ROBERTS 1973); these new theories are mainly based on what the various dormancy-breaking agents have in common; some of the latter play a role in the ecological situation, whereas others are in use only in laboratory experiments. In a series of papers Roberts and co-workers (for references see ROBERTS 1973) developed the hypothesis that stimulation of the activity of the pentose phosphate (PP) pathway leads to loss of dormancy. These authors classified the dormancy-breaking agents into a number of categories, and found a very close resemblance between the stimulation of germination and the possible stimulation of the PP pathway, which for the relevant ecological factors can be summarized as follows:

nitrate   is known to break seed dormancy in a large number of species
             (TOOLE *et al*. 1956; STEINBAUER & GRIGSBY 1957; WILLIAMS & HARPER
             1965) and stimulates hydrogen acceptance by intermediates of the
             PP pathway; it shares this property with agents such as nitrite,
             oxygen, and methylene blue, which have been shown to stimulate
             germination in certain cases;

temperature has a number of quite different effects on dormancy, depending on
             whether the seed is "dry" or "wet" (ROBERTS 1973); freshly shed
             seeds that have been kept dry show a rapid loss of dormancy at
             high temperatures (after ripening); when kept imbibed immediately
             after harvesting they germinate only within a small temperature
             range that is sometimes high (35-45°C) and in other cases low
             (3-7°C) (VEGIS 1964); another well known way of breaking dormancy
             is to keep the imbibed seeds for a certain time at low temperatures
             (3-5°C) (stratification treatment); in some cases fluctuating
             temperatures are required to stimulate germination. According to
             ROBERTS (1973), in all these treatments a stimulation of the PP
             pathway seemed to lead to a concomitant increase in the seed's
             ability to germinate;

light   has effects on dormancy breaking, invariably via the phytochrome
             system; light-sensitive seeds respond to red light and their
             germination is promoted, whereas far red counteracts this
             stimulation; there does not seem to be enough evidence yet to
             prove that the PP pathway is involved, but there are indications
             that this may be the case, e.g. the changes in respiration seen
             after exposure to light (EVENARI 1961) and the involvement of
             phytochrome in the stimulation of gibberellic acid synthesis
             (LOVEYS & WAREING 1971).

Although we still do not know the exact role of the PP pathway, the above-mentioned working hypothesis is rather attractive and may help to reconcile a number of experimental results from various sources. Nevertheless, differences between species and within species between populations will have to be quantitatively specified if we want to understand their behaviour in the field. As an example of such a detailed approach, mention should be made of the studies done on the germination pattern of winter annuals by JANSSEN (1973a, 1973b, 1974), who analyzed the effects of light and temperature on seed behaviour in combined laboratory and field experiments. These investigations included alleviation of dormancy, shifts in optimum and maximum temperatures from dissemination onward, and the interaction between light and temperature. By applying experimentally obtained parameters to the physical conditions of the microsites, the author was able to use computer simulation to predict the behaviour of the seeds. The different behaviour of representatives of two microsites on dry sandy soils in the coastal dunes near Oostvoorne (The Netherlands) was clarified in this way. Seeds of *Veronica arvensis* and *Myosotis ramosissima* appeared to be well adapted to their respective habitats.

PEGTEL (1976), who compared the ecological behaviour of two varieties of *Sonchus arvensis,* one a coastal and the other an arable type, demonstrated a difference between their germination responses to soil temperature and soil moisture regimes, the response fitting the situation in the respective habitats. Germination trials in the field confirmed the results obtained from laboratory experiments. Nevertheless, under natural conditions only a few seedlings were found,mainly because of the limited life-span of the achenes. Adaptation was assumed to be of little ecological value in this case, particularly since field tests showed that sown seeds of both varieties ultimately germinated in both habitats.

2.4. GERMINATION

Viable, non-dormant seeds germinate if the environment is suitable. The essential environmental factors are an adequate supply of water, a suitable temperature, an adequate supply of oxygen, and in some species either the presence or absence of light. According to most definitions, a seed may be considered to have germinated when the radicle breaks through the seed coat. Seedlings germinating on the ground become fixed in the soil by subsequent root growth. It is obviously important for this process to occur quickly, particularly at sites where conditions are subject to rapid changes. Seeds germinating within the soil must complete emergence before their reserves are exhausted. In any case, the germination process represents a risky period in the life-cycle of plants in the field.

Hence a seed's germination rate should be high in order to ensure rapid attachment to the soil and to diminish the risk. In most of the literature on germination the percentage of germinated seeds is plotted against time. A hundred per cent germination can be reached within one or two days or may take much longer even though the germination rate of each individual seed is still fast. In this way the germination rate of the population is indicated and the curves show seed polymorphism (RORISON 1973). A steep slope means that everything is staked on one throw ("gamblers") and is therefore rather risky (JANSSEN 1973b). The risks are mainly determined by changes in water availability, since the most sensitive tissue, the growing root-tip, is confined to the upper soil layer which is bound to follow changing weather conditions quite rapidly. In bare soil changes in temperature and water content are much more pronounced than when there is a cover, particularly at high levels of irradiance.

Young root parts of various species do not differ according to their function, which means that differences in sensitivity between species are mainly determined by either the rate of root differentiation or the capability of the root to resume growth upon alleviation of stress. These predominantly morphogenetic properties determine the changes of survival in environments with changing degrees of adversity (MILTHORPE & MOORBY 1974).

Both the temperature and the water content of the soil are important in determining the rate of germination and the germination percentage ultimately reached. At constant temperatures the proportion of seeds that germinate tends to increase with rising temperature up to an optimum and then decreases at higher temperatures. The rate of germination follows the same pattern, but the optima frequently occur at slightly different temperatures (Fig. 2, GULLIVER & HEYDECKER 1973). In the ascending part of the curve the rate of germination appears to be almost linearly related to temperature. HEGARTY (1973) showed that diurnal fluctuations within this temperature range can be simply handled by using the average value. Complications are encountered when non-linear parts of the curve have to be included.

GULLIVER & HEYDECKER (1973) also reviewed the effects of water supply on germination. The response pattern to an increase in water supply is rather similar to the pattern for temperature (Fig. 3), despite the obvious dissimilarity of these factors.

An optimum curve has generally been demonstrated when the percentage germination was plotted against the water status of the substrate. High levels of water supply may eliminate some of the seeds, thus reducing the percentage germinated although still favouring the rate of germination of the surviving seeds. The most likely assumption is that an excessive water supply affects the emergence of seedlings by reducing oxygen availability. Due to subsequent crusting of the soil under field conditions, this situation may even continue for a certain time after a return to a lower water level.

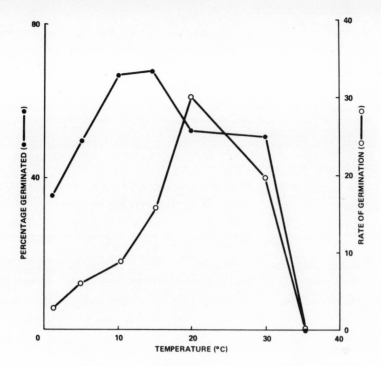

FIG. 2.   *Influence of temperature on the rate of germination and on the percentage of seeds ultimately germinating* (GULLIVER & HEYDECKER 1973)

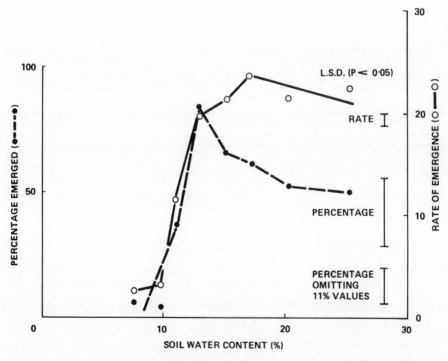

FIG. 3.   *Effect of the water content of the soil on the rate of emergence and percentage of seedlings ultimately emerging* (GULLIVER & HEYDECKER 1973)

2.5. ESTABLISHMENT OF SEEDLINGS

Reserves present in the seed are limited. Hence, the seedling has in due course to supply itself with all the essentials for growth. Soil factors affect seedling growth via their influence on root growth and root activities. The latter can be divided into two categories, *viz.* absorption of water and minerals, on the one hand, and, on the other, the synthesis of substances which are essential for shoot performance and are not (or insufficiently) synthesized in the shoot itself. In the past, most attention was paid to the meaning of the absorptive capacities of the roots for the performance of the whole plant, but more recently the emphasis has shifted to the role of root-borne growth essentials (e.g. hormones) in determining such processes like shoot growth, green leaf area duration, chlorophyll formation, and even the rates of photosynthesis and transpiration. The relative importance of these control systems may alter from case to case. As a whole, however, all these activities will depend on the supply of energy from the shoot. It has been shown repeatedly that there is an accurate control mechanism which regulates the feed back between root and shoot activities (BROUWER 1963; BROUWER & KLEINENDORST 1967).

Since the basic problem of the young seedling is the energy supply, its success will depend on illumination conditions. If it is situated in a standing crop that absorbs most of the light, the energy supply will be poor. Consequently, the over-all performance of the seedling will be bad and root growth will be more restricted than shoot growth. However, seedlings can perhaps survive for a rather long time in such a situation, because, to a certain extent, respiration losses are coupled to the available reserve of carbohydrate (McCREE 1970; PENNING DE VRIES 1975). Nevertheless, some maintenance respiration will go on and the seedlings will require light intensities for themselves well above the compensation point. As shown by LAZENBY (1955) for *Juncus effusus*, the surrounding vegetation can seriously limit seedling survival. Slow growth may also affect the incidence of pathogens (HARPER *et al.* 1955).

3. VEGETATIVE GROWTH

Within a vegetation much depends on the morphological characters and spatial distribution of shoot parts. A rosette plant is deemed to have fewer chances to reach the required level of illumination than tall species or plants with more vertically directed leaves. VAN DOBBEN (1967) and HARPER (1965) discussed the strategical implications on the basis of detailed comparison of the density-dependent mortality of broad-leaved horizontally developed plants and narrow-leaved upright plants.

When leaves reach a more exposed position, the energy supply improves but at the same time there is a greater need for root activities. In fact, the seedlings are now in the same situation as those growing from the very beginning on exposed areas. Freely growing plants and plants partially exposed in a vegetation are both affected by physical soil conditions, so it is not necessary to distinguish between the growth response in early and more mature phases of vegetative development: in free-growing plants the growth distribution follows a quite regular pattern from seedling emergence until flowering (VAN DOBBEN 1962; TROUGHTON 1974). Complications may occur if, in the course of the plant's development, the nature of the processes that limit whole-plant growth changes as a result of interactions between individual plants (interspecific as well as intraspecific competition). Hence, in any study of the effects of physical soil factors on the vegetative development attention should be paid to: (a) differences in response between freely growing plants and plants in dense vegetations; and (b) differences in response due to interactions with other factors in the environment.

The influence of the soil temperature on bean plants in successive developmental stages may serve as example here. During germination the optimum temperature is quite high, the rate of germination being determined mainly by the biochemical reactions by which the reserves in the seeds are converted into structural tissues in the seedlings.

Once the shoots have emerged, there is a distinct drop in optimum temperature. In this phase the absorption processes determine the growth rate. The temperature curve is a reflection of the root activities and is determined by both the quantity of roots present and the absorption per gram of roots.

At a later stage, when the plants have developed a closed-crop surface, the rate of dry-matter production is the same over a large part of the temperature range. Only at very low temperatures at which the leaves are partly wilted does root temperature cause a measurable reduction in dry-matter production.

The dotted line in the top part of Fig. 4 shows the effect of the root temperature on plant height. The course of this line is also indicative of leaf area development. In such situations taller plants are favoured in that they overshadow smaller plants. The results presented in Fig. 4 suggest the following interpretation. Root-zone temperatures primarily affect root growth and development. In addition, they determine the water permeability of the root tissue (KRAMER 1949). As a consequence of the latter, the water supply to the shoot will be affected and this in turn will cause differences in leaf water potential and hence leaf-area development.

At very low root temperatures (5 and 10 °C) there is hardly any extension growth, because leaf water potentials are very low. Despite the markedly reduced increase in leaf area, the photosynthetic activity is much less affected (GROBBELAAR 1963). This leads to an accumulation in the plants of reserves that remain partially available for growth resumption as soon as conditions improve. The relative insensitivity of the photosynthetic process

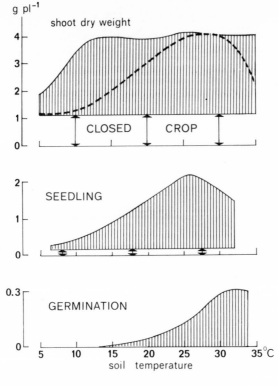

FIG. 4. *Effect of soil temperature on growth of bean seedlings in various stages of development.*
Bottom: germination during the first 10 days after sowing.
Middle: weight increase during 8 days after the start of treatment of plants pregrown uniformly until they reached the weight indicated by the arrows.
Top : weight increase during 8 days after the start of treatment of plants pregrown uniformly until they reached the weight indicated by the arrows (closed green-leaf surface)

explains why the effects of adverse soil conditions are less harmful once a closed-crop surface has been attained. All light is then captured and diverted into dry matter. This course of events illustrates a phenomenon known from agricultural experience, namely that deficiencies in the root medium cause a reduction in yield mainly by reduction in leaf growth, thus reducing leaf area (WATSON 1947).

This also means that the degree of adversity of a given soil factor depends largely on the density of the vegetation. In a dense vegetation, light is the limiting factor and thus diminishes the importance of soil factors. In open vegetations soil factors are more important. This might partly explain why annuals or biennials may occur in a relatively open vegetation whereas dense vegetations are almost completely composed of perennials with different growth forms (PEGTEL, personal communication).

The responses of bean plants to soil temperature, as discussed above, are representative for those plants in which the growing points of the shoot are situated well above ground level. In rosette plants and in grasses in the vegetative stage these growing points are located in or near the soil surface, and are therefore influenced by the soil temperature. As a result, the growth of such plants is more profoundly affected by soil temperature (Fig. 5), since not only root growth and root activity but also shoot development, e.g. leaf appearance (BROUWER et al. 1973), are directly controlled by this factor.

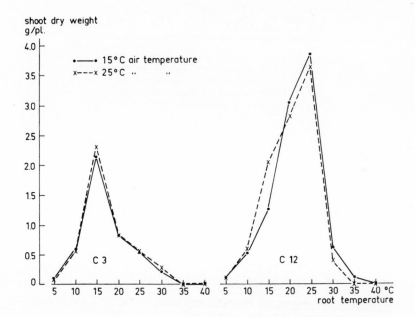

FIG. 5.   *Dry weights of shoots of two ryegrass* (Lolium perenne) *clones grown for 9 weeks on nutrient solutions kept at the indicated root temperatures and two different air temperatures* (KLEINENDORST & BROUWER 1965)

KLEINENDORST & BROUWER (1967) showed in climate-room experiments that air temperature was of negligible importance in perennial ryegrass, the growth rate being governed solely the the temperature of the root medium (Fig. 6). PEACOCK (personal communication) found that the temperature at a height of 0.5 cm above the soil surface determined the growth of perennial ryegrass in the field. In addition to leaf-area development, soil factors -including soil temperature- may affect leaf orientation. This further complicates a quantitative evaluation of the significance of soil temperature when plants are grown in competition. The results of these studies suggest that the degree of adversity is also dependent on other external conditions to which the plants are exposed (interaction). This seems to be supported by the results

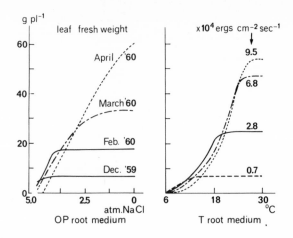

FIG. 6.   *Effects on the growth of bean plants*
of light intensity interacting with
osmotic concentration and temperature
of the root medium

shown in Fig. 6 concerning the interaction between light level and the osmotic
concentration as well as the temperature of the root medium. The response to a
reduced water supply induced by either an enhanced osmotic concentration or by
low root temperatures is of minor importance at low light intensities. The
effects of a lower water potential, which persists under such conditions of
low light supply, are completely obscured. At increasing light intensities the
growth rate of the controls increased correspondingly, and the effect of the
adverse factor became manifest at decreasing levels. It is interesting to see
that in the range of maximally tolerable adversity a shift also occurred.
Plants which had survived the stress treatments at low light intensities died
when treated in the same way at high light intensities. Under low root
temperatures this mortality was due to the low water potential which developed
as a consequence of the high irradiance. At high NaCl concentrations the death
of the plants might have been caused by the toxic effects of ion accumulation
in the tissue. An analysis of the components responsible for these growth
responses showed that both morphological (or phenotypical) adaptations (leaf
area ratio) and physiological processes (net assimilation rate) are involved
(Fig. 7).

A comparison of the effects of temperature and osmotic potential has shown
that the physiology of the response is rather similar. Both factors influence
shoot growth through their effect on the water balance of the plant. The
interaction with light intensity can be explained on this basis. However, the
water balance is only one of the several possible ways in which adverse root
treatments can influence plant performance (BROUWER 1973). Other root
activities, for instance mineral absorption or hormone production, might have

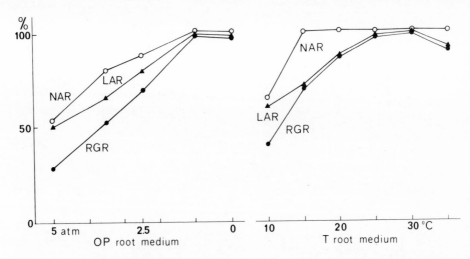

FIG. 7. *Effects of osmotic concentrations and suboptimal temperatures of the root medium on the relative growth rate of bean plants, exerted via morphological (leaf area ratio) and physiological (net assimilation) characters*

been reduced too. Roots have been reported (LIVNE & VAADIA 1972; WIEDENROTH 1974) to produce gibberellic acid and cytokinin-like substances contributing to leaf elongation, stomatal control, chlorophyll formation, etc. It has been shown that this aspect was not critical in the experiments in question (BROUWER & KLEINENDORST 1967). Under other conditions,however, these aspects may become growth limiting. The same holds for possible changes in biochemical pathways in the roots, i.e., changes whose effects on the overall plant performance are overshadowed by the effects of the water potential.

We find a different situation when we consider another factor in the root environment, namely gas-exchange restrictions. Much research has been done on the effects of poorly aerated soil on plant growth and performance (ARMSTRONG 1974). Comparison of plants grown in well-aerated soils (well supplied with oxygen, no accumulation of CO_2) with plants grown in soils with reduced gas exchange invariably shows reduced root growth. Quantitatively, considerable differences between species occur (Fig. 8). The penetration of roots into anaerobic layers depends on a number of internal and external factors. A high level of metabolism, high temperatures, and high organic nitrogen content all reduce maximum penetration considerably, e.g. in our experiments with water-logged sandy soils, maize roots penetrated to a depth of 75 cm below the ground-water table at 15°C and low nitrogen nutrition. The depth of penetration was reduced to 20-25 cm at 25°C and a higher nitrogen nutrition level.

In their response to anaerobiosis, maize plants occupy an intermediate position between plant species whose roots hardly penetrate the anaerobic

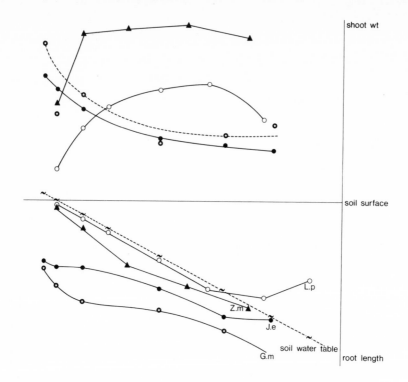

FIG. 8. *Rooting depth and shoot weight of various species*
as affected by the height of the soil water-table
(constantly maintained at levels between +5 and
-80 cm relative to the soil surface).
G.m. = Glyceria maxima, J.e. = Juncus effusus,
L.p. = Lolium perenne, Z.m. = Zea mays

regions in the soil (*Phaseolus vulgaris*) and others whose roots are much less
affected than maize roots (*Oryza sativa, Phragmites australis*). The marked
divergency of the responses alone suggests that at least in some species there
has been a considerable degree of adaptation to the reduced gas exchange
properties of the soil. With respect to the physiological aspects of this
adaptation, various plant properties are involved because anaerobiosis is
generally accompanied by changes in other soil properties, *viz.*: low partial
pressures of oxygen, accumulation of carbon dioxide, a lowered oxidation-
reduction (redox) potential, and the accumulation of reduced forms of carbon,
nitrogen, sulphur, iron, and manganese, some of which may reach toxic levels.
ARMSTRONG (1975) discussed the degree to which tolerance is achieved on the
basis of the following plant properties: (a) the ability to exclude or tolerate
soil-borne toxins, (b) the development of air-space tissue, (c) the ability to
metabolize anaerobically and tolerate an accumulation of anaerobic metabolites,
and (d) the ability to respond successfully to periodic inundation.

It has long been known that roots of intact plants can oxidize reduced substances in the root medium (MOLISCH 1888). FUKUI (1953) associated this ability to oxidize the rhizosphere with the ability to penetrate reduced paddy soils. The oxygen diffuses from the atmosphere, via intercellular spaces in shoots and roots, to the root surface. The limited size of the normally present intercellular cavities, cannot even satisfy the oxygen demands of normal root metabolism, let alone provide sufficient rhizosphere oxygenation to avoid intoxication by reduced soil-borne substances. Most plants respond to reduced oxygen tension by enhancing the porosity of the cortical tissue: values of about 5 per cent in well-aerated soils can increase to between 15 and 85 per cent, depending on the species (Fig. 9) and the degree of anaerobiosis. It is evident that such large cavities could considerably reduce

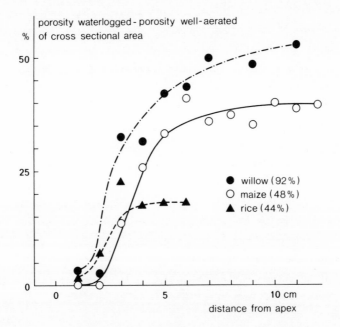

FIG. 9.   *Increase in porosity (expressed as percentage of cross-sectional area) of the roots of willow, rice, and maize plants in waterlogged soil and soil at field capacity. (Numbers between parentheses represent the highest porosity values found in waterlogged soils)*

diffusion resistance, but whether this completely explains their existence is still an open question (WILLIAMS & BARBER 1961; ARMSTRONG 1971). Perhaps plants adapt more readily when less tissue has to be maintained. In this respect it is noteworthy that the development of similar cavities in the root cortex is induced by a deficiency of various mineral elements.

The adequacy of the oxygen supply to the active sites determines whether or not normal aerobic respiration can proceed. If it cannot, most plants shift over to anaerobic dissimilation processes, and this leads to the accumulation in the plant root of toxic products such as ethanol and lactic acid. Adapted species have been shown to accumulate less toxic substances instead, for instance shikimic acid and malic acid (ARMSTRONG 1974). In addition to bio-chemical and anatomical adaptations, mention should be made of adaptations of the morphology of the root system. In wet soils an O_2 gradient will generally be maintained from the soil surface to the lower soil layers. Strongly reduced root growth in deeper soil layers leads to enhanced induction and growth of (adventitious) roots at the stem base, where conditions are less adverse.

It is generally rather difficult to decide which factor limits root development in a particular case. As already mentioned, radial oxygen losses enable some plants at least to improve conditions in the direct vicinity of the roots. Differences between plant species include differences in sensitivity to a number of factors acting at the same time (SHEIKH 1970). Whereas root growth responses of various species to waterlogging tend qualitatively in the same direction although they differ greatly, the concomitant shoot responses show even greater deviations quantitatively. Fig. 8   shows two typical response patterns. In *Glyceria maxima* and *Juncus effusus* the shoot weights are highest at very high ground-water levels, and decrease at lower water tables. *Zea mays* and *Lolium perenne* show distinct optima. In our case the experiment was started with established seedlings. LAZENBY (1955), who worked with variations in water level immediately after sowing, found a similar descending curve for *Juncus effusus*. Water levels lying more than 15 cm below the surface dit not stop growth of the pregrown plants, but germination did not occur at water tables deeper than 10 cm. KLEINENDORST & BROUWER (1967) found that aeration of the nutrient solution affected rooting and subsequent growth of various clones of perennial ryegrass in a quite different way, since aeration proved to be more essential for root initiation than for root elongation. They also observed distinct clonal differences.

The effect of a transition from favourable to adverse conditions depends on the nature of adaptive differences. The time required by plants to adapt them-selves to the new situation determines their chances of survival (BROUWER & WIERSUM 1977). Rapid responses are required to meet a suddenly inadequate water supply or increased transpiration rate, because water in the tissue has to be replaced continuously. With respect to the mineral supply, much greater variation is acceptable. The same holds for the carbohydrate supply, since a rather efficient feedback between photosynthesis and respiration is thought to exist. These characteristics vary according to the species.

The differences in response determine whether one species will win or hold its ground at one place and others will not. The responses of beans (*Phaseolus*) and willow cuttings (*Salix* spec.) to flooding may be mentioned as examples of such differential behaviour. After a period of growth in well-aerated soil, root elongation in both species is stopped by flooding. Shoot growth, however, is affected only in beans, which indicates that root functioning is reduced more in beans than in willows. This is confirmed by the time course of transpiration before and after flooding. Both species respond with an enhanced root growth in the soil layer near the surface (less adverse). After some time this leads in bean plants to a partial recovery of transpiration and growth. Coincident with the reduction in transpiration of the *Phaseolus* plants, their leaves show chlorophyll breakdown as well as other symptoms of ageing, which can be ascribed to a decreased synthesis of cytokinins by the treated roots (REID 1976).

Initiation of the recovery of such bean plants requires the growth of adventitious roots near the soil surface, the induction of which takes some time. If climatic conditions favour high transpiration rates, the plants will die before they reach this stage. When the plants are in a growth stage during which new root initiation normally does not occur (generative stage), recovery is almost impossible. Here again, the response depends strongly on the prevailing conditions and the stage in the life-cycle.

The local situation also determines whether reduced root growth affects shoot growth via the hormone balance, via the water requirements, or via the mineral absorption. BANNISTER (1964) demonstrated that waterlogging affected the distribution of *Calluna vulgaris* and *Erica cinerea* and *E. tetralix* via differences in its effect on the water balance of these species. In other cases unfavourable aeration conditions in the soil could be compensated for an additional supply of fertilizer (HAMMOND et al. 1955; SIEBEN 1963; BROUWER 1977).

It should be mentioned again that the response depends to a great extent on the situation in the vegetation. The relative importance of a given soil factor for growth is greater for individuals in open vegetations than for those in a closed vegetation, since in the latter light is more likely to be the ultimate limiting factor. In such a situation, however, rather small differences in root response may have important consequences for the whole plant, since being a little ahead in the beginning will have considerable advantages in the competition for light as well as in the exploration of soil-bound resources (ELLERN et al. 1970).

4. GENERATIVE STAGE

Despite considerable differences in the reproduction pattern (GRUBB 1977) of various species, some general remarks can be made. In plants in which the life-cycle is completed with flowering and subsequent seed formation, conditions during the vegetative stage determine both the size of the machinery for photosynthesis and the number of "sites" (seeds) that will be filled.

Usually there is a good correlation between the size of the plant and the number of seeds initiated, since both are mainly determined by the amount of light energy that has been captured (a certain degree of phenotypic plasticity can be observed here). High fertility levels tend to favour vegetative growth rather than seed initiation; low radiation levels frequently tend to postpone the onset of flowering and, even more so, seed setting. Both phenomena can be ascribed to the low carbohydrate level prevailing under these conditions. Therefore, in determinate plants, which complete their life-cycle with seed ripeness, the amount of light intercepted before flowering determines the number of seeds, whereas the amount of light intercepted after flowering determines the total ultimate seed weight, provided no calamities occur in the interval. Such calamities may cause serious disturbance of the normally rather precise correspondence between the number of flowers and viable seeds. Pollination is a particularly sensitive stage of the life-cycle, and is adversely affected by soil drought or by other soil conditions, such as flooding, which induce internal water deficits.

High air temperatures tend to have adverse effects as well. The success of seed-setting depends mainly on the balance between the levels of reserve carbohydrates and hormones produced by the developing fruits. Thus, in fertile soils or at low light intensities moderate drought may be favourable, since it tends to enhance the carbohydrate content. In determinate plants, after the onset of flowering, almost the entire dry-matter production is drained into the inflorescences. The first sign that this has started is that root growth is considerably reduced and fairly soon afterwards stops completely. This means that the roots produce less cytokinins and less of these substances is transported to the shoot. The resulting low cytokinin level in the leaves is the main reason for their senescence, which in turn is accentuated by unfavourable soil conditions (LIVNE & VAADIA 1972). It is clear that the maintenance of a green, light-absorbing surface (leaf-area duration; WATSON 1947) after seed-setting will determine the mass and possibly also the quality of seeds. Very little is known, however, about the effect of physical soil conditions on seed quality.

Although an appreciable number of herb species reach the end of their life-cycle very soon after flowering, many species flower in successive waves, and in these species vegetative growth and reproduction to on simultaneously for a considerable time. In a sense this behaviour tends to spread the risk of complete failure. Reduction of seed-setting due to adverse soil conditions during one wave may be at least partially compensated for by the next wave. Evidence concerning this phenomenon is, however, very scarce in the ecological literature.

Spreading of the production of viable seeds has also been observed as a year-to-year variation in perennial species (GRUBB 1977), but very little is known about the external factors that determine either failure or abundance.

5. REFERENCES

ABDALLA, F.H. & E.H. ROBERTS, 1968 - Effects of temperature, moisture and oxygen on the induction of chromosome damage in seeds of barley, broad beans, and peas during storage. *Ann. Bot.*, 32, 119-126.

ARMSTRONG, W., 1971 - Radial oxygen losses from intact rice roots as affected by distance from the apex, respiration and waterlogging. *Physiol. Pl.*, 25, 192-197.

ARMSTRONG, W., 1975 - Waterlogged soils. *In:* I. R. ETHERINGTON (Editor), *Environment and plant ecology*, Wiley, London, p. 181-219.

BANNISTER, P., 1964a - Stomatal responses of heath plants to water deficits. *J. Ecol.*, 52, 151-158.

BANNISTER, P., 1964b - The water relations of certain heath plants. I. Introduction, germination and establishment. *J. Ecol.*, 52, 423-432.

BANNISTER, P., 1964b - The water relations of certain heath plants. II. Field studies. *J. Ecol.*, 52, 468-481.

BANNISTER, P., 1964b - The water relations of certain heath plants. III. Experimental studies: General conclusions. *J. Ecol.*, 52, 499-510.

BARTON, L.V., 1961 - *Seed preservation and longevity.* Leonard Hill, London.

BERRIE, A.M.M. & D.S.H. DRENNAN, 1971 - The effect of hydration-dehydration on seed germination. *New Phytol.*, 70, 135-142.

BRADSHAW, A.D., 1965 - Evolutionary significance of phenotypic plasticity in plants. *Adv. Gen.*, 13, 115-155.

BROUWER, R., 1963 - Some physiological aspects of the influence of growth factors in the root medium on growth and dry matter production. *Jaarb. I.B.S. (Wageningen) 1963,* 11-30.

BROUWER, R., 1973 - Dynamics of plant performance. *Proc. int. Soc. Hort. Sci.*, 38, 31-49.

BROUWER, R., 1974 - A comparison of the effect of drought and low root temperatures on leaf elongation and photosynthesis in maize. *Proc. int. Soc. Hort. Sci.*, 39, 141-152.

BROUWER, R. & L.J. WIERSUM, 1977 - Root aeration in relation to crop growth. *In:* U.S. GUPTA (Editor), *Crop Physiology*, Oxford IBH Pub. Co. Janpath, New Delhi, India, p. 157-201.

BROUWER, R., 1977 - Root functioning. *In:* J.J. LANDSBERG & C.V. CUTTING (Editors), *Environmental effects on crop physiology,* Academic Press, New York, p. 229-245.

BROUWER, R. & A. KLEINENDORST, 1967 - Responses of bean plants to root temperatures. III. Interactions with hormone treatments. *Jaarb. IBS (Wageningen) 1967,* 11-28.

BROUWER, R., KLEINENDORST, A. & J.T. LOCHER, 1973 - Growth responses of maize plants to temperature. *In: Plant response to climatic factors.* Proc. Uppsala Symp. UNESCO, Paris, p. 169-174.

CARBON, B.A., 1973 - Diurnal water stress in plants grown on a coarse soil. *Austr. J. Soil Res.*, 24, 33-42.

CARCELLOR, M.S. & A. SORIANO, 1972 - Effect of treatments given to the grain, on the growth of wheat roots under drought conditions. *Can. J. Bot.*, 50, 105-108.

CURRIE, J.A., 1973 - The seed-soil system. *In:* W. HEYDECKER (Editor), *Seed ecology*, Butterworths, London, p. 463-480.

DOBBEN, W.H. VAN, 1962 - Influence of temperature and light conditions on dry-matter distribution, development rate and yield in arable crops. *Neth. J. agr. Sci.*, 10, 377-389.

DOBBEN, W.H. VAN, 1967 - Physiology of growth in two *Senecio* species in relation to their ecological position. *Jaarb. IBS (Wageningen) 1967,* 75-83.

EAVIS, B.W., TAYLOR, H.M. & M.G. HUCK, 1971 - Radicle elongation of pea seedlings as affected by oxygen concentration and gradients between shoot and root. *Agron. J.*, 63, 770-772.

EAVIS, B.W., 1972 - Soil physical conditions affecting seedling root growth. *Pl. Soil.*, 36, 613-622.

ELLERN, S.J., HARPER, J.L. & G.R. SAGAR, 1970 - A comparative study of the distribution of the roots of *Avena fatua* and *A. strigosa* in mixed stands using a 14C-labelling technique. *J. Ecol.*, 58, 865-868.

EVENARI, M., 1961 - A survey of the work done in seed physiology by the Department of Botany, Hebrew University, Jerusalem, Israel. *Proc. int. Seed Test. Ass.*, 26, 597-657.

FUKUI, J., 1953 - Studies on the adaptability of green manure and forage crops to paddy field conditions. *Proc. Crop Sci. Soc. Japan*, 22, 110-112.

GORDON, A.G., 1973 - The rate of germination. *In:* W. HEYDECKER (Editor), *Seed ecology*, Butterworths, London, p. 391-419.

GROBBELAAR, W., 1963 - Responses of young maize plants to root temperatures. *Meded. Landb. Hogesch. Wageningen*, 63, 1-71.

GRUBB, P.J., 1977 - The maintenance of species-richness in plant communities. The importance of the regeneration niche. *Biol. Rev.*, 52, 107-145.

GULLIVER, R.L. & W. HEYDECKER, 1973 - Establishment of seedlings in a changeable environment. *In:* W. HEYDECKER (Editor), *Seed ecology*, Butterworths, London, p. 433-462.

HAMMOND, L.C., ALLAWAY, W.H. & W.E. LOOMIS, 1955 - Effects of oxygen and carbon dioxide levels upon absorption of potassium by plants. *Pl. Physiol.*, 30, 155-161.

HARPER, J.L., 1965 - Establishment, aggression and cohabitation in weedy species. *In:* H.G. BAKER & G.L. STEBBINS (Editors), *The genetics of colonising species*, Academic Press, New York, p. 243-268.

HARPER, J.L., LANDRAGIN, P.A. & J.W. LUDWIG, 1955 - The influence of environment on seed and seedling mortality. II. The pathogenic potential of the soil. *New Phytol.*, 54, 119-131.

HARPER, J.L. & J. OGDEN, 1970 - The reproductive strategy of higher plants. I. Concept of strategy with special reference to *Senecio vulgaris* L. *J. Ecol.*, 58, 681-698.

HARRISON, B.J., 1966 - Seed deterioration in relation to storage conditions and its influence upon germination, chromosomal damage and plant performance. *J. natn. Inst. agric. Bot.*, 10, 644-649.

HEGARTY, T.W., 1970 - The possibility of increasing field establishment by seed hardening. *Hort. Res.*, 10, 59-64.

HEGARTY, T.W., 1973 - Temperature relations of germination in the field. *In:* W. HEYDECKER (Editor), *Seed ecology*, Butterworths, London, p. 411-432.

HEMBERG, T., 1949 - Significance of growth inhibitory substances and auxins for the rest period of the potato tuber. *Physiologia Pl.*, 2, 24-36.

HEYDECKER, W., 1973 - Seed ecology. *In:* W. HEYDECKER (Editor), *Seed ecology*, Butterworths, London, p. 1-5.

HURD, E.A., 1974 - Phenotype and drought tolerance in wheat. *Agric. Meteorol.*, 14, 39-55.

JANSSEN, J.G.M., 1973a - The relations between variation in edaphic factors and microdistribution of winterannuals. *Acta bot. neerl.*, 22, 124-134.

JANSSEN, J.G.M., 1973b - Effects of light, temperature and seed age on the germination of the winter annuals *Veronica arvensis* L. and *Myosotis ramosissima* Rochel ex Schult. *Oecologia*, 12, 141-146.

JANSSEN, J.G.M., 1974 - Simulation of germination of winter annuals in relation to microclimate and microdistribution. *Oecologia*, 14, 197-228.

KABRA, Y.P., 1971 - Application of split-root technique in orthophosphate absorption experiments. *J. agric. Sci.*, 77, 77-81.

KAYS, S.J., NICKLOW, C.W. & D.H. SIMONS, 1974 - Ethylene in relation to the response of roots to physical impedance. *Pl. Soil*, 40, 565-571.

KLEINENDORST, A. & R. BROUWER, 1967 - Responses of two different clones of perennial ryegrass to aeration of the nutrient solution. *Jaarb. IBS (Wageningen) 1967*, 29-38.

KLEINENDORST, A. & R. BROUWER, 1970 - The effect of temperature of the root medium and of the growing point of the shoot on growth, water content and sugar content of maize leaves. *Neth. J. agric. Sci.*, 18, 140-148.

KOLLER, D., MAYER, A.M., POLJAKOFF-MAYBER, A. & S. KLEIN, 1962 - Seed germination. *A. Rev. Pl. Physiol.*, 13, 437-464.

KRAMER, P.J., 1949 - *Plant soil water relationships*. Mac Graw Hill, New York.

LAZENBY, A., 1955 - Germination and establishment of *Juncus effusus* L. I. The effect of different companion species and of variation in soil and fertility conditions. *J. Ecol.*, 43, 103-119.

LAZENBY, A., 1955 - Germination and establishment of *Juncus effusus* L. II. The interaction effects of moisture and competition. *J. Ecol.*, 43, 595-605.

LIVNE, A. & Y. VAADIA, 1972 - Water deficits and hormone relations. *In:* T.T.
    KOZLOWSKI (Editor), *Water deficits and plant growth.* III, p. 255-275.
LOVEYS, B.R. & P.F. WAREING, 1971 - The red light production of gibberellin
    in etiolated wheat leaves. *Planta,* 98, 109-116.
LUXMOORE, R.J., STOLZY, L.H. & J. LETEY, 1970 - Oxygen diffusion in the
    soil-plant system. *Agron. J.,* 62, 317-332.
McCREE, K.J., 1970 - An equation for the rate of respiration of white clover
    plants grown under controlled conditions. *In: Prediction and measurement
    of photosynthetic productivity, (Proc. IBP/PP Technical Meeting, Trebon),*
    Pudoc, Wageningen, p. 221-229.
MILTHORPE, F.L. & J. MOORBY, 1974 - *An introduction to crop physiology.*
    University Press, Cambridge, 195 p.
MOLISCH, H., 1880 - Über Wurzelausschiedungen und deren Einwirkung auf
    organischen Substanzen. *Sber. Akad. Wiss. Wien Math. Nat. Kl.,* 96, 84-89.
PASSIOURA, J.B., 1972 - The effect of root geometry on the yield of wheat
    growing on stored water. *Austr. J. agric. Res.,* 23, 745-752.
PEGTEL, D.M., 1976 - On the ecology of two varieties of *Sonchus arvensis.Ph.D.
    Thesis Groningen,* 148 p.
PENNING DE VRIES, F.W.T., 1975 - Use of assimilates in higher plants. *In:
    Photosynthesis and productivity in different environments.* I.B.P., 3,
    459-480, University Cambridge Press.
PONS, T.L., 1976 - An ecophysiological study in the field layer of ash
    coppice. I. Field measurements. *Acta bot. neerl.,* 25, 401-416.
PONS, T.L., 1977 - An ecophysiological study in the field layer of ash
    coppice. II. Experiments with *Geum urbanum* and *Cirsium palustre* in
    different light intensities. *Acta bot. neerl.,* 26, 29-42.
PONS, T.L., 1977 - An ecophysiological study in the field layer of ash
    coppice. III. Influence of diminishing light intensity during growth
    on *Geum urbanum* and *Cirsium palustre.* *Acta bot. neerl.,* 26, 251-263.
RALSTON, D.S. & W.H. DANIEL, 1972 - Effect of temperatures and water table
    depth on the growth of creeping bentgrass roots. *Agron. J.,* 64, 709-713.
REID, D.M., 1976 - Crop responses to waterlogging. *In:* O. GUPTA (Editor), *Crop
    physiology I.* Oxford & IBH Pub. Co. Janpath New Delhi, p. 157-201.
ROBERTS, E.H., 1973 - Oxidative processes and the control of seed germination.
    In: W. HEYDECKER (Editor), *Seed ecology,* Butterworths, London, p. 189-218.
RORISON, J.H., 1973 - Seed ecology - Present and Future. *In:* W. HEYDECKER
    (Editor), *Seed ecology,* Butterworths, London, p. 497-517.
RUTTER, A.J., 1955 - The composition of wet-heath vegetation in relation to
    the water table. *J. Ecol.,* 43, 507-543.
SAGAR, G.R. & J.L. HARPER, 1964 - Biological flora of the British Isles
    (*Plantago major* L., *P. media* L. and *P. lanceolata* L.). *J. Ecol.,* 52,
    189-221.
SHEIKH, K.H., 1970 - The response of *Molinia caerulea* and *Erica tetralix* to
    soil aeration and related factors. *J. Ecol.,* 58, 141-154.
SIEBEN, W.H., 1964 - Invloed van de ontwateringstoestand op stikstofhuis-
    houding en opbrengst. *Landbouwk. Tijdschr.,* 76, 784-802.
SIMON, E.W., 1974 - Phospholipids and plant membrane permeability. *New
    Phytol.,* 73, 377-420.
STEINBAUER, G.P. & B. GRIGSBY, 1957 - Interaction of temperature, light and
    moistening agents in the germination of weed seeds. *Weeds,* 5, 175-182.
THOMPSON, P.A., 1970 - Characterization of germination response to temperature
    of species and ecotypes. *Nature,* 225, 827-831.
THOMPSON, P.A., 1973 - Geographical adaptation of seeds. *Proc. XIXth. Easter
    School Univer. of Nottingham,* 31-58.
TOOLE, E.H., HENDRICKS, S.B., BORTHWICK, H.A. & V.K. TOOLE, 1956 - Physiology
    of seed germination. *A. Rev. Pl. Phys.,* 7, 299-324.
TROUGHTON, A., 1974 - The growth and function of the root in relation to the
    shoot. *In:* J. KOLEK (Editor), *Structure and function of primary root
    tissues,* Bratislava, p. 153-164.
VEGIS, A., 1964 - Climatic control of germination, bud break and dormancy.
    In: G.C. EVANS (Editor), *Environmental control of plant growth,* Academic
    Press, New York, p. 265-287.
VILLIERS, T.A., 1973 - Ageing and the longevity of seeds in field conditions.
    In: W. HEYDECKER (Editor), *Seed ecology,* Butterworths, London, p. 265-288.

WAREING, P.F., 1965 – Endogenous inhibitors in seed germination and dormancy. *In:* W. RUHLAND (Editor), *Encycl. Pl. Physiol.,* 15 - 2, p. 909-924.

WAREING, P.F., STADEN, J. VAN & D.B. WEBB, 1973 – Endogenous hormones in the control of seed dormancy. *In:* W. HEYDECKER (Editor), *Seed ecology,* Butterworths, London, p. 145-155.

WATSON, D.J., 1947 – Comparative physiological studies on the growth of field crops. *Ann. Bot. N.S.,* 11, 41-76.

WIEDENROTH, E.M., 1974 – Die Bedeutung der Wurzel für die Chlorophyllbildung und den Gaswechsel in den oberirdischen Organen höherer Pflanzen. *Wiss. Z. Humboldt-Univ. Math. Nat. Reihe XXIII,* 6, 631-640.

WILLIAMS, J.T., & J.L. HARPER, 1965 – Seed polymorphism and germination. I. Influence of nitrates and low temperatures on the germination of *Chenopodium album. Weed Res.,* 5, 141-150.

WILLIAMS, W.T. & D.A. BARBER, 1961 – The functional significance of aerenchyma in plants. *In:* F.L. MILTHORPE (Editor), *Mechanisms in biological competition.* (S.E.B. Proc. XV) University Press, Cambridge, p. 132-144.

WIT, C.T. DE, 1960 – On competition. *Versl. landbouwk. Onderz.,* 66 - 8.

WIT, C.T. DE & J.P. VAN DEN BERGH, 1965 – Competition between herbage plants. *Neth. J. agric. Sci.,* 13, 212-221.

WOODRUFF, D.R., 1969 – Studies on presowing drought hardening of wheat. *Austr. J. agric. Res.,* 20, 13-24.

6. DISCUSSION

On the question of the physical condition resulting in yellowing of shoots, BROUWER pointed out that this effect could be caused by low temperature as well as by oxygen deficiency. The process of yellowing associated with ageing could be stopped by a relatively rapid change in the stress situation if the air humidity is not too low.

ROZEMA (Amsterdam) then asked about the causal background of the increase in biomass production of flooding-tolerant plants under anaerobic conditions. BROUWER thought that the effect may be due to an unaffected assimilation rate, an enhancement of the leaf area, or a reduced respiration.

WENT (Nevada) drew attention to the ecological importance of the experiments with varying levels of the water-table, because they will explain the distribution of plants in the field. BROUWER commented that his experiments have already been started with established plants or tillers, whereas the critical ecological point may be the germination stage. As LAZENBY (1955) has demonstrated for *Juncus effusus,* the germination of this species is quite sensitive to lowering of the water-table, whereas the established plants can also thrive well at rather low water-tables.

CHAPHEKAR (Bangor) asked about the influence of a high water-table or soil compactness on the acceleration of flowering. BROUWER replied that most of his experiments were restricted to the vegetative state, but due to a correlation between the size of plants and the possibility for flowering and seed production he believed that the response will be the same as for vegetative growth.

VAN DER MEIJDEN (Leiden) asked about the ecological advantage of a high temperature demand for germination and the lower temperature optimum for growth. BROUWER confirmed the more general phenomenon of the need for high temperature at the early growth stages. At the moment, however, the ecological importance remains uncertain.

Mechanisms of adaptation to physical and chemical factors in plants

1. INTRODUCTION

The mechanisms by which plants adapt to environmental conditions have been studied almost as long as plants themselves have been an object of scientific investigation. Early workers observed a very high frequency of stomata in the leaves of xerophytes and immediately wondered whether such a system could help the plant to survive a period of drought. Experimentation soon showed that under conditions of ample water supply xerophytes had a very high rate of transpiration (per unit leaf area) but under limitation of soil water the numerous stomata allowed the plant to regulate its water loss in a much more precise way than would be possible in leaves with fewer stomata (for a review of the older literature, see CRAFTS et al. 1949).

The experimental approach also proved valuable in studies on the frost resistance of plants and as early as 1912, MAXIMOV contributed experimental evidence supporting the hypothesis that a high sugar content of plant cells formed part of the protective mechanism against frost damage: epidermal strips of cabbage leaves floating on a sucrose solution were more resistant to sub-zero temperatures than strips on salt solutions or distilled water or the intact tissue itself. Slowly, the increasing knowledge about the "hardiness of plants", as LEVITT (1956) described this field of research, gave rise to specialized research areas according to the environmental factor under investigation, including resistance to freezing (cryobiology), chilling, high temperature, salinity and drought, to mention the most important areas. LEVITT (1972) gave a synthesis of the knowledge gained by this work in a book entitled "Responses of plants to environmental stresses", which reflects the specialization in this field in a time when the majority of the plant physiologists are working with plant material grown under controlled conditions guaranteeing optimal growth.

However, optimal conditions for growth are not necessarily the main factors to be investigated by ecophysiologists, because the conditions encountered are usually suboptimal. Therefore, the establishment and maintenance of populations of a given plant species in natural vegetations will depend on many other factors, related to the physiology of the individual plant besides optimal conditions for growth.

At this point it may be useful for the experimental approach of the problem
how to investigate the adaptation mechanisms, to classify the vegetations
composed of these plants into the following groups. It should be kept in mind
here that such a classification is a prerequisite for any experimental approach.

There are three main types of vegetation:

Type 1. Vegetations consisting of plant species occurring under conditions in
which a single environmental factor or a cluster of environmental factors are
decisive and limiting for the establishment or maintenance of populations of
the species concerned. Such vegetations may be relatively poor with respect to
the number of species of (higher) plants and consequently may have a monotonous
appearance. Examples of this type include vegetations of salt marshes
(*Salicornia* spp.; the habitat being characterized by flooding with sea water);
peat bogs (*Sphagnum* spp.; permanently wet, the water having a low pH value and
a low mineral content); dry heath areas (*Calluna vulgaris*; habitat relatively
dry with a low mineral level and exposed to grazing); and submerged aquatic
vegetations (*Potamogeton* spp.; water with a high pH value and relatively strong
movement). Many of the pioneer vegetations belong to this category.

Type 2. This is a variant of type 1, but the environmental factor (or cluster
of factors) acting as key factor strongly promotes a high growth rate of the
individual plants of the population concerned. This type of vegetation is
often highly dependent upon human activities. Dutch arable fields with their
high yields are examples of such vegetations, but vegetations less directly
dependent on human interference also belong to this type for instance
vegetations composed of reed (*Phragmites australis*) and reed-grass (*Glyceria
maxima*) along the banks of eutrophic lakes or meadows with a cover of English
ryegrass (*Lolium perenne*). The species concerned often have a relatively high
tolerance for the effect of foreign substances (pollution), treading, tillage,
weeding, and other human activities. It should be stressed that most of our
knowledge about plant physiology concerns species originating from vegetations
of this type.

Type 3. Vegetations of greater diversity than those of type 1, composed of
several to many species often representing a wide variety of life forms. In
well-developed situations the floristic composition of such vegetations is
very rich and includes many rarely encountered species. The diversity of the
vegetation is also expressed in a marked seasonal variation. Competition
between the individual plants seems to be much lower in vegetations of type 3
than in those of type 2, which is undoubtedly due to the lower growth rate of
the individual plants and the low or very low productivity of such vegetations.

Vegetations of type 3 are found in gradient situations of environmental
conditions, e.g. on the borderline between areas differing in topography, soil
type, hydrology, salinity, or the level of mineral nutrients. It has been
suggested by VAN LEEUWEN (1966) and WESTHOFF *et al.* (1970) that the more
complex the system of gradients in environmental conditions, the more intricate

the pattern of the existing vegetation, especially under typical conditions. Type 3 vegetations are often difficult to preserve in densely populated areas characterized by vegetations of type 2. The physiology of species characteristic for such vegetations has hardly been studied at all, but intuitively one would suggest that the individual plants in such an environment must have evolved mechanisms to maintain proper conditions for the physiological functioning of their tissues and organs under the fluctuating environmental conditions. If these three types of vegetation are considered in relation to GRIME's three primary strategies (see this volume) it is evident that the stress-tolerant strategy concerns vegetations of type 1 and 3 and the ruderal and competitive strategy concerns vegetations of type 2, distinguishing between productive vegetation types of severely disturbed and of relatively undisturbed environments.

2. METHODOLOGY

In the following sections examples will be presented of adaptations by plants to extreme environmental conditions (such as prevail in vegetations of type 1) as well as responses of plants to fluctuations in environmental conditions, i.e., responses possibly having adaptive value for species that are characteristic for more diverse vegetations (type 3). Because of the scarcity of experimental data on the latter subject, caution must be applied in the interpretation of the results and generalization is not permissible.

Adaptations of plants occur at several levels of organization, from the intact plant organ down to the cells and cell organelles. Photosynthesis provides the source of energy for the formation of all the structures required to cope with extreme or fluctuating environmental conditions. Measurements on respiration, and more specifically on the efficiency of respiration, can provide information on the utilization of metabolic energy in growth, for instance in species of high-yield vegetations (type 2) and in the formation of the structures needed to cope with extreme environmental conditions (type 1) or with conditions prevailing in the diverse vegetations (type 3). On the cellular and subcellular levels striking alterations are observed as a response to different environmental stresses such as prevail in vegetations of types 1 and 3. Examples of such alterations will be presented on the level of the biomembranes which regulate so many physiological reactions. Specifically, the effect of the environment on the lipid composition of biomembranes and on the functioning of lipid-dependent membrane enzymes will be discussed in more detail.

The role of membranes in responses to environmental conditions can be studied in several ways. In the first place, comparison can be made between species or varieties, preferably grown under identical environmental conditions. The species are arranged in order of adaptation to environmental factors or stresses (e.g. salinity, low temperature), and correlation with membrane properties, such as membrane enzyme activity or membrane lipid composition is sought. Secondly, an environmental condition can be varied and species differing in ecological adaptation to this environmental factor can be compared. A particularly useful experimental approach is to introduce an abrupt environmental change and follow the ensuing changes in membrane properties as a function of time. Such observations may indicate, for instance, the specific role of a membrane enzyme or membrane lipid in relation to the structure and transport properties of the biomembrane in question, enabling it to function properly under the altered environmental conditions. Specifically, plant species that have to cope with fluctuating environmental conditions (type 3) may yield information on adaptation on the membrane level when the last of these experimental approaches is used.

3. PHOTOSYNTHESIS

3.1. ADAPTATION TO SUN AND SHADE

As is evident from a recent review on environmental control of photosynthesis (MARCELLE 1975), photosynthetic efficiency is highly dependent on environmental conditions. In a densely shaded environment it is the efficiency with which the plant is able to absorb and utilize light of low intensity together with a minimal investment in constituents of the chloroplasts that determines the efficacy of the adaptation to shade. In *Alocasia macrorhiza,* a shade plant of rain forests, the chloroplast grana are extremely well developed and oriented in all directions, resulting in a very high chlorophyll concentration per chloroplast as well as per cell (BJÖRKMAN 1975). Such an extreme shade plant shows, on a chlorophyll basis, very low levels of chloroplast soluble protein, chloroplast ribulose diphosphate carboxylase, and several carriers of the electron transport chain such as plastoquinone, cytochrome f, and cytochrome b_6 (BJÖRKMAN *et al.* 1972).

The reverse holds for a species with a preference for the sun, *Atriplex hastata,* which is a beach plant (C_3) characterized by high levels of chloroplast soluble protein, chloroplast RuDP-carboxylase, and photosynthetic electron transport carriers, by less well developed grana, and by the occupation of a larger volume by the stroma fraction, thus permitting the plant to reach a much higher level of photosynthesis under light-saturation conditions. Other *Atriplex* species belonging to the same environment i.e., *Atriplex glabriuscula* (C_3) and *Atriplex sabulosa* (C_4), show photosynthesis curves with similar responses to temperature, even though the rate of photosynthesis of the C_4 species is consistently higher than that of the C_3 species. In both species photosynthesis decreases at temperatures above 35°C, whereas in *Tidestromia oblongifolia*, a C_4 species of the hot desert photosynthesis reaches its temperature optimum above 45°C. With respect to photosynthesis, the C_4 species are clearly at an advantage in bright sunlight. In that situation ATP production by chloroplasts does not limit CO_2 reduction even though the number of ATP molecules required to reduce a CO_2 molecule is considerably higher than for the chloroplasts of C_3 plants (BLACK 1973).

It should be noted that physiological factors that may depend indirectly upon the C_4 mechanism of photosynthesis, may give C_4 plants and advantage over C_3 plants. HOFSTRA & STIENSTRA (1977) observed that in dry and open fields in Indonesia the C_4 grasses occurring there, *Axonopus compressus* and *Setaria plicata*, showed a higher initial growth rate and a lower shoot/root ratio than the investigated C_3 grass (*Oplismenus compositus*), which grows in the shade. These factors may have competitive value and explain why C_3 grasses were absent in the open fields.

3.2. PHOTOSYNTHESIS OF AQUATIC PLANTS

Another interesting adaptation of photosynthesis occurs in submerged aquatic plants, which may absorb a considerable quantity of bicarbonate in addition to dissolved carbon dioxide gas. The distribution of the carbon dioxide components (CO_2, HCO_3^-, and CO_3^{2-}) is pH-dependent, and at pH 9.5 calculation gives a saturation concentration of bicarbonate of 20 mM, corresponding to a saturated CO_2 concentration at pH 7.0. Such high concentrations will cause a flux of HCO_3^- into the leaf that is almost equal to the rate of uptake of CO_2, assuming a CO_2 concentration of 75 ppm inside the leaf (HELDER & ZANSTRA 1977). The absorbed bicarbonate is utilized in photosynthesis, and part of the solar energy is used to pump the split hydroxyl ions ($HCO_3^- \rightarrow CO_2 + OH^-$) back into the medium. The hydroxyl efflux system is located in peripheral bands on *Chara* cells (LUCAS 1975, 1976).

In *Potamogeton* leaves a polar transport occurs, bicarbonate being absorbed at the lower surface of the leaf and hydroxyl ions expelled at the upper surface (HELDER 1975).

4. RESPIRATION

4.1. MITOCHONDRIAL RESPIRATION

Dark respiration seems to be regulated by various factors. Several authors found dark respiration of photosynthetic tissue to be reduced or even completely inhibited in the light (MANGEL *et al.* 1974; CHEVALIER & DOUCE 1976; RAVEN 1972, 1976), whereas in other experiments no effect of light was measurable (RAVEN 1972; CHAPMAN & GRAHAM 1974).

HEICHEL (1971) compared two corn varieties with significantly different rates of dark respiration. The variety with the lowest dark respiration showed the highest growth rate. Thus, it seems that as far as growth is concerned, part of the dark respiration can be considered wasteful (ZELITCH 1975). Part of the mitochondrial respiration occurs along an alternative pathway which produces only one-third as much ATP for each pair of hydrogen ions oxidized as does the conventional pathway, and sometimes no ATP is produced at all. The alternative pathway is insensitive to cyanide and antimycin but is specifically inhibited by the salicylhydroxamic acid (SCHONBAUM et al. 1971), a compound which does not affect the conventional pathway. In leaves, at least half of the mitochondrial respiration is cyanide-insensitive (BONNER & WILDMAN 1946), but insensitivity to cyanide has also been observed in potato tubers (VAN DER PLAS 1977) and in the sunk cabbage spadix (BAHR & BONNER 1973). It has been suggested that in the latter organ cyanide-insensitive respiration might contribute by heat production to the well-known elevated temperature characteristic for the developing spadix of Araceae.

The alternative pathway may function as a regulator of the redox state of the cell under conditions of excess production of reducing power (NADH), e.g. when the plant is exposed to conditions limiting growth of the plant. It can be important that respiration continues under such "stress" conditions, which limit growth, because at the same time the plant may need respiratory activity for the synthesis of cell material needed to cope with the newly developed "stress" condition (LAMBERS, personal communication). In this connection it is important to note that during ageing the phospholipid level in sweet potato tuber mitochondria is lowered, while concomitantly the cyanide-insensitive respiration of these mitochondria is increased due to a change in the mitochondrial membrane structure (NAKAMURA & ASAHI 1976). Regulation of respiration and its response to various environmental conditions is still a highly neglected field of research.

The alternative pathway is easily inhibited by exposure of plant roots to anaerobiosis (SOLOMOS 1977). It is conceivable that the stimulation of growth of the flooding-insensitive Senecio aquaticus (LAMBERS 1976) and other swamp plants observed under anaerobiosis is attributable to inhibition of non-phosphorylating oxidase (LAMBERS & SMAKMAN 1977). Under anaerobic conditions growth respiration of the roots of this Senecio species in only one-third of that of aerobically grown plants, a phenomenon which again underlines the importance of inhibition of this "wasteful" respiration by flooding (LAMBERS & STEINGRÖVER 1977). The significance of this oxidase for environmental adaptations remains an open question, but the results obtained so far indicate a possible role of this enzyme in adaptation to flooding.

5. PHYSIOLOGY OF MEMBRANES: EFFECTS OF FREEZING AND LOW TEMPERATURES

5.1. SUPERCOOLING

Freezing resistance in higher plants has been discussed in relation to many cellular characteristics. A first prerequisite for resistance is supercooling

of the intracellular water (GEORGE et al. 1974) to prevent intracellular
damage due to freezing. The role of supercooling and nucleation of ice was
studied by RASMUSSEN et al. (1975) in single cells, whereas BERVAES et al.
(1977) attempted to study this phenomenon in the more complicated system of
higher-plant tissues. To this end, the kinetics of freezing damage were
investigated in apple and pine trees. As expected, the killing rate associated
with freezing was lowest in cold-acclimated trees. When this rate is plotted
against the physical supercooling parameter, $1/T^3 \cdot (\Delta T)^2$, the results indicate
that in cold-acclimated trees supercooling is indeed part of the mechanism of
frost protection.

GEORGE et al. (1974), BURKE et al. (1975), and GEORG & BURKE (1977) had
suggested that deep supercooling has an important effect in the xylem of many
trees. It is of interest that in North America many tree species have a
northern distribution limit which is characterized by the rarity of minimum
temperatures below -40°C in any year. This temperature limit has physical
significance, since the limit for the supercooling of water is about -41°C,
and calorimetric and nuclear magnetic resonance studies have indeed shown that
in such species ice formation in the xylem starts in the region of -30°C to
-40°C (GEORGE & BURKE 1977).

5.2. THE PLASMA MEMBRANE

Besides supercooling, the chemical and physical condition of the plasma
membrane is crucial for frost resistance: after thawing, damage due to frost
becomes visible as a loss of turgor caused by destruction of the plasma
membrane structure. Upon lowering of the temperature, the lipid matrix of the
plasma membrane starts to crystallize, saturated lipid molecules first,
followed by less saturated lipid molecules at lower temperatures. The membrane
proteins tend to aggregate in the remaining area of non-crystalline lipids
(called the liquid-crystalline phase), and finally denaturation of membrane
proteins takes place by oxidation of sulfhydryl groups of protein molecules
in such close contact that disulfide bridges are formed and restoration of the
original state of the plasma membrane after the temperature rises is no
longer possible (LEVITT 1969).

5.3. LIPIDS AND FROST RESISTANCE

Plants whose winter hardiness is increased by low temperatures during growth
are characterized by a high total lipid content. Under these conditions there
is an appreciable increase in the level of two lipids viz. phosphatidyl
choline and phosphatidyl ethanolamine (alfalfa leaves, KUIPER 1970; poplar
bark, YOSHIDA 1974; black locust bark, SIMINOVITCH et al. 1975; wheat
seedlings, DE LA ROCHE et al. 1972, 1973, 1975, and WILLEMOT 1975; rape leaves,
SMOLENSKA & KUIPER 1977). The elevated level of these lipid fractions is
accompanied by a depressed level of phosphatidyl glycerol. When the plant

tissue is damaged by freezing, the level of phosphatidic acid rises sharply, even if the tissue remains frozen (WILSON & RINNE 1976).

In many studies on frost hardiness an increase in lipid unsaturation, and more specifically an increased level of linolenic acid, has been found. From a study done in wheat varieties differing in their ability to harden to cold, DE LA ROCHE *et al.* (1975) concluded that the observed elevated levels of linolenic acid in plants grown at low temperature (2°C) reflected only a response to the low temperature growth condition without a direct connection with frost-hardening itself. In agreement with this conclusion is the observation made in the bark of poplar and black locust trees that no change in unsaturation occurred during the entire year, even though the frost sensitivity of the bark tissue in the winter differs widely from that in the summer. The assumption that an elevated level of linolenic acid is a factor in the functioning of plants at low temperatures is supported by the finding that the survival of cotton seedlings at 8°C was strongly reduced when the seedlings had been treated with a specific inhibitor of linolenic acid synthesis (HILTON *et al.* 1971; ST. JOHN & CHRISTIANSEN 1976).

5.4. FLEXIBILITY OF MEMBRANES AND CYCLIC ACIDS

Mitochondria rich in unsaturated lipids show a higher degree of flexibility and permeability to water than mitochondria containing a higher proportion of saturated lipids (LYONS & RAISON 1970). In pine trees whose winter hardiness can be increased by a low temperature treatment as well as by a short-day treatment, specific effects of these environmental factors on lipid composition are noteworthy: behenic acid was exclusively synthesized upon transfer of the trees to low temperature, and a cyclic acid was suggested upon exposure of the plants to short-day conditions (BERVAES *et al.* in preparation). Such cyclic fatty acids were found in early spring plants (KUIPER & STUIVER 1972). Like polyunsaturated fatty acids, they guarantee a high flexibility of the involved membrane at low temperatures. Large quantities of cyclopropane fatty acids were observed in the sulfolipid fraction of the snow drop and of *Anthriscus sylvestris*, provided the plants were collected early in the spring. Cyclic acids were absent in flowering *Anthriscus* plants at the end of May. Compared with polyunsaturated acids, cyclic acids are less susceptible to photo-oxidation at (high) day-time temperatures and thus the production of cyclic acids might be a mechanism used by species exposed to extreme daily temperature fluctuations. Plant species of widely different habitats occurring in The Netherlands were screened for cyclic acids and, interestingly enough, two grasses from inland sand dunes (*Ammophila arenaria* and *Corynephorus canescens*) showed cyclopropane fatty acids in the phosphatidyl choline fraction of the leaves (KUIPER & STUIVER 1972).

A similar observation was made by DERTIEN *et al.* (1977) in lichens of sand-dune areas. Tree-growing species like *Evernia prunastri*, *Parmelia saxatilis*, and *Hypogymnia physodes* were characterized by high levels of the polyunsaturated linoleic and linolenic acids, whereas the terrestrial species from the sand-dune area, *Cetraria islandica* and *Cladonia impexa,* contained large quantities of cyclic acids (Table 1). In the latter species an analogue of nephromopsic acid containing a lactone ring was indicated by mass spectrometry.

TABLE 1. *Fatty acid composition of tree-growing and terrestrial lichens expressed as percentage of total fatty acids. The numbers of the fatty acids refer to the number of C-atoms and of double bonds, respectively (after* DERTIEN *et al.* 1977)

Fatty acids	Tree-growing			Terrestrial	
	Evernia prunastri	*Parmelia saxatilis*	*Hypogymnia physodes*	*Cetraria islandica*	*Cladonia impexa*
Saturated (16:0 + 18:0)	13.0	20.4	17.2	12.9	15.6
Monoenoic (18:1)	14.3	17.8	20.6	20.8	17.4
Dienoic (18:2)	20.3	34.5	34.9	25.2	19.1
Trienoic (18:3)	43.9	18.3	12.9	9.9	9.2
20 or more C-atoms*	7.0	13.8	14.8	26.3	38.1

* including cyclic lichen acids

Species adapted to extremely high day-time temperatures may contain large quantities of non-oxidazible lipids, and the organ-pipe cactus (*Lemairocereus thurberii*) contains sterols in levels up to 50% of the dry weight of the plant (KIRCHER & BIRD 1976).

5.5. PHOTO-OXIDATION AT LOW TEMPERATURES

Damage of chloroplasts by photo-oxidation at low temperatures (1°C) is sometimes observed in chilling-sensitive plants. In *Cucumis* leaves the chloroplast envelope ruptured and vesicles formed in the thylakoids at 1°C in the light (VAN HASSELT 1974a). Under the same conditions *Cucumis* leaves showed rapid degradation of linolenic acid (VAN HASSELT 1974b). The linolenic acid level of the dark control at 1°C was not affected. Blue light was especially effective for the photo-oxidation of unsaturated fatty acids, which indicates that in addition to chlorophyll, carotenoids contributed to the photo-oxidation of unsaturated fatty acids. DE KOK & KUIPER (1977) showed that specifically monogalactose diglyceride, which lipid complexes with chlorophyll, was degraded in the photo-oxidative process. In all probability, photo-oxidation of monogalactose diglyceride -which is mainly esterified with linolenic acid-can be prevented by tocopherol, and the level of this compound was found to be very low in *Cucumis* (VAN HASSELT, in preparation). This anti-oxidant was virtually absent in the lichen species studied by DERTIEN *et al.* (1977). Lichens with a large quantity of polyunsaturated fatty acids may therefore be sensitive to photo-oxidation. In this connection it is of interest that lichen species with the highest sensitivity to SO_2 pollution show visible symptoms of photo-oxidation (bleaching) (see also Table 2). SO_2 stimulates photo-oxidation, because at low pH the undissociated H_2SO_4 functions as a rather strong oxidator.

TABLE 2. *Fatty acid composition of tree-growing lichens (expressed as % of total fatty acids) as related to their sensitivity to air pollution. High figures (BARKMAN 1958; HAWKWORTH & ROSE 1970) refer to high sensitivity of the lichens; the latter increases also from A upwards (DE WIT 1976)*

Species	Fatty acids			Sensitivity to air pollution		
	dienoic (mainly linoleic acid)	trienoic (mainly linolenic acid)	20 or more C-atoms (saturated and cyclic acids)	BARKMAN (1958)	DE WIT (1976)	HAWKWORTH & ROSE (1970)
Ramalina fastigiata	11.8	54.9	10.4	8	F	7
Evernia prunastri	20.3	43.9	7.0	8	C	5
Parmelia saxatilis	34.5	18.3	13.8	10	F	5
Ramalina farinacea	31.1	15.0	18.3	7	D	5
Parmelia sulcata	20.9	17.0	34.2	6	B	4

A preliminary survey of species of *Ramalina, Evernia,* and *Parmelia,* which differ greatly in sensitivity to SO_2, indeed showed that the level of polyunsaturated fatty acid was directly related to the SO_2 sensitivity of the species (DE KOK, unpublished experiments).

5.6. TERPENES

A very interesting problem is presented by the chloroplasts of pine needles, which are known to continue photosynthesis even at sub-zero temperatures. Unlike the situation in most other higher plants, the monogalactose diglyceride fraction of these chloroplasts is esterified not with linolenic acid but with more saturated fatty acids. BERVAES *et al.* (1972) showed that upon dehardening of pine trees, terpene components (denoted as "extra-long-chain-fatty acids") in the chloroplasts moved from the digalactosyl diglyceride fraction to the monogalactosyl diglyceride fraction, as showed by DEAE-cellulose column chromatography. Extraction from the above fractions by

thin-layer chromatography, followed by n.m.r. and mass-spectrometry, indicated
that the chemical structure of these components was that of a monomethylester
of a cyclic diterpene dicarboxylic acid: pinifolic acid (BERVAES, in
preparation). Interconversion of the two forms takes place by internal
rearrangement of the molecule, i.e., by ring opening and closing, the open
ring being characteristic for hardened pine trees (Fig. 1). The chemical
nature of pinifolic acid is also very intriguing. Like sterols, this compound

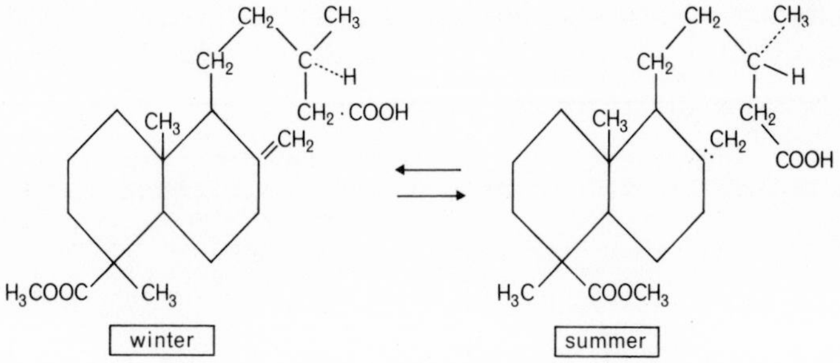

FIG. 1.   *Interconversion of two forms of pinifolic acid
monomethyl ester which was extracted from pine
needles in various times of the year* (BERVAES,
unpublished data). *The winter form is characterized
by a vinyl group in the ring structure. By charge
transfer within the molecule the summer form
attains a higher degree of electron mobility
in the ring structure*

functions as a membrane stabilizer and reduced temperature effects: the
melting point of the relatively unsaturated egg phosphatidyl choline rises
from 14° to 20°C (cholesterol: 18°C) and the melting point of the saturated
dipalmitoyl phosphatidyl choline, drops from 41° to 24°C (cholesterol: 32°C),
which provides a uniform flexibility of the lipid matrix of the biomembrane
over a wide range of temperatures (STULEN & BERVAES, in preparation).

5.7. ATP-ASES AND COLD-SENSITIVITY

Membrane ATPases and soluble ATPases may be sensitive to cold or freezing
temperatures (RACKER 1959; McCARTY & RACKER 1966; KUIPER 1971). Sensitivity
of the membrane protein to frost depends on the lipid environment. LIVNE &
RACKER (1969) showed that addition of lipids, with sulfolipid as the most
efficient type, are among the factors that give stability to the chloroplast
ATPase. ATPases of plant tissues required phosphatidyl choline and sulfolipid
for proper functioning (KUIPER 1972), but many other membrane enzymes require

a lipid matrix as well. For example, LYONS & RAISON (1970) observed that in plants sensitive to chilling succinate oxidation of mitochondria showed a non-uniform temperature dependence resulting in broken curves when the logarithm of the rate of oxidation was plotted against temperature. At low temperatures the rate of oxidation is strongly temperature dependent, whereas at higher temperatures a much smaller temperature response was observed. The break in such temperature curves can be attributed to a phase shift of the mitochondrial membrane lipid matrix from the crystalline phase to the liquid-crystalline phase at rising temperatures. Such breaks are absent when the mitochondria originate from chilling-resistant plants, because the increased degree of unsaturation of the lipid matrix in such mitochondria lowers the temperature threshold at which crystallization occurs. Similar responses have also been seen in chloroplasts (NOBEL 1974).

5.8. CONCLUDING REMARKS

The foregoing illustrates that with respect to temperature, not only membrane structure but also physiological functions as dependent on that structure may show adaptations to low temperatures (polyunsaturated fatty acids), freezing (zwitterionic phospholipids), high temperatures (sterols), and extreme daily temperature variations (aliphatic cyclic acids) in sand dune grasses and terrestrial lichens, as well as a more uniform temperature response over a wide temperature range (terpenes in pine needle chloroplasts) and the need of an anti-oxidant (tocopherol) to prevent photo-oxidation when the enzymatic reactions are reduced at unfavourable temperatures. A careful evaluation of the environmental temperature regime to which an individual plant is exposed is needed to recognize the "solution" the species has selected to cope with the different demands put by various temperature regimes on the organism, either for survival alone or for functioning in general.

6. PHYSIOLOGY OF MEMBRANES: EFFECTS OF SALINITY

6.1. OSMOREGULATION

Osmoregulation is obligatory for the maintenance of turgor pressure under saline conditions. Most higher plants utilize ion transport to build up the required osmotic pressure, i.e. by pumping ions into the vacuole. Others excrete accumulated salts from the leaves or the roots back into the environment. To protect the cytoplasm from salt damage, high levels of non-electrolytes are often observed in the cells: polyalcohols (glycerol, galactoglycerol, and sugars) and amino acids (mainly proline). For further details the reader is referred to POLJAKOFF-MAYBER & GALE (1975).

6.2. BIOMEMBRANES AND SALINITY

As regards the effects of salinity on membranes, energy is needed for the active transport of ions across the ectoplast and tonoplast to build up the required turgor pressure by the pumping of ions into the vacuole; cation-specific ATPases indicate the presence of such ion pumps. Secondly, in salt-sensitive species the membrane structure may be damaged by salt, resulting in the release of protein and the suppression of the active uptake mechanism for phosphate and glucose (NIEMAN & WILLIS 1971). Furthermore, under such conditions membrane-stabilizing divalent cations such as Mg^{2+} and Ca^{2+} are replaced by Na^+. Damage done by high salt concentrations is clearly visible under the electron microscope (POLJAKOFF-MAYBER 1975). Thirdly, sensitivity to chloride and other anions depends on the anion permeability of the lipid matrix of the biomembranes. Biomembranes containing (acidic) phospholipids tend to be virtually impermeable to cations and relatively more permeable to anions. For this reason, a distinction should be made between the effects of cations (Na^+) and of anions (Cl^-) as factors in salinity.

6.3. SENSITIVITY TO Na^+

A distinction can be made between natrophilic and natrophobic species. In natrophilic species Na^+ is taken up readily and is uniformly distributed in the plant. In some of these species a ($Na^+ + K^+$)-activated ATPase has been isolated and correlated with transport of these ions as a response to a saline environment (sugar beet, HANSSON & KYLIN 1969; *Avicennia,* KYLIN & GEE 1970). Stimulation of the enzyme by $Na^+ + K^+$ proved to be dependent on the sulfolipid level, and the removal of phosphatidyl choline -the other lipid found to be effective in the reconstitution of the activity of lipid-depleted plant-root ATPases (KUIPER 1972)- did not result in loss of activity. A similar ($Na^+ + K^+$)-ATPase preparation of animal origin, the salt gland of the duck has also been related to the level of an acidic sulfur-containing lipid, viz. sulphatide (KARLSSON *et al.* 1971). Roots of salt-tolerant *Plantago* species (*P. maritima* and *P. coronopus*) show higher levels of sulfolipid than roots of species from non-saline habitats (D.KUIPER, unpublished results). This acidic lipid might play an important role in the regulation of Na^+ transport in natrophilic plants.

A natrophobic species like *Phaseolus vulgaris* shows limited uptake of Na^+ by the roots together with effective excretion mechanisms that prevent any accumulation of Na^+ in the leaves (MARSCHNER 1974). When leaf segments of bean plants are exposed to NaCl, Na^+ accumulation can no longer be prevented, and this results in leakiness of membranes, loss of K^+, and damage to the chloroplasts. Sugar beet leaf discs exposed to NaCl do not show any of these symptoms, which indicates differences in membrane structure between bean and sugar beet leaves. In the sugar beet and cotton, growth is even stimulated by NaCl and in the latter species, under saline conditions phosphate is preferentially incorporated into phospholipid (TWERSKY & FELHENDLER 1973).

6.4. SENSITIVITY TO Cl⁻

KUIPER (1968a) compared the lipids of the roots of five varieties of grapes differing markedly in the translocation of Cl^- to the leaves. When these roots were exposed to moderate salt stress, the most salt-sensitive variety accumulated 15 times more Cl^- in the leaves than the most resistant variety. The monogalactosyl diglyceride content was directly related to Cl^- transport to the leaves, and this lipid was found to be the most efficient of the Cl^- transporters tested in a transport model (KUIPER 1968b). The phosphatidyl ethanolamine and phosphatidyl choline levels were inversely related to Cl^- accumulation. The roots of the most salt-sensitive variety had a very low sterol content. The charged phospholipids contributed strongly to the low Cl^- transport to the leaves. The effect on Cl^- transport of the addition of lipid to the root environment was also studied (KUIPER 1969). Galactolipids added to the root environment increased Cl^- transport to the roots, stem, and leaves of bean plants, whereas similarly supplied phosphatidyl choline was only absorbed by the roots; no transport to the stem and leaves could be dectected. When glycophosphoryl choline, a precursor of phosphatidyl choline, was added to the root environment, this substance was transported to the stem and leaves and incorporated into these tissues as phosphatidyl choline. The addition of glycerophosphoryl choline to the roots greatly reduced Cl^- transport to the leaves, thus demonstrating the importance of phosphatidyl choline in the regulation of Cl^- transport to the leaves when the plant is exposed to saline conditions.

7. PHYSIOLOGY OF MEMBRANES: RELATION TO MINERAL NUTRITION

7.1. INTRODUCTION

Higher plants have developed various mechanisms for regulating the uptake of mineral nutrients. Under limitation of the supply, species may develop a specific mechanism to guarantee sufficient uptake of an essential nutrient. MARSCHNER (1975) gives the example that under iron deficiency, sunflower roots lower the pH of the root environment. This results in leakage of reducing substances from the roots, which in turn reduce the local Fe^{3+} ion, which is not available for the plant, to Fe^{2+}. The reduced Fe^{2+} ions guarantee iron nutrition of the plants until the pH value of the root environment rises again, an iron deficiency develops, and the whole cycle is repeated.

A relationship between adaptation to variations in the availability of nutrients on the one hand and membrane properties on the other hand seems appropriate for plants, which unlike animals are unable to move away when changes occur in the environmental conditions of the habitat. The plasma membrane of the outer root cells is exposed to every fluctuation of the soil conditions and, to guarantee the required levels of uptake of essential nutrients, plants have had to adapt the properties of their cell membranes to a considerable extent. As an example, English ryegrass roots were found to have four different isozymes of ATPase that were stimulated by various combinations of concentrations of Na^+ and K^+. The activity of these isozymes varied along the root axis and was strongly affected by the level of the available nutrients. When the salt concentration was lowered from 35 to 0.75 mM, an isozyme specifically stimulated by Na^+ was detected that could have a function in osmotic adaptation by regulating the efflux of Na^+ (NELSON & KUIPER 1975). There are many other examples, but this discussion will be restricted to two crop plants (wheat and oat) and to species of *Plantago*, which will be discussed with respect to Ca^{2+} and Mg^{2+} ions.

7.2. Ca^{2+} AND Mg^{2+} IN WHEAT AND OAT

Wheat roots showed a high proportion of Ca^{2+}-stimulated ATPase activity, whereas in oat roots ATPase stimulation by Mg^{2+} was dominant (KYLIN & KÄHR 1973; KÄHR & KYLIN 1974; KÄHR & MAX MØLLER 1976). These species also differed in their response to the nutritional level of the substrate on which the plants were grown. Low-salt roots of oats showed a higher activity of divalent cation-stimulated ATPase than did high-salt roots. The reverse was found for wheat roots (KYLIN & KÄHR 1973; KÄHR & KYLIN 1974). These observations on the ATPase activity correlated well with field observations on the nutrient demands of these species. Oat is a crop of acid soils with a low mineral content, whereas wheat prefers calcium-rich soils with a high content of mineral nutrients.

The lipids of oat roots are more unsaturated than those of wheat roots, which is consistent with the habitat of these crops, oat having a preference for lower soil temperatures than wheat (KÄHR *et al*. 1976). When wheat and oat plants were grown at 18° and 25°C and different levels of nutrition, the highest activity of oat roots (in the presence of Mg^{2+}) was found in plants grown at 18°C and a low salt level (KÄHR & MAX MØLLER 1976).

7.3. Ca^{2+} AND Mg^{2+} IN *PLANTAGO*

The ATPase activity of the microsomal membrane fractions of the roots of several plantain species of various habitats was tested (D. KUIPER in preparation). The highest Ca^{2+}- and Mg^{2+}-stimulated ATPase activities were observed in species from relatively nutrient-poor environments (*Plantago lanceolata, Plantago coronopus,* and *Plantago media*), and species from relatively nutrient-rich environments (*Plantago major* and *Plantago maritima*) showed much lower activities (Fig. 2). A similar distinction between the plantain species could be made for the affinity of the ATPases for Mg^{2+} and Ca^{2+}. *Plantago major* and *Plantago maritima* shared other characteristics of their Ca^{2+}- and Mg^{2+}-stimulated ATPases. Besides an optimum in activity at pH 6.5 observed in all of the *Plantago* species tested, a second optimum was

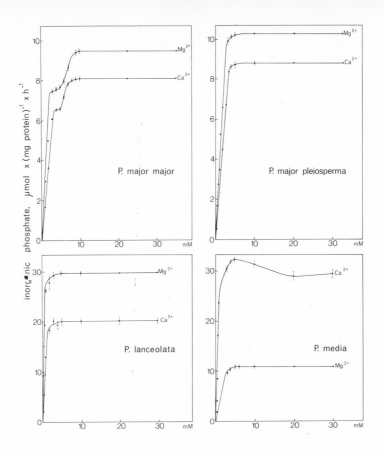

FIG. 2. *Effect of Ca^{2+} and Mg^{2+} ions on the ATPase activity
of microsomal fraction from the roots of various*
Plantago *species from different ecological habitats:*
Plantago major *spp.* major *and P. major* spp. pleiosperma,
relatively nutrient-rich habitats; P. lanceolata *and*
P. media, *relatively nutrient-poor habitats and from
rather acidic and alcaline soils respectively
(unpublished data of* D. KUIPER)

observed at pH 8.0. Also, the Mg^{2+}- and Ca^{2+}-stimulated ATPase activities in
both species showed a biphasic response when activity was plotted against
ionic concentration, but only in plants grown on nutrient-rich solution. For
this reason it is suggested that the high-affinity ATPase system of *Plantago
major* and *Plantago maritima* is located at the plasma membrane and the low-
affinity system at the tonoplast. Biphasic ion-uptake curves have been
reported by PITMAN (1976) for Ca^{2+} (corn, cotton) and Mg^{2+} (barley).

In *Plantago lanceolata*, Mg^{2+} stimulated the ATPase activity more than Ca^{2+}
did; in *Plantago major* and *Plantago maritima* the effect of these ions was
virtually equal; and in the species occurring on calcareous soil (*Plantago*

media) Ca^{2+} had a much higher stimulatory effect than Mg^{2+}. In this respect *Plantago lanceolata* resembles oat and *Plantago media* wheat, except that in both *Plantago* species stimulation by Mg^{2+} (*Plantago lanceolata*) and by Ca^{2+} *Plantago media* is most dramatic in the roots of plants grown under low-salt conditions.

With respect to the ATPase activity of the entire root system, *Plantago coronopus* showed almost no response to fluctuations in the level of nutrients, a slight response was detectable in *Plantago lanceolata* and *Plantago media*, and *Plantago major* and *Plantago maritima* were severely affected by a drop in the nutrient level. The affinity of the Ca^{2+}- and Mg^{2+}-stimulated ATPase in the roots increased after transfer of the plants to low-salt conditions, which partly compensated for the lowered capacity of the ATPase. The well-regulated ATPase activity in *Plantago coronopus* is consistent with an ecological adaptation of the species, which enables it to grow under conditions fluctuating between saline and nutrient-poor.

7.4. CONCLUDING REMARKS

For salinity as well as for mineral nutrition, parallels can be drawn between biochemical properties of root-cell membranes and the specific demands the root environment makes on the individual plant. The various examples concerning salinity and membrane properties as well as mineral nutrition and ATPase activity in plants from different habitats clearly show the usefulness of this approach. As already mentioned, the right connection must be made between the levels of energy production and utilization on the one hand and the responses of the membranes involved on the other hand, and it is obvious that only an approach making use of several methods can provide a sound basis for an understanding of the relationship between membrane properties, physiological responses, and ecological adaptations. The *Plantago* research mentioned above forms part of such a project which was initiated by the Institute for Ecological Research and various university departments in The Netherlands.

8. ACKNOWLEDGEMENTS

I would like to thank J.C.A.M. Bervaes (A.E.), Dr. Ph.R. van Hasselt, Dr. R.J. Helder, L.J. de Kok, D. Kuiper, and J.T. Lambers for permission to use their research results.

9. REFERENCES

BAHR, J.T. & W.D. BONNER, 1973 - Cyanide-insensitive respiration. I. The steady states of skunk cabbage and bean hypocotyl mitochondria. *J. biol. Chem.*, 248, 3441-3445.

BARKMAN, J.J., 1958 - *Phytosociology and ecology of cryptogamic epiphytes*. Van Gorcum, Assen, The Netherlands, 628 p.

BERVAES, J.C.A.M., KETCHIE, D.O. & P.J.C. KUIPER, 1977 - Kinetics of freezing damage in apple bark and pine needles. *Physiologia Pl.*, 40, 35-38.

BERVAES, J.C.A.M., KUIPER, P.J.C. & A. KYLIN, 1972 - Conversion of digalactosyl diglyceride (extra long carbon chain conjugates) into monogalactosyl diglyceride of pine needle chloroplasts upon dehardening. *Physiologica Pl.*, 27, 231-235.

BJÖRKMAN, O., 1975 - Environmental and biological control of photo-synthesis: inaugural address. *In*: R. MARCELLE (Editor), *Environmental and biological control of photosynthesis*. W. Junk, The Hague, The Netherlands, p. 1-16.

BJÖRKMAN, O., BOARDMAN, N.K., ANDERSON, J.M., THORNE, S.W., GOODCHILD, D.J. & N.A. PYLIOTIS, 1972 - Effect of light intensity during growth of *Atriplex patula* on the capacity of photosynthetic reactions, chloroplast components and structure. *Carnegie Institute, Washington, Year Book*, 71, 115-135.

BLACK, C.C., 1973 - Photosynthetic carbon fixation in relation to net CO_2 uptake. *A. Rev. Pl. Physiol.*, 24, 253-286.

BONNER, J. & S.G. WILDMAN, 1946 - Enzymatic mechanisms in the respiration of spinach leaves. *Archs. Biochem.*, 10, 497-518.

BURKE, M.J., BRYANT, R.G. & C.J. WEISER, 1974 - Nuclear magnetic resonance of water in cold acclimating red osier dogwood stem. *Pl. Physiol.*, 54, 392-398.

CHAPMAN, E.S. & D. GRAHAM, 1974 - The effect of light on the tricarboxylic acid cycle in green leaves. I. Relative rates of the cycle in the dark and the light. *Pl. Physiol.*, 53, 879-898.

CHEVALIER, D. & R. DOUCE, 1976 - Interactions between mitochondria and chloroplasts in cells. I. Action of cyanide and of 3-(3,4 - dichlorophenyl)-1, 1-dimethylurea on the spore of *Funaria hygrometrica*. *Pl. Physiol.*, 57, 400-402.

CRAFTS, A.S., CURRIER, H.B. & C.R. STOCKING, 1949 - *Water in the physiology of plants*. Chron. Bot. Ca. USA, 240 p.

DERTIEN, B.K., KOK, L.J. DE & P.J.C. KUIPER, 1977 - Lipid and fatty acid composition of tree-growing and terrestrial lichens. *Pl. Physiol.*, 40, 175-180.

GEORGE, M.F. & M.J. BURKE, 1977 - Cold hardiness and deep supercooling in xylem of shagbark hickory. *Pl. Physiol.*, 59, 319-325.

GEORGE, M.F., BURKE, M.J., PELLET, H.M. & A.G. JOHNSON, 1974 - Low temperature exotherms and woody plant distribution. *Hort. Science*, 9, 519-522.

HANSSON, G. & A. KYLIN, 1969 - ATPase activities in homogenates from sugar beet roots. *Z. Pflanzenphysiol.*, 60, 270-275.

HASSELT, Ph. R. VAN, 1974 - Photooxidative damage to the ultrastructure of *Cucumis* chloroplasts during chilling. *Proc. K. ned. Akad. Wet. (Amsterdam) Series* C 77, 50-56.

HASSELT, Ph. R. VAN, 1974 - Photooxidation of unsaturated fatty acids in *Cucumis* leaf discs during chilling. *Acta bot. neerl.*, 23, 156-166.

HAWKSWORTH, D.L. & F. ROSE, 1970 - Quantitative scale for estimating sulphur dioxide air pollution in England and Wales using epiphytic lichens. *Nature*, 227, 145-148.

HEICHEL, G., 1971 - Confirming measurements of respiration and photosynthesis with dry matter accumulation. *Photosynthetica*, 5, 93-98.

HELDER, R.J., 1975 - Flux-ratios and concentration-ratios in relation to electrical potential differences and transport of rubidium ions across the leaf of *Potamogeton lucens* L. *Proc. K. ned. Akad. Wet. (Amsterdam) Series* C 38, 376-388.

HELDER, R. & P. ZANSTRA, 1977 - Changes of the pH at the upper and lower surface of bicarbonate assimilating leaves of *Potamogeton lucens*. *Proc. K. ned. Akad. Wet. (Amsterdam)*, in press.

HILTON, J.L., ST. JOHN, J.B., CHRISTIANSEN, M.N. & K.H. NORRIS, 1971 – Interactions of lipoidal materials and a pyridazinone inhibitor of chloroplast development. *Pl. Physiol.*, 48, 171-177.

HOFSTRA, J.J. & A.W. STIENSTRA, 1977 – Growth and photosynthesis of closely related C3 and C4 grasses as influenced by light intensity and water supply. *Acta bot. neerl.*, 26, 63-72.

KÄHR, M., BERVAES, J.C.A.M., KYLIN, A. & P.J.C. KUIPER, 1976 – Influence of mineral nutrition on ATPase activities and relation saturated to unsaturated fatty acids in ATPase preparations from wheat and oats. *In: Proc. Transmembrane Ionic Exchanges in Plants,* Rouen, in press.

KÄHR, M. & A. KYLIN, 1974 – Effects of divalent cations and oligomycin on membrane ATPases from roots of wheat and oat in relation to salt status and cultivation. *In:* U. ZIMMERMAN & J. DAINTY (Editors), *Membrane transport in plants,* Springer, Berlin, p. 321-325.

KÄHR, M. & I.MAX MØLLER, 1976 – Temperature response and effect of Ca^{2+} and Mg^{2+} on ATPases from roots of oats and wheat as influenced by growth temperature and nutritional status. *Physiologia Pl.*, 38, 153-158.

KARLSSON, K.A., SAMUELSSON, B.E. & G.O. STEEN, 1971 – Lipid pattern and Na^+-K^+-dependent adenosine triphosphatase activity in the salt gland of duck before and after adaptation to hypertonic saline. *J. Membrane Biol.*, 5, 169-184.

KIRCHER, H.A. & H. BIRD, 1976 – Lipids of organ pipe cactus, *Lemairocereus thurberii*. *In: Lipids and Lipidpolymers in higher Plants,* Abstr. Symp. Bot. Inst. Univ. Karlsruhe, West Germany, 68-69.

KOK, L.J. DE & P.J.C. KUIPER, 1977 – Glycolipid degradation in leaves of the thermophilic *Cucumis sativus* as affected by light and low-temperature treatment. *Physiologia Pl.*, 39, 123-128.

KUIPER, P.J.C., 1968a– Lipids in grape roots in relation to chloride transport. *Pl. Physiol.*, 43, 1367-1371.

KUIPER, P.J.C., 1968b– Ion transport characteristics of grape root lipids in relation to chloride transport. *Pl. Physiol.*, 43, 1372-1374.

KUIPER, P.J.C., 1969 – Effect of lipids on chloride and sodium transport in bean and cotton plants. *Pl. Physiol.*, 44, 968-972.

KUIPER, P.J.C., 1970 – Lipids in alfalfa leaves in relation to cold hardiness. *Pl. Physiol.*, 45, 684-686.

KUIPER, P.J.C., 1971 – Potato ATPase: sensitivity to hydrostatic pressure of a cold-labile form. *Biochim. biophys. Acta*, 250, 443-445.

KUIPER, P.J.C., 1972 – Temperature response of adenosine triphosphatase of bean roots as related to growth temperature and to lipid requirement of the adenosine triphosphatase. *Physiologia Pl.*, 26, 200-205.

KUIPER, P.J.C. & C.E.E. STUIVER, 1972 – Cyclopropane fatty acids in relation to earliness in spring and drought tolerance in plants. *Pl. Physiol.*, 49, 307-309.

KYLIN, A. & R. GEE, 1970 – Adenosine triphosphatase activities in leaves of the mangrove *Avicennia nitrida* Jacq. *Pl. Physiol.*, 45, 169-172.

KYLIN, A. & M. KÄHR, 1973 – The effect of magnesium and calcium ions on adenosine triphosphatases from wheat and oat roots at different pH. *Physiologia Pl.*, 28, 452-457.

LAMBERS, H., 1976 – Respiration and NADH-oxidation of the roots of flood-tolerant and flood-intolerant *Senecio* species as affected by anaerobiosis. *Physiologia Pl.*, 37, 117-122.

LAMBERS, H. & G. SMAKMAN, 1977 – Respiration of the roots of flood-tolerant and flood intolerant *Senecio* species: affinity for oxygen and resistance to cyanide. (in press).

LAMBERS, H. & E. STEINGRÖVER, 1977 – Efficiency of root respiration of a flood-tolerant and a flood intolerant *Senecio* species as affected by low oxygen tension. *Physiologia Pl.*, (in press).

LEEUWEN, C.G. VAN, 1966 – A relation-theoretical approach to pattern and process in vegetation. *Wentia,* 15, 25-46.

LEVITT, J., 1956 – *The hardiness of plants.* Acad. Press, New York, USA, 278p.

LEVITT, J., 1969 – Growth and survival of plants at extremes of temperature – a unified concept –. *Symp. Soc. Exp. Biol.*, 23, 395-448.

LEVITT, J., 1972 – *Responses of plants to environmental stresses.* Acad. Press, New York, USA, 697 p.

LIVNE, A. & E. RACKER, 1969 - Partial resolution of the enzymes catalyzing photophosphorylation. V. Interaction of coupling factor 1 from chloroplasts with ribonucleic acid and lipids. *J. biol. Chem.*, 244, 1332-1338.

LUCAS, W.J., 1975a- Photosynthetic fixation of ^{14} carbon by internodal cells of *Chara corallina*. *J. exp. Bot.*, 26, 331-346.

LUCAS, W.J., 1975b- The influence of light intensity on the activation and operation of the hydroxyl efflux system of *Chara corallina*. *J. exp. Bot.*, 26, 347-360.

LUCAS, W.J., 1976a- Plasmalemma transport of HCO_3^- and CH^- in *Chara corallina*: non-antiporter systems. *J. exp. Bot.*, 27, 19-31.

LUCAS, W.J., 1976b- The influence of Ca^{2+} and K^+ on $H^{14}CO_3^-$-influx in internodal cells of *Chara corallina*. *J. exp. Bot.*, 27, 32-42.

LYONS, J.M. & J.K. RAISON, 1970 - Oxidative activity of mitochondria from plant tissues sensitive and resistant to chilling injury. *Pl. Physiol.*, 45, 386-389.

MANGAT, B.S., LEVIN, W.B. & R.G.S. BIDEWELL, 1974 - The extent of dark respiration in illuminated leaves and its control by ATP levels. *Can. J. Bot.*, 52, 673-682.

MARCELLE, R., 1975 - *Environmental and Biological Control of Photosynthesis*. Junk, The Hague, 408 p.

MARSCHNER, H., 1974 - Mechanisms of regulation of mineral nutrition in higher plants. *In*: R.L. BIELESKI, A.R. FERGUSON & M.M. CRESWELL (Editors), *Mechanisms of regulation of plant growth*. R. Soc. N.Z., 12, p. 99-109.

MAXIMOV, N.A., 1912 - Chemische Schutzmittel der Pflanzen gegen Erfrieren. I. *Ber. dt. bot. Ges.*, 30, 52-65.

MAXIMOV, N.A., 1912 - Chemische Schutzmittel der Pflanzen gegen Erfrieren. II. Die Schutzwirkung von Salzlösungen. *Ber. dt. bot. Ges.*, 30, 293-305.

MAXIMOV, N.A., 1912 - Chemische Schutzmittel der Pflanzen gegen Erfrieren. III. Über die Natur der Schutzwirkung. *Ber. dt. bot. Ges.*, 30, 504-516.

McCARTY, R.E. & E. RACKER, 1966 - Effect of a coupling factor and its antiserum on photophosphorylation and hydrogen ion transport. *In: Energy conversion by the photosynthetic apparatus. Brookhaven Symp. Biol.*, 29, 202-212.

NAKAMURA, K. & T. ASAHI, 1976 - Changes in properties of the inner mitochondrial membrane during mitochondrial biogenesis in aging sweet potato tissue slices in relation to the development of cyanide-insensitive respiration. *Archs. Biochem. Biophys.*, 174, 393-401.

NELSON, P.V. & P.J.C. KUIPER, 1975 - Properties of (sodium + potassium)-stimulated ATPases in English ryegrass roots and implications in ion transport. *Physiologia Pl.*, 35, 263-268.

NIEMAN, R.H. & C. WILLIS, 1971 - Correlation between the suppression of glucose and phosphate uptake and the release of protein from viable carrot root cells treated with monovalent cations. *Pl. Physiol.*, 48, 287-293.

NOBEL, P.S., 1974 - Temperature dependence of the permeability of chloroplasts from chilling-sensitive and chilling-resistant plants. *Planta*, 115, 369-372.

PITMAN, M.G. ANDERSON, W.P. & U. LÜTTGE, 1976 - Transport processes in roots. *In*: U. LÜTTGE & M.G. PITMAN (Editors), *Transport in plants*. Encycl. Plant Physiol. II part B New Series vol. 2, Springer Berlin, p. 57-151.

PLAS, L.H.W. VAN DER, LOBSE, P.A. & L.D. VERLEUR, 1976 - Cytochrome-c dependent, antimycin-A resistant respiration in mitochondria from potato tuber (*Solanum tuberosum* L.). *Biochim. biophys. Acta*, 430, 1-12.

PLAS, L.H.W. VAN DER, SCHOENMAKER, G.S. & S.J. GERBRANDY, 1977 - CN-resistant respiration in a *Convolvulus arvensis* L. cell structure. *Plant Sci. Letters*, 8, 31-33.

POLJAKOFF-MAYBER, A., 1975 - Morphological and anatomical changes in plants as a response to salinity stress. *In*: A. POLJAKOFF-MAYBER & J. GALE (Editors) *Plants in saline environments*. Springer, Berlin, p. 97-117.

RASMUSSEN, D.M., MACAULY, M.N. & A.P. MACKENZIE, 1975 - Supercooling and nucleation of ice in single cells. *Cryobiology*, 12, 328-339.

RAVEN. J.A., 1972 - Endogeneous inorganic carbon sources in plant photo-
    synthesis. I. Occurrence of the dark respiration pathways in illuminated
    green cells. *New Phytol.*, 71, 227-247.

RAVEN, J.A., 1976 - The rate of cyclic and non-cyclic photophosphorylation and
    oxidative phosphorylation and regulation of the rate of ATP consumption
    in *Hydrodictyon africanum*. *New Phytol.*, 76, 205-212.

ROCHE, I.A. DE LA & C.J. ANDREWS, 1973 - Changes in phospholipid competition
    of a winter wheat cultivar during germination at 2C and 24C. *Pl. Physiol.*,
    51, 468-473.

ROCHE, I.A. DE LA, ANDREWS, C.J., POMEROY, M.R., WEINBERGER, P. & M. KATES,
    1972 - Lipid changes in winter wheat seedlings (*Triticum aestivum* L.) at
    temperatures inducing cold hardiness. *Can. J. Bot.*, 50, 2401-2409.

ROCHE, I.A. DE LA, POMEROY, M.K. & C.J. ANDREWS, 1975 - Changes in fatty acid
    composition in wheat cultivars of contrasting hardiness. *Cryobiology*, 12,
    506-512.

SCHONBAUM, G.R., BONNER Jr., W.D., STOREY, B. & J.T. BAHR, 1971 - Specific
    inhibition of the cyanide-insensitive respiratory pathway in plant
    mitochondria by hydronamic acids. *Pl. Physiol.*, 47, 124-128.

SIMINOVITCH, D., SINGH, J. & I.A. DE LA ROCHE, 1975 - Studies on membranes in
    plant cells resistant to extreme freezing. I. Augumentation of
    phospholipids and membrane substance without changes in unsaturation of
    fatty acids in hardening of black locust bark. *Cryobiology*, 12, 144-153.

SMOLENSKA, G. & P.J.C. KUIPER, 1977 - Effect of low temperature upon lipid and
    fatty acid composition of roots and leaves of winter rape plants.
    Physiologia Pl., (in press).

SOLOMOS, T., 1977 - Cyanide-resistant respiration in higher plants. *A. Rev.
    Pl. Physiol.*, 28, 279-297.

ST. JOHN, J.B. & M.N. CHRISTIANSEN, 1976 - Inhibition of linolenic acid
    synthesis and modification of chilling resistance in cotton seedlings.
    Pl. Physiol., 57, 257-259.

TWERSKY, M. & R. FELHENDLER, 1973 - Effect of water quality on relationship
    between cationic species and leaf lipids at two different development
    stages in cotton. *Physiologia Pl.*, 29, 396-401.

WESTHOFF, V., BAKKER, P.A., LEEUWEN, C.G. VAN & E.E. VAN DER VOO, 1970-1973 -
    Wilde Planten. Ver. Beh. Natuurmon. in Ned., The Netherlands, part 1-3,
    (Dutch).

WILLEMOT, C., 1975 - Stimulation of phospholipid biosynthesis during frost
    hardening of winter wheat. *Pl. Physiol.*, 55, 356-359.

WILSON, R.F. & R.W. RINNE, 1976 - Effect of freezing and cold storage on
    phospholipids in developing soybean cotyledons. *Pl. Physiol.*, 57,
    270-273.

WIT, T. DE, 1976 - Epiphytic lichens and air pollution in The Netherlands.
    Research Institute for Nature Management, The Netherlands, Verh. 8.

YOSHIDA, S., 1974 - Studies on lipid changes associated with frost hardiness
    in cortex in woody plants. *Contr. Inst. low Temp. Sci., Hakkaido Univ.*,
    B 18, 1-43.

ZELITSCH, I., 1975 - Environmental and biological control of photosynthesis:
    general assessment. *In:* R. MARCELLE (Editor), *Environmental and
    biological control of photosynthesis*. Junk, The Hague, p. 251-262.

10. DISCUSSION

QUESTION: Is there a correlation between the rate of fatty acid
substitution and the unpredictability of the environment in which species
live?

ANSWER: I think the category "fluctuations of the environment" will more
or less fit your "unpredictability". Terrestrial lichens, for instance, have
very widely fluctuating rates of synthesis and breakdown of lipids and lichen
acids, whereas the tree-growing lichens maintain a much more constant level

in these respects.

QUESTION: Halophytes, which are succulent, have a very high elasticity of the cells. Would you expect this phenomenon to be correlated with a biochemical change in the membrane?

ANSWER: For the sugar beet, some results are available. Nyctostatin, a compound which removes sterols from biomembranes and make them leaky to K^+, is ineffective in the sugar beet, which indicates that the sterols present in the sugar beet biomembranes are inaccessible to this compound, unlike the membranes of glycophytes. Work done by D. KUIPER in our laboratory showed that sulpholipid, which is essential for functioning of the $(Na^+ + K^+)$-ATPase observed in halophytes, is present in higher levels in *Plantago maritima* and *Plantago coronopus* than in the glycophytic *Plantago* species. Furthermore, a halophyte like *Plantago coronopus* is much better equipped than *Plantago major* to maintain a constant sterol level in its membranes, because the latter lacks a regulatory mechanism.

QUESTION: You started your lecture by mentioning that Professor MAXIMOV found that increased resistance to frost was correlated with increased sugar content in the plant. Do the sugars have other functions besides that of increasing the osmotic pressure of the cell?

ANSWER: Sugars are needed to prevent intracellular freezing damage to the cytoplasm.

The rhizosphere as part of the plant-soil system

1. INTRODUCTION

The organic materials formed by higher plants during the process of photosynthesis are decomposed mainly by microorganisms. These breakdown processes take place primarily in the soil. The constituents of the aerial parts of the plant become available to the microflora by volatilization from the leaves, leaching by rain, mist, or dew, and by the falling of leaves or branches. The subterranean parts provide the microorganisms with root excretions, sloughed-off cells, and dead roots.

The present discussion is mainly concerned with the processes occurring under the influence of the carbon provided by the living root system. It should be kept in mind, however, that compounds originating from intact aerial parts, i.e., leached and volatilized products, may considerably alter the surface layers of the soil around the plant. Many of these compounds are toxic to other plant species or inhibit the germination of their seeds. The important ecological effects of these allelochemicals supplied by the aerial parts will not be discussed here; the reader is referred to reviews by TURKEY (1969), WHITTAKER (1970, 1971), WENT (1970), and RICE (1974).

The present paper deals with the problem of whether the microbial processes going on in the environment of the root, i.e., the rhizosphere, are influenced by the plant in a specific way and thus contribute to niche differentiation. Special attention will also be given to the effects of the rhizosphere microflora on the availability of plant nutrients such as nitrogen and phosphorus. For a discussion of other aspects of rhizosphere research, the reviews given by ROVIRA (1974) and MANGENOT (1975) should be consulted.

For an understanding of microbial processes in the soil, a few points should be kept in mind. Due to the small dimensions of the microorganisms their environmental conditions differ widely from place to place and fluctuate strongly in time. In the surface layers appreciable differences in temperature occur, the availability of moisture and oxygen in the interior and on the surface of soil aggregates may be completely different, and the supply of energy and nutrients is highly variable in space and time. Consequently, microorganisms may easily loose their niche because of changes in environmental conditions and thus become subject to mineralization. Due to these variations in conditions on a millimeter scale, many microbial processes may proceed simultaneously at different places in the rhizosphere: for instance, mineralization of nitrogen may be accompanied by its immobilization, fixation of atmospheric nitrogen into organic compounds may coincide with the decomposition of nitrate into gaseous products.

Much of our insight into the processes going on in the rhizosphere is based on experiments done in arable and horticultural plant species, most of which are rapidly growing annuals. As far as perennial plants are concerned the data available derive almost exclusively from species of permanent grasslands. Our knowledge of these processes as they occur in natural vegetations is still very incomplete, and much work will have to be done before a more satisfactory picture can be obtained.

In such natural vegetations the primary production is often limited by a lack of nutrients or water, and a surplus of energy is available in the form of carbonaceous materials. For their nitrogen and phosphorus supply, these plants depend almost completely on the mineralization of organic compounds by microorganisms, whereas arable and horticultural crops, which are usually given fertilizers at the beginning of the growth season, have a surplus of nitrogen available during a great part of the year and energy is the factor limiting microbial activities. Therefore, it is hardly surprising that experiments with such crops have yielded different results from those performed in natural, unfertilized vegetation.

2. DESCRIPTION OF THE RHIZOSPHERE

The rhizosphere is composed of three zones. The rhizosphere in the strict sense which consists of the soil around the roots in which soluble and volatile compounds excreted by the roots diffuse. The rhizoplane, which comprises the root surface and the mucigel covering part of the roots behind the root cap. The endorhizosphere, which consists of epidermis and cortex cells invaded by saprophytic microorganisms. It should be noted that no microorganisms have been observed in the cells of the central cylinder.

The number of microorganisms in the rhizosphere (R) is much higher than in the soil further away from the root (S). R/S ratios of about 6 are usually found, but in poor soils the rhizosphere effect is much more pronounced, as can be seen from Table 1, which shows data obtained from arable crops by ROUATT & KATZNELSON (1961) and from species of poor dune soils by WEBLEY et al. (1951). High R/S values are found particularly in the rhizosphere of leguminous plants.

The rhizosphere effect can often be observed at considerable distances from the root. This is illustrated by data of PAPAVIZAS & DAVEY (1961) obtained with Blue Lupine (Table 2).

TABLE 1. *Colony counts of bacteria in root-free soil and in the rhizosphere of crop plants and wild species*

Species	Colony counts (10^6/g soil)		
	Rhizosphere	Root-free soil	R/S ratio
*Trifolium pratense**	3,260	134	24
*Avena sativa**	1,090	184	6
*Linum usitatissimum**	1,015	184	5
*Zea mays**	614	184	3
*Atriplex babingtonii***	23.3	0.016	1455
*Ammophila arenaria***	3.58	0.016	223
*Agropyron junceum***	3.56	0.016	222

* Data from ROUATT & KATZNELSON (1961)

** Data from WEBLEY *et al.* (1951)

TABLE 2. *Extent of the rhizosphere of Blue Lupine* (PAPAVIZAS & DAVEY 1961)

Distance from root (mm)	Microorganisms (10^3/g oven-dry soil)		
	Bacteria	Streptomycetes	Fungi
0	159,000	46,700	355
0- 3	49,000	15,500	176
3- 6	38,000	11,400	170
9-12	37,400	11,800	130
15-18	34,170	10,100	117
>80	27,300	9,100	91

TABLE 3. *Composition of the microflora of the rhizosphere of a 2-year-old sand culture of perennial ryegrass, of permanent grassland on sand and of sandy arable soil, respectively* (WOLDENDORP 1963)

	Rhizosphere	Grassland	Arable soil
% of total number			
Bacteria	75	70	90
Streptomycetes	25	30	9
Fungi	0	0	1
% of bacteria			
G(-) rods	60	58	11
Bacilli	3	10	14
Coryneforms	37	32	75

Fungi and streptomycetes were stimulated mainly in the rhizoplane, the effects on bacterical numbers occurred over greater distances. From these findings it can be concluded that under a permanent cover of plants the microbial processes in the upper layer of the soil are under the influence of the roots. Consequently, the composition of the microflora of grassland has much more in common with that of the rhizosphere soil (Table 3).

In the rhizosphere the microorganisms are not evenly distributed over the root surface. The region of the newly formed root-cap cells is generally free of bacteria and fungal hyphae, but the mucilaginous sheath around the roots soon becomes colonized by bacteria, some of which are able to lyse the mucigel (BREISCH et al. 1975). The older root parts are often densely colonized by many layers of bacterial cells, but the bacteria occur commonly in colonies with only a limited number of cells. These colonies are particularly common at the junction of cell walls and at places where lateral roots emerge. This uneven distribution of the microorganisms over the root surface was also shown by NEWMAN & DOWEN (1974), who used the techniques developed by GREIG-SMITH (1964) to study patterns in vegetation. Although some parts of the root are densely colonized, most of the root surface is devoid of microorganisms. At most 15 per cent of the root surface is covered by microorganisms, but ROVIRA et al. (1974) found lower values for a number of grassland species (Table 4).

TABLE 4. *Percentage of root surface covered by bacteria in grassland species* (ROVIRA *et al.* 1974)

Species	% cover	Species	% cover
Hypochaeris radicata	9.3	*Cynosurus cristatus*	6.4
Holcus lanatus	8.6	*Rumex acetosa*	6.3
Lolium perenne	8.4	*Trifolium repens*	6.2
Anthoxanthum odoratum	8.1	*Plantago lanceolata*	4.7

Our knowledge of the distribution of microorganisms on the root surface has been greatly increased by studies done with transmission and stereoscan electron microscopy (JENNY & GROSSENBACHER 1963; DART & MERCER 1964; GREAVES & DARBYSHIRE 1972; ROVIRA *et al.* 1974; BREISCH *et al.* 1975; OLD & NICHOLSON 1975). For such investigations use should be made of soil-grown roots, because it has been found that the rhizoplane populations of these roots are many times smaller than those of roots grown on agar or water culture.

From the foregoing it is clear that microorganisms must be taken into consideration in studies on plant physiology, because of their presence in appreciable areas of the root surface.

3. ROOT EXCRETIONS AND THEIR EFFECTS ON THE RHIZOSPHERE

3.1. QUANTITATIVE ASPECTS

Initially, the quantitative aspects of root excretions were investigated in sterile (axenic) plants grown in water culture. As a rule, only the compounds with a low molecular weight were analysed. Under such conditions, small quantities were found to be excreted.

Later it was found, however, that considerably higher quantities were supplied to the rhizosphere by plants grown in soil. This difference from water cultures has been ascribed to the greater damage undergone by the roots in soil (for a review of this subject, see HALE *et al.* 1971). The quantities of excreted products were also found to be higher for non-sterile plants (see e.g. BILES & CORTEZ 1975). To explain this difference it has been suggested that the rhizosphere microflora may affect root exudation by altering the metabolism and permeability of the root. Finally, the soluble compounds constitute only a small fraction of the total amounts supplied by the living plant to the soil. Gaseous products and non-soluble compounds of high molecular weight are quantitatively of much greater importance. The ratio between water soluble compounds, water isoluble compounds, and volatile compounds, has been estimated to be 1 : 3-5 : 8-10 (ROVIRA & DAVEY 1974). Seventy to eighty per cent of the volatile compounds consists of carbon dioxide originating from respiration by the root and microorganisms (ROVIRA 1972).

When these effects are taken into account, it is not surprising that much higher quantities of compounds deriving from the living root system were recently found in experiments with soil-grown, non sterile plants (Table 5). The data in Table 5 are to be regarded as minimum values, since some of the labelled carbon dioxide originating from respiration may have escaped.

TABLE 5. *Reported quantities of carbon supplied to the soil by living root systems*

Reference	Conditions	Quantities of carbon
VANCURA (1964)	Water stress, 10 days	7-10% of biomass
SHAMOOT *et al.* (1968)	14 CO_2, whole growth period	9-42 g/100 g of roots
MARTIN (1973)	14 CO_2, whole growth period	10-38 g/100 g of roots 6% of photosynthesis
BREISCH (1974)	14 CO_2, 1 day	10-20% of photosynthesis
WAREMBOURG (1975)	14 CO_2, 3 days	12-16% of photosynthesis
BILLES *et al.* (1975)	Leached sugars, 35 days	4.5-8% of biomass
BALANDREAU & HAMED-FARES (1975)	Calculated from N_2 fixation	10-20% of biomass

A large share of this carbon is contributed by sloughed-off rootcap cells. According to CLOWES (1971), these cells have a lifespan of only one day. For maize, the amount of sloughed-off rootcap cells has been estimated at 10 tons of dry matter per hectare (SAMTSEVITCH 1971).

The carbon dioxide evolved in the rhizosphere is only partially derived from root respiration. A roughly equal part originates from microorganisms in the rhizosphere (Table 6).

TABLE 6. *Contribution of the microflora to CO_2 production in the rhizosphere. The data were obtained by comparing sterile and non-sterile plants*

LUNDEGÅRDH (1927)	33%	REUSZER (1949)	60%
WAKSMAN & STARKEY (1931)	45%	NILOVSKAYA (1970)	35%
STILLE (1938)	>90%	TROLLDENIER (1972)	50%
BARKER & BROYER (1942)	50%		

More or less similar results were obtained by WOLDENDORP (1963), who compared the oxygen uptake of sterile and non-sterile root systems of pea plants and found that about 40 per cent of the total respiration could be ascribed to the microflora. However, these results should be considered with some caution, because it is well known that microorganisms influence the morphology and the metabolism of the root system considerably, which complicates comparison between sterile and non-sterile plants. Nevertheless, it is evident that the rhizosphere is the scene of considerable metabolic activity giving rise to locally high concentrations of carbon dioxide and low concentrations of oxygen. That the rhizosphere in permanent grasslands is the main site of respiratory activities in the soil is also indicated by data of WOLDENDORP (1963), who found that in grassland about 90 per cent of the total soil respiration was due to the living root system.

3.2. QUALITATIVE ASPECTS

In the study of the composition of root excretions, it is essential to avoid not only modification of these compounds by microorganisms but also their contamination by either leached or volatilized products from the aerial parts or root-decomposition products. Consequently, most investigations have been carried out in axenic plants grown in water or in sand. In such studies many different water-soluble compounds were found to be excreted. As an example, compounds reported in the literature on axenic wheat roots are cited in Table 7. Many of these compounds are involved in the normal metabolism of the plant, but many of the secondary metabolic products could also be detected in small amounts in the medium in which axenic roots were grown. Among these products are phenols and terpenes, which are toxic to bacteria, as well as compounds which inhibit or stimulate the growth of fungi and compounds which are inhibitors or attractants of nematodes. Furthermore, compounds supplied to the leaves, such as herbicides (see HALE *et al.* 1971), were found to be excreted in the root medium. For instance, an inhibitor of nematodes isolated from the rhizosphere of *Asparagus* was found to inhibit nematodes in the rhizosphere of tomato plants (ROHDE & JENKINS 1958) when applied to their leaves.

TABLE 7. *Compounds excreted by axenic wheat roots*

Volatile compounds	Low-molecular compounds	High-molecular compounds
CO_2	Sugars (10)	Polysaccharides
Ethanol	Amino acids (27)	Enzymes
Isobutanol	Vitamines (10)	
Isoamylalcohol	Organic acids (11)	
Acetoin	Nucleotides (4)	
Isobutyric acid		

The polysaccharides (Table 7) isolated from the rhizosphere are thought to originate from the mucigel formed by the Golgi apparatus of the plant cells. The composition of the mucigel resembles that of the sugars and uronic acids of the cell wall (BURKE *et al.* 1974). In this respect there are considerable differences between monocotyles, dicotyles, and gymnosperms. Among the enzymes found in the rhizosphere, the plant origin of peroxidase, nuclease, and invertase has been established (COLLET 1975). Other enzymes, e.g. urease, ATPase, and phosphohydrolases, are not excreted but are localized close to the root surface.

From the published work done in sterile root systems it is difficult to establish essential differences between plant species. This is due to the divergence between the conditions of cultivation, the age of the plants and the methods of analysis used by the various investigators. As SCHROTH & HILDEBRANDT (1964) showed, four investigators who studied the excretion of amino acids by oats obtained completely different results.

Only in a limited number of cases have specific differences in the nature and quantities of the excreted compounds been definitely established. A number of investigators have shown that legumes in general excrete higher amounts and relatively more nitrogenous compounds than other plant species (VANCURA & HANZLIKOVA 1972), and a comparative study on varieties of rice which were either resistant or susceptible to root disease showed quantitative and qualitative differences in excreted products (MACRAE & CASTRO 1967).

Under non-sterile conditions it is even more difficult to establish the specific nature of the excreted products. Not only are the quantities altered by microorganisms but their composition is modified as well. In non-sterile plants MARTIN (1971) found that 45 per cent of the water-soluble compounds had a molecular weight higher than 10,000 and 75 per cent higher than 1,000, whereas the compounds excreted by axenic roots were of low molecular weight. Moreover, the extent and the composition of the mucigel is influenced by microorganisms.

Some evidence that different plant species have a specific rhizosphere has been obtained by analysis of their microflora. It was found by NEAL et al. (1970, 1973) that about 20 per cent of the bacteria in the rhizosphere of the resistant wheat variety *Apex* were antagonistic to the root-rot fungus *Cochliobolus sativus.* (Table 8).

TABLE 8.   *Microorganisms in the rhizosphere of one susceptible and two root-rot resistant varieties of wheat* (NEAL et al. 1970)

Variety	Bacteria (x10^6)	Fungi (x10^3)	Root-rot antagonists(%)
Apex, resistant	251.5	352.9	20
S-616, susceptible	576.8	123.7	0
S-A5B*, resistant	266.5	81.5	19
Non-rhizosphere soil	44.9	119.5	7

* susceptile variety S-615 on which resistance was conferred by substitution of *Apex* chromosome 5B

In the rhizosphere of the susceptible variety S 615 such antagonistic bacteria were absent. However, when resistance was conferred on the susceptible variety by substitution of the *Apex* resistance chromosome, 20 per cent of the bacteria in the rhizosphere became antagonistic to the plant disease. The composition of the microflora of the susceptible variety differed from that of the resistant ones in other respects too (see Table 8).

Another example of such specific rhizosphere effects was given by COLEY-SMITH & KING (1970), who studied the fungus *Sclerotium cepivorum*. This fungus, which can persist for many years in the soil as sclerotia, is pathogenic only to *Allium* species. Under natural soil conditions, the sclerotia will germinate only in the rhizosphere of *Allium*. Germination was also obtained with root extracts of *Allium* species, whereas extracts of other plant species had no effect. The *Allium* extracts proved to contain antibiotic substances (alkylthiosulphinates). Moreover, it was found that other organic

compounds in sterilized soils also induced germination of the sclerotia. The
results were explained by assuming that components of the microflora normally
suppress germination in the soil. The *Allium* species excrete antibiotic
substances which inhibit these components of the microflora, thus releasing
the fungistatic action. But the concentration of the antibiotic compounds
responsible for the germination is so low that no effect on the bacteria
could be expected. But since no suppression of the bacterial flora in the
rhizosphere of *Allium* species has been found either, the excretion of specific
antibiotic compounds by the roots of *Allium* species still cannot be considered
to have been established beyond doubt. Other examples from the realms of
phytopathology point similarly to the plant-specific nature of the rhizosphere,
but in no case has it been demonstrated that such differences resulted from
specific root excretions rather than from different proportions of the same
substances (LOUVET 1975).

It is not clear whether the allelochemicals accumulating in arid regions
under vegetations like the chaparral in California or the *Rosmarino-Ericion*
in the Mediterranean are excreted by the roots or originate from leaching and
volatilization from the aerial parts or from the decomposition of dead plant
materials. For instance, the latter have been seen as the source of the
cyanide that accumulates in soils under a monoculture of peaches. The compound
is thought to be formed by the decomposition of the roots, which can
accumulate amygdaline in concentrations up to 6 per cent of their dry weight
(JUSTE 1975). The cyanide accumulation prevented replanting of peaches. Many
other plant species also contain cyanogenic glucosides that give rise to
cyanide, but in soils with a normal moisture level the compound is decomposed,
which results in an accumulation of cyanide-resistant or decomposing microorganisms. For instance, from the rhizosphere of flax a strain of *Bacillus
pumilus* was isolated which was extremely resistant to cyanide (STROWZONSKI &
STROBEL 1969) and from that of tapioca strains of *Streptomyces* sp. and
Aspergillus sp. which used HCN as a nitrogen source (SADASIVAM 1974).

The accumulation of cumarine-decomposing microorganisms in the rhizosphere
of *Anthoxanthum odoratum* can be similarly explained. Cumarine decomposition
in soil must be rather complete, because a suppressing effect of *A. odoratum*
on other plant species has not been found.

Other examples of soil fatigue and also the suppression of wheat by poppies
or of annual weeds by buckwheat, are in all likelihood to be ascribed likewise
to the decomposition of root remains rather than to the excretion of toxic
compounds by the living root system. The same explanation may also hold for
the effects of legumes on the nitrogen supply of other plant species.

Even though their origin is not definitely established, there is ample
chemical evidence that in arid regions compounds toxic to other plant species
may accumulate in the soil around plants. In the temperate regions the
existence of specific rhizosphere effects is suggested by the composition of

the microflora, but these effects can often be explained in terms of detoxification instead of accumulation of toxic compounds. Much more research is required before it can be concluded that plants influence each other by means of their root excretions and thus contribute to niche differentiation.

3.3. FACTORS AFFECTING EXCRETION

3.3.1. General remarks

The effects of environmental factors on excretions and the rhizosphere micro-flora have been the subject of many studies recently reviewed by HALE *et al.* (1971), ROVIRA & DAVEY (1974), LESPINAT & BERLIER (1975), and TROLLDENIER (1975). A few comments will suffice here.

It should be noted that the effects of these factors on the composition of the excretions have been studied in experiments with sterile plants in which only the water-soluble compounds were analysed. As has been discussed above (see 3.1.), such studies give an incomplete picture of the real level of excretion. The effects of the environmental factors on the composition of the microflora and on respiration in the rhizosphere by the microflora were investigated with non-sterile plants. But to obtain a correct picture of these effects, studies must be done with both sterile and non-sterile plants.

3.3.2. Plant age and stage of development

In studies on the effect of plant age it has been found (e.g. HAMLEN *et al.* 1972) that the excretions per gram of roots were highest in the first weeks after germination. These studies were all performed with annual crops and the decreasing excretion may be due to the tendency of maturing annuals to transport the assimilates from the leaves to their seeds rather than to the roots. In perennials a different picture is to be expected. In experiments with perennial ryegrass, WOLDENDORP (1963) observed that toward the end of the growing season considerable quantities of nitrogen were transported back to the root system and also to the soil.

With respect to the composition of the microflora, appreciable changes were found to occur during the plant development. Initially, Gram-negative rods predominate but these are gradually replaced by Coryneform bacteria, as shown for instance by the author (Table 9). In experiments with a mixture of pure cultures of an *Arthrobacter* and a *Pseudomonas* strain added to axenic flax plants, a similar replacement was found (WOLDENDORP 1975). This replacement was not necessarily caused by a change in the composition of the excretion products.

A similar replacement was found when the mixture of the two bacteria was cultivated in a dialysis culture continuously supplied with low quantities of mineral salts and glucose. When glutamic acid was used as carbon source, the replacement occurred much more slowly. The results could be explained by

TABLE 9.   *The influence of plant age on the proportions of G(-)rods and*
           Coryneforms in the rhizosphere of flax (WOLDENDORP 1975)

Plant age (weeks)	G(-)rods (%)	Coryneforms (%)	Other microorganisms (%)
1	90	4	6
4	81	14	5
8	60	32	8
12	49	48	3

assuming that initially the roots are colonized by rapidly growing
Gram-negative rods. When further increases in the number of microorganisms are
impossible because of a shortage of nutrients, the more efficient Coryneforms
take over. This hypothesis was confirmed by experiments in which the growth
rate and the maintenance requirement of a *Pseudomonas* and a *Arthrobacter*
strain were determined (Table 10).

TABLE 10.   *The influence of glucose and glutamic acid on the generation time*
            and maintenance requirement of Pseudomonas *str 64 and* Arthrobacter
            str 356

Strain	Substrate	Generation time (min)	Maintenance requirement (mg/g dr.m./h)
P 64	Glucose	30	12
A 356	Glucose	56	4
P 64	Glutamic acid	21	9
A 356	Glutamic acid	78	7

These results also indicate that the stimulation of Gram-negative rods by
amino acids is caused by an increased growth rate and is not due to a specific
requirement for these compounds, as suggested by many investigators.

3.3.3. Mineral nutrition

It has been shown that the quantities of excreted products are increased by
nutrient deficiencies. COLLINS & RAILLY (1968) found that higher amounts of
amino acids and organic acids were excreted when sulphate was replaced by
chloride in the culture solution. Similar increases were found by SHAY & HALE
(1973) and by EL-KHAB-BASHA & BEKHASI (1973) under calcium deficiency
conditions. TROLLDENIER (1971a) observed an increased excretion of amino acids
by potassium deficient wheat plants (Table 11).

TABLE 11.   *Amino acids and amides (10^{-9} Mole/g root dr. m.) in the*
             rhizosphere of wheat as a function of the K supply
             (TROLLDENIER 1971a)

Amino acid	K supply	
	7 meq	28 meq
Alanine	149.3	19.1
Glutamic acid	72.7	5.4
Arginine	56.8	62.8
Proline	53.9	60.8
Glucine	65.2	26.6
Aspartic acid	40.6	18.8
Threonine	31.0	0.0
Asparagine	28.6	0.0
γ-aminobutyric acid	25.2	7.7
Serine	11.4	12.7
Lysine	7.1	4.4
Total	541.8	218.1

Working with non-sterile plants, this author found higher numbers of microorganisms under a potassium shortage, and the contribution of the microorganisms to root respiration increased from 57 per cent in the presence of sufficient potassium to 66 per cent under a deficiency of this element (TROLLDENIER 1971b).

However, the reverse effect has been observed for nitrogen, i.e., the amounts of excreted products increased with increasing levels of this element (see e.g. MACURA 1961). The number of the bacteria and the proportion of Gram-negative rods were also higher the higher the nitrogen level (WOLDENDORP 1963).

3.3.4. Defoliation

According to a number of authors, defoliation stimulates excretion by the root system (VANCURA 1969; MARTIN 1971; HAMLEN *et al*. 1972), but in experiments with perennial ryegrass given labelled nitrogen, WOLDENDORP (1963) and HUNTJENS (1972) found that after defoliation nitrogen was transported to regrown tops and more soil nitrogen was taken up than before. These results point to a decreased excretion of carbonaceous compounds, as will be discussed below (see section 5). The discrepancy might be explained by assuming that in the former experiments part of the root system was killed by the defoliation and then decomposed. Since defoliation by mowing and grazing are important measures in nature management, more insight is needed into the effects on the accumulation of soil organic matter and recycling of nutrients.

3.3.5. Moisture stress

Under conditions of moisture stress, increased excretion has been observed. This occurred particularly in wilting plants (VANCURA 1964; VANCURA & GARCIA 1969; REID 1974). In such dry soils a partial sterilization takes place, which kills off part of the microflora. Remoistening of the soil leads to a rapid regrowth of usually a restricted number of microbial species, which use the accumulated excretion products of the plants and the killed microorganisms as a source of food. This gives rise to a flush in the mineralization of nutrients to which specialized plant species, such as winter annuals, are adapted.

3.3.6. Other factors

The light intensity and the temperature at which plants grow affect the amount and composition of the excreted products. At high light intensities a positive effect was observed, and no decrease occurred during the night (LESPINAT & BERLIER 1974). However, it has been claimed that the production of respiratory carbon dioxide is related to the photoperiod (OSMAN 1971). Under anaerobic conditions, particularly at high CO_2 concentrations in the atmosphere of the rhizosphere, there is also a quantitative and qualitative shift in the excreted compounds (RITTENHOUSE & HALE 1971). When the carbon dioxide level in the atmosphere reached 30 per cent, the roots showed an ethanol content of 300 mg per gram of dry matter (SMUCKER 1971).

3.3.7. Conclusions

From the data presented above it is evident that the quantities and the nature of the root excretions are considerably influenced by environmental factors. These effects do not run parallel with dry matter production. Some of them, for instance wilting, anaerobic conditions, and Ca and K deficiencies, which affect dry-matter production adversely, have a stimulating effect on the excretions, and the same holds for other factors such as a high light intensity or an ample supply of nitrogen, which influence general productivity favourably.

As far as the excretion of primary metabolic products is concerned, our insight into the combined effects of these factors on processes going on in the rhizosphere of non-sterile plants is very incomplete, and nearly nothing is known about plant-specific effects.

In the foregoing, ways in which living plant roots influence their environment have been discussed. Conversely, the enhanced microbial activity induced by the root excretions affects the growth of the plant in various ways. Due to the inhomogeneity of the rhizosphere, particularly in the rhizoplane, some of these effects occur very locally.

The growth and morphology of the root system are considerably changed by microorganisms. Generally, the root length was found to be decreased by the microflora, the roots being stunted, yellowish-brown, and limp. Sometimes they were more branched and the root hairs better developed (BOWEN & ROVIRA 1961; BLONDEAU 1970; DARBYSHIRE & GREAVES 1970). Other investigators, however, observed increased root growth under non-sterile conditions (MILLER & CHAN 1970).

When pure cultures of microorganisms or extracts of such cultures were added to sterile plants, conflicting results were also obtained. SOBIEZCZANSKI (1966) divided his collection of rhizosphere bacteria into three groups: strains without an effect on root growth, strains which stimulated root growth, and strains which inhibited root growth.

Although it is not yet possible to reconcile these conflicting results, two aspects may play a role here. Firstly, stimulation under non-sterile conditions was found mainly when the plants were cultivated in sterilized soil which had been inoculated with a soil suspension. In such experiments the sterilization procedure may have led to increased mineralization of nutrients. Secondly, inhibition or stimulation by the same compounds is often a question of concentration, which was not known in these experiments. Furthermore, stimulatory and inhibitory substances may cancel each other's effects. In any case, it is clear that the rhizosphere microflora has a marked influence on the growth and development of the root system and thus also on its respiration and nutrient uptake.

The effects of the microflora on the morphology of the root system are probably to be ascribed to phytohormones such as auxins, gibberellins, and cytokinins. Many species of microorganisms produce such compounds in pure culture (for a review, see BLONDEAU 1975), but the quantities formed under normal soil conditions are unknown. Therefore, this subject will not be discussed here.

Other effects of the rhizosphere microflora on plant growth, such as protection against pathogens, detoxification of secondary plant metabolites, and effects on the rhizosphere atmosphere, have already been referred to above (see 3.1. and 3.2.). The profound effects of the microorganisms on the availability of nutrients will be treated below in more detail.

5. EFFECTS OF THE RHIZOSPHERE MICROFLORA ON THE AVAILABILITY OF NITROGEN

5.1. GENERAL REMARKS

Apart from small inputs by rain and electrical discharge, the plants of natural vegetations are completely dependent on the activities of microorganisms for their nitrogen supply. The microorganisms provide the plants with nitrogen by the fixation of atmospheric nitrogen and the mineralization of organic nitrogenous compounds. In the various steps of the nitrogen cycle (Fig. 1), bacteria in particular play a major role.

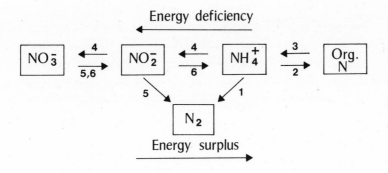

FIG. 1.   *Schematic representation of the nitrogen cycle*

1. Nitrogen fixation   *4. Nitrification*
2. Ammonium immobilization   *5. Denitrification*
3. Nitrogen mineralization   *6. Nitrate assimilation*

This cyclic chain of nitrogen transformations is kept going by the continuous input of solar energy (MOROWITZ 1966, WOLDENDORP 1975). This means that the nitrogen cycle is closely related to the carbon cycle.

When sufficient energy is available in the form of organic compounds, the energy-demanding steps of the nitrogen cycle predominate (i.e., nitrogen fixation, nitrogen assimilation, denitrification); consequently, nitrogen is found in a reduced state, i.e. as organic nitrogen. In the absence of energy, nitrification is the predominant process, which leads to the most oxidized state of nitrogen, viz. nitrate.

Form the foregoing it is clear that the state in which nitrogen occurs in the soil is to a high degree determined by the presence or absence of organic carbon. Therefore, plants have a profound influence on the nitrogen transformations in the soil, because they provide carbonaceous matter. Under those natural conditions where nitrogen and not photosynthesis is the limiting factor in the productivity of plants, immobilization prevails and nitrogen remains predominantly in the reduced state. The conditions for the fixation of atmospheric nitrogen are in principle likewise favourable, but nitrogen-fixing microorganisms have to compete with other microorganisms for the organic compounds. The process of denitrification will be stimulated when nitrate is present, but because the conditions for nitrification are generally not favourable in natural vegetations, this will usually not be the case. As a rule, plants have to compete with the nitrogen-immobilizing microorganisms for mineralized ammonium to satisfy their nitrogen demand. It should be kept in mind here that due to the inhomogeneity of the rhizosphere, locally nitrogen is not a limiting factor; at such spots, conditions may permit nitrification.

The various nitrogen transformations going on in the rhizosphere will be discussed below in more detail.

5.2. MINERALIZATION AND IMMOBILIZATION OF NITROGEN

It has been found that a constant mineralization of soil-organic nitrogen occurs in all soils. The energy liberated during this breakdown process is used by the microflora to re-immobilize part of the mineralized nitrogen; this continuous mineralization-immobilization cycle has been called nitrogen turnover (JANSSON 1958). A relative shortage of carbonaceous compounds leads to an increase of the total mineral nitrogen, a surplus to a decrease of the total inorganic nitrogen level (ammonium plus nitrate). The existence of nitrogen turnover in the soil can be easily demonstrated by the addition of ^{15}N-labelled ammonium: under all conditions the label will be incorporated into the organic matter in the soil, as shown in Table 12. In this example 6 mg labelled ammonium was incorporated into soil organic matter after 27 days and, in addition, 3 mg was chemically fixed into clay minerals.

TABLE 12.   *Nitrogen turnover after the addition of 30 mg $^{15}NO_3^-$ and 28 mg $^{15}NH_4^+$-N to grassland soil* (WOLDENDORP 1963)

Treatment	Days	NO_3^--N		NH_4^+-N		Organic
		$14 + 15_N$	15_N	$14 + 15_N$	15_N	15_N
Control	0	0	0	3	0	0
"	27	63	0	9	0	0
NH_4^+-N	0	0	0	30	27	3
"	27	73	10	18	9	9
NO_3^--N	0	30	30	3	0	0
"	27	92	29	9	0.5	0.5

However, hardly any of the added nitrate was immobilized. This is due to the preference of microorganisms for ammonium over nitrate in the immobilization process. Therefore, nitrate immobilization will occur only when sufficient ammonium nitrogen is not available. It is also clear that the process of nitrification (which requires the presence of ammonium) will not occur simultaneously with nitrate immobilization.

Studies with labelled nitrogen have shown that only part of the organic nitrogen (about 5-10 per cent) is involved in nitrogen turnover (JANSSON 1958), the rest forming a passive pool (Fig. 2).

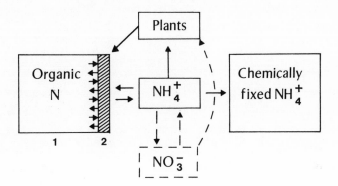

FIG. 2.   *Nitrogen turnover in soil and competition for ammonium between immobilization, nitrification, and plants. 1= Passive organic nitrogen pool, 2= Active organic nitrogen pool*

The small active nitrogen pool proved to consist of the soil microflora and their direct decomposition products. The equilibrium in the microflora is easily disturbed, which results in the death of some species and their replacement by others. Such disturbances as mixing of a grassland soil by ploughing or activity of animals (e.g. moles), cutting of trees, changes in pH (lime), drying and remoistening, and changes from anaerobic to aerobic conditions, result in considerable flushes of nitrogen. These flushes are ecologically very important, and plant species such as biennials and winter annuals are specially adapted to them.

Plants have to compete with heterotrophic microorganisms for the mineralized ammonium (Fig. 2). From the above considerations it can be concluded that the extent of the turnover cycle rather than its net result governs nitrogen uptake by plants.

In most vegetations where plant growth is not limited by a lack of moisture, there is a surplus of carbonaceous matter. RICHARDSON (1938), who studied the availability of nitrogen in grassland soils, invariably found low levels of mineral nitrogen. According to WOLDENDORP (1963), even these low levels are artefacts resulting from analytical errors. HUNTJENS (1972), who made an extensive study of the availability of mineral nitrogen in grassland soils, has shown that a gradual accumulation of organic nitrogen takes place under the influence of the living root system. It was indeed the living root system that suppressed mineralization, because killing of the grass resulted in a rapid nitrogen mineralization.

Under a permanent cover of plants, therefore, ammonium will be the principal source of nitrogen to plants. It is only on partially bare soils, after a spell of dry weather or some other disturbance of the equilibrium in the microflora, that mineralization may exceed the demands of the microflora and plants, and nitrification can take place. In this respect the present author agrees with Jansson (1958), who considered the nitrate fraction in soil as a storage pool formed by the surplus nitrogen not used in immobilization, uptake by plants, or chemical fixation.

5.3. NITRIFICATION

In the foregoing it has been shown that in vegetations where productivity is limited by the nitrogen supply, nitrification is of minor importance. Since the 1938 publication of RICHARDSON, who found no nitrate in unfertilized grassland, nitrification in the vicinity of plant roots has been the subject of many studies. A number of investigators also came to the conclusion that the absence of nitrification in grassland and forest soils is due to the unavailability of ammonium for the process, because of the predominance of ammonium immobilization and plant uptake (see e.g. GORING & CLARK 1948; PURCHASE 1974).

However, in 1951 THERON advanced another theory. This author ascribed the suppression of nitrification in the vicinity of roots to ˄ toxic effect of root excretions. This hypothesis, which has been investigated by a number of authors (MUNRO 1966a, b; NEAL 1969; RICE 1974; MOORE & WAID 1971), is consistent with the susceptibility of the chemolithotrophic nitrifiers to organic compounds. Plant species of climax ecosystems were thought to inhibit the nitrification more than those of pioneer vegetations. According to RICE (1974), the suppression of nitrification should have a selective advantage for the plant, because nitrogen losses by the leaching of nitrate are avoided and no energy is necessary for the reduction of nitrate to ammonia in the plant. Therefore, by using ammonium instead of nitrate, species of climax ecosystems can reach a higher efficiency of nitrogen utilization than pioneer plants. It has been shown that plant roots indeed contain compounds -such as phenols, terpenes, and alkaloids- which are toxic to nitrifying bacteria.

Although the hypothesis that plants of climax vegetations suppress nitrification by means of toxic compounds is attractive, the evidence put forward to support it is not completely convincing. There is no proof that under natural conditions the toxic compounds in the rhizosphere reach levels at which nitrification is suppressed. The results of experiments in which the process was inhibited by root extracts or leachates which were obtained under non-sterile conditions and whose composition was insufficiently characterized, cannot be considered conclusive evidence. Furthermore, certain facts are not in accordance with the theory. For instance, NAKOS (1975) found that nitrification could not be restored by leaching of the soil and the addition of nitrifying bacteria. However, nitrification started immediately after the removal of grasses or trees (WOLDENDORP 1963; BRAR & GIDDENS 1968; HUNTJENS 1972). In some grassland soils which normally showed no nitrification, the process was found to occur when rain followed a prolonged dry period.

Most of the advocates of the suppression of nitrification by toxic compounds do not mention the fact that THERON, who formulated the theory, withdrew it in 1963 in favour of the immobilization explanation. In the present state of our knowledge, the low rates of nitrification in the vicinity of plant roots must still be ascribed at least partially to the absence of ammonium.

A good qualitative impression of the process of nitrification can be obtained by counting nitrifying bacteria, since these organisms are dependent on the oxidation of inorganic nitrogen. Rhizosphere counts have given conflicting results. For the rhizosphere of wheat, for example, lower (KATZNELSON *et al*. 1956), similar (MOLINA & ROVIRA 1963), and higher RIVIÈRE 1960) numbers of nitrifying organisms as compared with those in non-rhizosphere soil have been reported. Plants of different ages had different effects on nitrifiers (ROVIRA 1965). In the presence of mature plants, low numbers of nitrifying organisms were generally found. This was also the case for permanent grassland soils (ROBINSON 1963; MEIKLEJOHN 1968).

It should be kept in mind that the absence of nitrate in the rhizosphere is not necessarily an indication of the absence of nitrification. Nitrate has been found in grasstops even though the compound could not be detected in the soil (VAN BURG 1962).

In conclusion it can be stated that several different approaches are required for the study of nitrate supply to plants in natural vegetations. Such investigations should include simultaneous determination of the number of nitrifying bacteria, the rate of formation of nitrate after the removal of the living plants, the accumulation of nitrate inside the plants, and the level of nitrate reductase in various parts of the plants.

5.4. DENITRIFICATION

The denitrification process is performed by bacteria that use nitrate instead of oxygen as a terminal electron acceptor. In this process, which can only occur under anaerobic conditions, nitrate is reduced to gaseous nitrogen and nitrous oxide. For the occurrence of denitrification in the soil, which has been discussed in detail by WOLDENDORP (1968, 1975), two conditions must be met, viz. a low level of oxygen in the soil solution and the presence of hydrogen donors, in the form of organic compounds, to reduce the nitrate. In soils without a cover of plants, denitrification occurs only under more or less waterlogged conditions. But in the rhizosphere these two prerequisites are at least partially fulfilled under soil conditions which are considered to be well aerated.

When labelled nitrate was applied to sods of permanent grassland, 10 to 20 per cent was lost by denitrification (Table 13).

TABLE 13.   *Denitrification losses after the addition of labelled nitrate to grassland sods* (WOLDENDORP 1963)

Soil type	Herbage	Roots	Soil	Loss
Sandy soil	65%	10%	10%	15%
Clay	57%	11%	7%	25%
Peat	57%	2%	22%	19%

Sods with killed root systems showed much lower losses, thus demonstrating the influence of the living root system on the process (WOLDENDORP 1963). These results were fully confirmed by Australian investigators (STEVENSON 1972, 1973).

The utilization of root excretions in the denitrification process was shown in experiments in which nitrate was supplied to perennial ryegrass and peas planted in soil devoid of organic matter (WOLDENDORP 1963). The rate of the process was influenced by the age and species of the plants. Root excretions from pea plants led to a higher denitrification rate than those from grass plants. This difference was traced to the excretion of relatively more amino acids by the legume.

In the above-mentioned experiments external nitrate was added to the soil. As mentioned under 5.3., in many natural vegetations little if any nitrate is formed. This raises the question of whether denitrification is an important process in such vegetations. To the best of the author's knowledge, there are no publications that deal with this question adequately, but the experiments performed by NOMMIK (1961) may offer an indirect answer. This author grew oats in soils containing different amounts of straw and labelled nitrate. When straw was added, a rapid immobilization of nitrate occurred. Subsequently, the nitrogen was partly remineralized and taken up by the oats. Treatments without added organic material led to a deficit in the nitrogen balance amounting to between 10 and 20 per cent. This result can be explained by assuming that nitrogen mineralized from soil organic matter is not lost by volatilization in the rhizosphere. This nitrogen was probably taken up by the plant in the ammonium form.

In the present author's opinion, the tentative conclusion to be drawn is that due to a low nitrification rate and nitrogen uptake by the plant in the form of ammonium, losses by denitrification are of minor importance in natural vegetations. On the other hand, there is no reason to suppose that any nitrate formed in the soil would behave differently than added nitrate. Therefore, denitrification losses could occur.

5.5. NITROGEN FIXATION

Under natural vegetations considerable stores of organic nitrogen often occur. It has been claimed that these stores derive partially from the fixation of atmospheric nitrogen by microorganisms. Such fixation processes are said to occur particularly in the tropics. Similar gains in organic soil nitrogen were also observed in vegetations which were virtually devoid of leguminous plants. MOORE (1966), who discussed the older literature on the subject, concluded that short-term field trials are too inaccurate to permit any conclusions concerning gains due to nitrogen fixation. Since lysimeters permit much more accurate nitrogen balances be made, WOLDENDORP (1968) analysed the published data obtained in lysimeter experiments, and found that in planted lysimeters receiving no nitrogen or only small amounts, small nitrogen gains occurred, whereas lysimeters given higher additions of nitrogen showed losses. In more recent experiments the nitrogen-fixing capacity of soils was investigated with the $^{15}N_2$ and acetylene reduction techniques. The results show that only gains amounting at most to 10-20 kg N per hectare per year occurred unless an additional carbon source was supplied (CLARK & PAUL 1970). BALANDREAU (1975) too found similar gains under field conditions (Table 14).

TABLE 14. *Nitrogen fixation under field conditions* (BALANDREAU 1975)

Plant cover	N gains (kg/ha/year)
Savanna (Africa)	10
Rice, wet (Africa)	30
Maize (France)	1-3.5
Lolium perenne	6.3
Dactylis glomerata	5.6
Festuca elatior	5.0

Only in rice on waterlogged soils were higher values found. Low values (1-3 kg) were found for nitrogen fixation in the temperate regions also by KAPUSTKA & RICE (1976). In dune vegetations nitrogen fixation contributed only a few kilograms to the nutrition of *Ammophila arenaria* (ABDEL WAHAB 1969).

Nitrogen gains have been ascribed to fixation by free-living, nitrogen-
-fixing bacteria in the rhizosphere, which use plant excretion products as a
source of reduction power. In the rhizosphere of a number of grasses,
especially C_4-species, nitrogenase activity was found (DOMMERGUES et al.1973;
DÖBEREINER et al. 1972; DAY et al. 1975; DAY & DÖBEREINER 1976; DE-POLLI et al.
1977). Particularly the rhizosphere of Bahia grass (*Paspalum notatum*) showed
high values, which were ascribed to the presence of the free-living nitrogen
fixer *Azotobacter paspali*, which so far has only been isolated from the
rhizosphere of Bahia grass. *Spirillum lipoforum* was isolated from the
rhizospheres of *Digitaria decumbens* and maize.

The high value of 90 kg per hectare initially recorded by DÖBEREINER et al.
(1972) could not be substantiated in subsequent work and is possibly to be
ascribed to the long preincubation of their samples. At two recent symposia
(Salamanca 1976, Uppsala 1976) the problem of nitrogen fixation in the
rhizosphere was discussed in detail. It seems possible that nitrogen fixation
may only be of quantitative importance for plants grown in waterlogged soil,
for instance rice, *Juncus balticus,* and *Agrostis tenuis* (BALANDREAU 1975;
TJEPKEMA & EVANS 1976).

5.6. CONCLUSIONS

As shown in the foregoing, under a cover of vegetation the carbonaceous
compounds provided by the plants have a conservational effect on the nitrogen
contained in the system. Mineralized ammonium is either re-immobilized or
taken up by the plants. Nitrification is often limited, and is usually induced
by a disturbance of the soil ecosystem. Small increases in the total nitrogen
content of the plant-soil system may result from fixation of atmospheric
nitrogen.

6. EFFECTS OF THE RHIZOSPHERE MICROFLORA ON THE AVAILABILITY OF PHOSPHORUS

In natural vegetations phosphorus is often a limiting factor in plant growth.
There is a great deal of similarity between the phosphorous and nitrogen
transformations in the rhizosphere section. As shown above under the
influence of the living root system there is a surplus of carbonaceous
compounds in the rhizosphere. The rhizosphere microflora using these
compounds immobilizes soluble phosphates in microbial tissue. During phosphate
turnover, plants roots have to compete with the microorganisms for the
soluble phosphates. Due to the immobilization process, there is a gradual
accumulation of organic phosphate in the soil. In established soil profiles
more than half of the total phosphorous content is often in the organic form
(Table 15), the main components being phytine (myo-inositol-hexaphosphate),
glycerophosphates, nucleotides, and nucleic acids.

TABLE 15. *Phosphorus in components of a native grassland ecosystem (g P/m^2)*
          (HALM *et al.* 1972)

Green plant material	0.21
Standing dead plant material	0.21
Consumers	< 0.1
Litter	0.07
Roots + rhizomes	1.07
Soil fauna + microorganisms	1.98
Water soluble P	10^{-7}
Organic P	137.5
Inorganic P	153.8

When mineralization occurs, the phosphate may be re-immobilized in new microbial tissue, sorbed by plant roots, or revert to Al-, Fe-, or Ca-bound inorganic phosphates of varying solubility.

In grassland soils the organic-phosphorus fraction plays a vital role in the phosphorus supply of the plants (HALM *et al.* 1972). The organic compounds are hydrolysed by phosphatehydrolases originating from the rhizosphere bacteria, the mycorrhizas, and the plant roots. The role of each of these in the decomposition of organic phosphorus has yet not been established (see also the paper by Mosse in this volume). In this respect it has been claimed that organic phosphates such as phytine and lecithins are taken up directly by the root system (ROGERS *et al.* 1940; MARTIN 1973). It is the opinion of the present author that much work will have to be done on transformation of organic phosphorus in the rhizosphere before a more complete picture can be obtained.

Since GERRETSEN (1948) found that non-sterile plants take up more phosphate from insoluble rock phosphates than do sterile plants, the influence of rhizosphere organisms other than mycorrhiza on the solubilization process has been the subject of many investigations. Although GERRETSEN's original findings could not be confirmed, a number of investigators found an increase in the rhizosphere or organisms which dissolved insoluble phosphates in laboratory studies (KATZNELSON 1962). This result may have been due to the production of 2-ketogluconic acid by the bacteria. This compound, which is a chelating agent of Fe^{3+} and Ca^{2+} ions, is accumulated by Gram-negative rods, e.g. *Pseudomonas* species, during the breakdown of glucose via the Entner-Doudoroff pathway. Other compounds produced by fungi such as citric acid, are also chelating agents of Ca, Fe and Al, and thus may contribute to the solubilization of the phosphate salts of these ions. However, there is no indication that these compounds are produced by the organisms in question under rhizosphere conditions. The claims that H_2S and CO_2 affect the solubilization of rock phosphates in the rhizosphere could not be substantiated either.

The role played by the rhizosphere microflora in the uptake of soluble phosphate has been controversial too. BARBER and co-workers (see e.g. BARBER 1974) demonstrated trapping of phosphate by rhizosphere microorganisms. At low external phosphate concentrations in water culture the effect was of practical significance, because less phosphate was available to the plants. Since significant trapping occurred at concentrations of up to 3×10^{-5} M phosphate, the trapping process was considered to be of ecological significance. However, EPSTEIN (1972) found that Barber's conclusions had to be rejected. It should be noted that Barber's results are fully explained by the phosphorus turnover cycle in the soil, which leads to its immobilization. On the other hand, it seems likely that the phosphorous incorporated into the microorganisms is relatively more available to the plant than Ca, Fe, and Al phosphate or rock phosphate.

Therefore, in the present author's opinion, many aspects of the role of the rhizosphere microflora in the phosphorous supply of the plant are still obscure and no definite conclusions can be drawn.

7. GENERAL CONCLUSIONS

It is clear from the foregoing that plant species rarely change the surroundings of their root system to such an extent that the growth of other plant species is influenced. There is no direct proof that different species excrete different compounds under field conditions. Nevertheless, it has been shown that the rhizosphere microflora may react in a specific way, particularly as regards the occurrence of phytopathogens. This specific microflora often seems to be involved in the decomposition of specific secondary plant metabolites. Only when the rate of this decomposition process is too low, which can occur in arid regions or after abrupt killing of plants (e.g. by ploughing), may toxic compounds affect the germination or growth of other species. Future research will undoubtedly provide more examples of specific processes going on in the rhizosphere. Nevertheless, it seems unlikely that these processes will be found to contribute to the process of niche differentiation.

The rhizosphere microflora has a profound influence on the availability of nutrients. At low nutrient levels, considerable amounts are immobilized into soil organic matter by the microflora for which root excretions serve as a carbon source. In natural vegetations, plants are dependent on the activities of microorganisms for their nitrogen en phosphorous supply. Consequently, these activities must be included in investigations on the availability of nutrients to plants.

8. REFERENCES

ABDEL WAHAB, A.M., 1969 - Role of microorganisms in the nitrogen nutrition of *Ammophila arenaria*. *Ph. D. Thesis,* Aberyswyth.

BALANDREAU, J., 1975 - Activité nitrogenasique dans la rhizosphère de quelques graminées. *Thèse,* Nancy.

BALANDREAU, J. & I. HAMAD-FARES, 1975 - Importance de la fixation d'azote dans la rhizospère du riz. *Soc. bot. Fr., Coll. Rhizosphère,* 109-119.

BARBER, D.A., 1974 - The absorption of ions by microorganisms and excised roots. *New Phytol.,* 73, 91-96.

BARKER, H.A. & T.C. BROYER, 1942 - Notes on the influence of microorganisms on growth of squash plants in water culture with particular reference to manganese nutrition. *Soil Sci.,* 53, 467-477.

BILLES G. & J. CORTEZ, 1975 - Étude de la production de polysaccharides bactériens au niveau de la rhizosphère de *Festuca arundinacea. Soc. bot. Fr., Coll. Rhizosphère,* 41-45.

BLONDEAU, R., 1975 - Les synthèses bactériennes de substances trophiques, vitaminiques et phytohormonales dans la rhizosphère. *Soc. bot. Fr., Coll. Rhizosphère,* 145-155.

BOWEN, G.D. & A.D. ROVIRA, 1961 - The effects of microorganisms on plant growth. I. Development of roots and root hairs in sand and agar. *Pl. Soil,* 15, 166-188.

BRAR, S.S. & J. GIDDENS, 1968 - Inhibition of nitrification in Bladen grassland soil. *Soil Sci. Soc. Amer. Proc.,* 32, 821-823.

BREISCH, H., 1974 - Contribution à l'étude du rôle des exsudats racinaires dans les processus d'agrégation des sols. *Thèse Doct. Spéc. Univ.,* Nancy.

BREISCH, H. GUCKERT, A. & O. REISINGER, 1975 - Étude au microscope électronique de la zone apicale des racines de mais. *Soc. bot. Fr., Coll. Rhizosphère,* 55-60.

BURG, P.F.J. VAN, 1962 - Internal nitrogen balance, production of dry matter and ageing of herbage and grass. *Thesis,* Wageningen.

BURKE, D., KAUFMAN, P., McNEIL, M. & P. ALBERSHEIM, 1974 - The structure of plant cell walls. *Pl. Physiol. Wash.,* 54, 109-115.

CLARK, F.E. & E.A. PAUL, 1970 - The microflora of grassland. *Adv. Agron.,* 22, 375-435.

CLOWES, F.A.L., 1971 - The proportion of cells that divide in root meristems of *Zea mays* L. *Ann. Bot.,* 35, 249-261.

COLEY-SMITH, J.R. & J.E. KING, 1970 - Response of resting structures of root infecting fungi to host exudates: an example of specificity. *In:* T.A. TOUSSOUN *et al.* (Editors), *Root diseases and soil-borne pathogens.* Univ. California Press, p. 130-133.

COLLET, G.F., 1975 - Exsudations racinaires d'enzymes. *Soc. bot. Fr., Coll. Rhizosphère,* 61-75.

COLLINS, J.C. & E.J. REILLY, 1968 - Chemical composition of the exudate from excised maize roots. *Planta,* 83, 218-222.

DARBYSHIRE, J.F. & M.P. GREAVES, 1970 - An improved method for the study of the interrelationships of soil microorganisms and plant roots. *Soil Biol. Biochem.,* 2, 63-71.

DART, P.J. & F.V. MERCER, 1964 - The legume rhizosphere. *Archiv Mikrobiol.,* 47, 344-378.

DAY, J.M. & J. DÖBEREINER, 1976 - Physiological aspects of N_2-fixation by a *Spirillum* from *Digitaria* roots. *Soil. Biol. Biochem.,* 8, 45-50.

DAY, J.M., NEVEST, M.C.P. & J. DÖBEREINER, 1975 - Nitrogenase activity on the roots of tropical forage grasses. *Soil Biol. Biochem.,* 7, 107-112.

DE-POLLI, H., MATSUI, E., DÖBEREINER, J. & E. SALATI, 1977 - Confirmation of nitrogen fixation in two tropical grasses by $^{15}N_2$-incorporation. *Soil Biol. Biochem.,* 9, 119-123.

DÖBEREINER, J., DAY, J.M. & P.J. DART, 1972 - Nitrogenase activity and oxygen sensitivity of the *Paspalum notatum* - *Azotobacter paspali* association. *J. gen. Microbiol.,* 71, 103-116.

DOMMERGUES, Y., BALANDREAU, J., RINANDO, G. & P. WEINHARD, 1973 - Non-symbiotic nitrogen fixation in the rhizospheres of rice, maize and different tropical grasses. *Soil Biol. Biochem.,* 5, 83-89.

EL-KHABBASHA, K.M. & G. RANGASWAMI, 1973 - Excretion of some organic substances from bean roots under various conditions of temperature, illumination and calcium ion concentration of the nutrient medium. *Izv. timiryazev.sel'.- khoz. Akad.*, 4, 204-208.

EPSTEIN, E., 1972 - Ion absorption by roots: the role of microorganisms. *New Phytol.*, 71, 873.

GERRETSEN, F.C., 1948 - The influence of microorganisms on the phosphate intake by plants. *Pl. Soil*, 1, 51-81.

GORING, C.A.I. & F.E. CLARK, 1948 - Influence of crop growth on mineralization of nitrogen in the soil. *Proc. Soil Sci. Soc. Am.*, 13, 261-266.

GREAVES, M.P. & J.F. DARBYSHIRE, 1972 - The ultrastructure of the mucilaginous layer on plant roots. *Soil Biol. Biochem.*, 4, 443-449.

GREIG-SMITH, M.A., 1964 - *Quantitative plant ecology*. Butterworths, London, 256 p.

HALE, M.G., FOY, C.L. & F.J. SHAY, 1971 - Factors affecting root exududation. *Adv. Agron.*, 23, 89-109.

HALM, B.J., STEWART, J.W.B. & R.L. HALSTEAD, 1972 - The phosphorus cycle in a native grassland ecosystem. *In: Isotopes and radiation in soil-plant relationships including forestry.* Proc. of the symposium, Vienna, 13-17 December 1974, STI/PUB/292, Vienna, IAEA, p. 571-589.

HAMLEN, R.A., LUKEZIC, F.L. & J.R. BLOOM, 1972 - Influence of age and stage of development on the neutral carbohydrate components in root exudates from alfalfa plants grown in gnotobiotic environment. *Can. J. Pl. Sci.*, 52, 633-642.

HUNTJENS, J.L.M., 1972 - Immobilization and mineralization of nitrogen in pasture soil. *Thesis*, Pudoc, Wageningen.

JANSSON, S.L., 1958 - Tracer studies on nitrogen transformations in soil with special attention to mineralization - immobilization relationships. *K.LantbrHögsk. Annlr*, 24, 101-361.

JENNY, H. & K. GROSSENBACHER, 1963 - Root-soil boundary zones as seen in the electron microscope. *Proc. Soil Sci. Soc. Am.*, 27, 273-277.

JUSTE, C., 1975 - Substances physiologiquement actives exsorbées par les semences et les racines de certains végétaux. *Soc. bot. Fr., Coll. Rhizosphère*, 31-39.

KAPUSTKA, L.A. & E.L. RICE, 1976 - Acetylene reduction (N_2-fixation) in soil and old field succession in central Oklahoma. *Soil Biol. Biochem.*, 8, 497-503.

KATZNELSON, H., ROUATT, J.W. & T.M.B. PAYNE, 1956 - Recent studies on the microflora of the rhizosphere. *Trans. Intern. Congr. Soil Sci. 6th.*, Paris, vol. C., 151-156.

KATZNELSON, H., PETERSON, E.A. & J.W. ROUATT, 1962 - Phosphate-dissolving microorganisms on seed and in the root zone of plants. *Can. J. Bot.*, 40, 1181-1183.

LESPINAT, P.A. & Y. BERLIER, 1975 - Les facteurs externes agissant sur l'excrétion racinaire. *Soc. bot. Fr., Coll. Rhizosphère*, 21-30.

LOUVET, J., 1975 - L'activité des champignons phytopathogènes dans la rhizosphère. *Soc. bot. Fr., Coll. Rhizosphère*, 183-192.

LUNDEGÅRDH, H., 1927 - Carbon dioxide evolution of soil, and crop growth. *Soil Sci.*, 23, 417-430.

MAC RAE, I.C. & T.F. CASTRO, 1967 - Root exudates of the rice plant in relation to akagare, a physiological disorder of rice. *Pl. Soil*, 26, 317-323.

MACURA, J., 1961 - Bacterial flora of the root surface of wheat grown in nutrient solutions deficient in nitrogen and phosphorus. *Folia microbiol. Praha*, 6, 279-281.

MANGENOT, F., 1975 - *La rhizosphère.* Soc. bot. Fr., Coll. Rhizosphère, tome 122, 215 p.

MARTIN, J.K., 1971 - ^{14}C-labeled material leached from the rhizosphere of plants supplied with [14]CO_2. *Aust. J. biol. Sci.*, 24, 1131-1142.

MARTIN, J.K., 1975 - ^{14}C-labeled material leached from the rhizosphere of plants supplied continuously with [14]CO_2. *Soil Biol. Biochem.*, 7, 395-399.

MARTIN, J.K., 1973 - The influence of the rhizosphere microflora on the availability of ^{32}P-myoinositol hexaphosphate phosphorus to wheat. *Soil Biol. Biochem.*, 5, 473-483.

MEIKLEJOHN, J., 1968 - Numbers of nitrifying bacteria in some Rhodesian soils under natural grass and improved pasture. *J. appl. Ecol.*, 5, 291-300.

MILLER, R.H. & T.J. CHAU, 1970 - The influence of soil microorganisms on the growth and chemical composition of soybean. *Pl. Soil*, 32, 146-160.

MOLINA, J.A.E. & A.D. ROVIRA, 1963 - The influence of plant roots on autotrophic nitrifying bacteria. *Can. J. Microbiol.*, 10, 249-257.

MOORE, A.W., 1966 - Non-symbiotic nitrogen fixation in soil and soil-plant systems. *Soils Fertil.*, 29, 113-128.

MOORE, D.R.E. & J.S. WAID, 1971 - The influence of washings of living roots on nitrification. *Soil Biol. Biochem.*, 3, 69-83.

MOROWITZ, H.J., 1966 - Physical background of cycles in biological systems. *J. theoret. Biol.*, 12, 60-62.

MUNRO, P.E., 1966a - Inhibition of nitrite oxidizers by roots of grass. *J. appl. Ecol.*, 3, 227-229.

MUNRO, P.E., 1966b - Inhibition of nitrifiers by grass root extracts. *J. appl. Ecol.*, 3, 231-238.

NAKOS, G., 1975 - Absence of nitrifying microorganisms from a Greek forest soil. *Soil Biol. Biochem.*, 7, 335-336.

NEAL, J.L., 1969 - Inhibition of nitrifying bacteria by grass and forb root extracts. *Can. J. Microbiol.*, 15, 633-635.

NEAL, J.L., ATKINSON, T.G. & R.L. LARSON, 1970 - Change in the microflora of spring wheat induced by disomic substitution of a chromosome. *Can. J. Microbiol.*, 16, 153-158.

NEAL, J.L., LARSON, R.L. & T.G. ATKINSON, 1973 - Changes in rhizosphere populations of selected physiological groups of bacteria related to the substitution of special pairs of chromosomes in spring wheat. *Pl. Soil*, 39, 209-211.

NEWMAN, E.J. & H.J. BOWEN, 1974 - Patterns of distribution of bacteria on root surfaces. *Soil Biol. Biochem.*, 6, 205-209.

NILOVSKAYA, N.T., KOVALENKO, V.K. & V.V. HAPTEV, 1970 - Uptake and liberation of carbondioxide by plant and microorganisms under artificial environmental conditions. *Soviet Pl. Physiol.*, 17, 567-572.

NOMMIK, H., 1961 - Effect of the addition of organic materials and line on the yield and nitrogen nutrition of oats. *Acta Agric.scand.*, 11, 211-226.

OLD, K.M. & T.H. Nicolson, 1975 - Electron microscopical studies of the microflora of roots of sand dune grasses. *New Phytol.*, 74, 51-58.

OSMAN, A.M., 1971 - Root respiration of wheat plants as influenced by age, temperature, and shoot irradiation. *Photosynthetica*, 5, 107-112.

PAPAVIZAS, G.C. & C.B. DAVEY, 1961 - Extent and nature of the rhizosphere of Lupines. *Pl. Soil*, 13, 384-390.

PURCHASE, B.S., 1974 - Evaluation of the claim that root exudates inhibit nitrification. *Pl. Soil*, 41, 527-539.

REID, P., 1974 - Assimilation, distribution and root exudation of ^{14}C Ponderosa pine seedlings under induced water stress. *Pl. Physiol. Wash.*, 54, 44-49.

REUSZER, H.W., 1949 - A method for determining the carbon dioxide production of sterile and non-sterile root systems. *Soil Sci. Soc. Amer. Proc.*, 14, 175-179.

RICE, E.L., 1974 - *Allelopathy*. Acad. Press, New York, 353 p.

RICHARDSON, H.L., 1938 - The nitrogen cycle in grassland soils with special reference to the Rothamsted Park grass experiment. *J. agric. Sci. Camb.*, 28, 73-121.

RITTENHOUSE, R.L. & M.G. HALE, 1971 - Loss of organic compounds from roots. II. Effect of O_2 and CO_2 tension on release of sugars from peanut roots under axenic conditions. *Pl. Soil*, 35, 311-321.

RIVIÈRE, J., 1960 - Étude de la rhizosphère du blé. *Ann. agron.*, 11, 397-440.

ROBINSON, J.B., 1963 - Nitrification in a New Zealand grassland soil. *Pl. Soil*, 19, 173-183.

ROGERS, H.T., PEARSON, H.W. & W.H. PIERRE, 1940 - Absorption of organic phosphorus by corn and tomato plants and the mineralizing action of exo-enzyme systems of growing roots. *Soil Sci. Soc. Amer. Proc.*, 5, 285-291.

ROHDE, R.A. & W.R. JENKINS, 1958 - Basis of resistance of *Asparagus officinalis* var. *altius* L. to the Stubby root nematode *Trichodorus christiei*. *Univ. Maryland Agr. Exp. Sta. Bull.*, A97, 19 p.

ROUATT, J.W. & H. KATZNELSON, 1961 - A study of the bacteria on the root surface and in the rhizosphere soil of crop plants. *J. appl. Bact.*, 24, 164-171.

ROVIRA, A.D., 1965 - Interactions between plant roots and microorganisms. *A. Rev. Microbiol.*, 19, 242-266.

ROVIRA, A.D., 1972 - Studies on the interactions between plant roots and microorganisms. *J. Austr. Inst. agric. Sci.*, 9, 91-94.

ROVIRA, A.D. & C.B. DAVEY, 1974 - Biology of the rhizosphere. *In:* E.W. CARSON (Editor), *The plant root and its environment*. Univ. Press of Virginia, Charlottesville, 153-204.

ROVIRA, A.D., NEWMAN, E.J., BOWEN, H.J. & R. CAMPBELL, 1974 - Quantitative assessment of the rhizoplane microflora by direct microscopy. *Soil Biol. Biochem.*, 6, 211-216.

SADASIVAM, K.V., 1974 - Cyanide-tolerant microorganisms in the rhizosphere of tapioca. *Soil Biol. Biochem.*, 6, 203.

SAMTSEVITCH, J.A., 1971 - Root excretions of plants. An important source of humus formation in soil. *Humus et Planta*, 5, 147-154.

SCHROTH, M.N. & D.C. HILDEBRANDT, 1964 - Influence of plant exudates on root infecting fungi. *Ann. Rev. Phytopathol.*, 2, 101-132.

SHAMOOT, S., McDONALD, L. & M.V. BARTHOLEMEA, 1960 - Rhizo-deposition of organic debris in soil. *Soil Sci. Soc. Amer. Proc.*, 32, 817-820.

SHAY, F.J. & M.G. HALE, 1973 - Effects of low levels of calcium on exudation of sugars and sugar derivatives from intact peanut roots under axenic conditions. *Pl. Physiol.*, 51, 1061-1063.

SIMPSON, J.R., 1962 - Mineral nitrogen fluctuations in soils under improved pasture in New South Wales. *Austr. J. agric. Res.*, 13, 1059.

SMUCKER, M., 1971 - Anaerobiosis as it affects the exudation and infection of pea roots grown in an aseptic mist chamber. *Ph. D. Thesis*, Michigan State University, Hickory Corners, U.S.A.

SOBIESZCZANSKI, J., 1966 - Studies on the role of microorganisms in the life of cultuvated plants. III. Origin of bacterial substances stimulating the growth of plants. *Acta microbiol. pol.*, 5, 71-79.

STEVENSON, R.C., 1972 - Soil denitrification in sealed soil-plant systems. *Pl. Soil*, 33, 113-140.

STEVENSON, R.C., 1973 - Evolution patterns of nitrous oxide and nitrogen in sealed soil-plant systems. *Soil Biol. Biochem.*, 5, 167-169.

STILLE, B., 1938 - Untersuchungen über die Bedeutung der Rhizosphäre. *Arch. Mikrobiol.*, 9, 477-485.

STROWRONSKI, B. & B.A. STROBEL, 1969 - Cyanide resistance and cyanide utilization by a strain of *Bacillus pumilus*. *Can. J. Microbiol.*, 15, 93-98.

THERON, J.J., 1951 - The influence of plants on the mineralization of nitrogen and maintenance of organic matter in the soil. *J. agric. Sci. Camb.*, 41, 289-296.

THERON, J.J., 1963 - The mineralization of nitrogen in soils under grass. *S. Afr. J. agric. Sci.*, 6, 155-164.

TJEPKEMA, J.D. & H.J. EVANS, 1976 - Nitrogen fixation associated with *Juncus balticus* and other plants of Fregon wetlands. *Soil Biol. Biochem.*, 8, 505-509.

TROLLDENIER, G., 1971a - Einfluss der Kalium- und Stickstoffernährung von Weizen auf die Bakterienbesiedlung der Rhizosphäre. *Landw. Forsch.*, 26, *Sonderh.*, 37-46.

TROLLDENIER, G., 1971b - Einfluss der Kalium- und Stickstoffernährung von Weizen sowie der Sauerstoffversorgung der Wurzeln auf Bakterienzahl, Wurzelatmung und Denitrifikation in der Rhizosphäre. *Zentbl. Bakt. ParasitKde*, Abt. II, 126, 130-141.

TROLLDENIER, G., 1972 - L'influence de la nutrition potassique de haricots nains sur l'exsudation de substances organiques marquées au ^{14}C, le nombre de bactéries rhizosphériques et la respiration des racines. *Rev. Ecol. Biol. Sol.*, 9, 595-603.

TROLLDENIER, G., 1975 - Influence de la fumure minérale sur l'équilibre biologique dans la rhizosphère. *Soc. bot. Fr., Coll. Rhizosphère,* 157-167.

TURKEY, H.B., 1969 - Implications of allelopathy in agricultural plant science. *Bot. Rev.,* 35, 1-16.

VANCURA, V., 1964 - Root exudates of plants. I. Analysis of root exudates of barley and wheat in their initial phase of growth. *Pl. Soil,* 21, 231-248.

VANCURA, V. & J.L. GARCIA, 1969 - Root exudates of reversibly wilted millet plants (*Panicum milliaceum* L.). *Oecol. Plant.,* 4, 93-98.

VANCURA, V. & A. HANZLIKOVA, 1972 - Root exudates from plants. IV. Differences in chemical composition of seed and seedling exudates. *Pl. Soil,* 36, 271-282.

WAKSMAN, S.A. & R.L. STARKEY, 1931 - *Soil and microbes.* John Wiley & Sons, New York, 260 p.

WAREMBOURG, F.R., 1975 - Le dégagement de CO_2 dans la rhizosphère des plantes. *Soc. bot. Fr., Coll. Rhizosphère,* 77-87.

WEBLEY, D.M., EASTWOOD, D.J. & C.H. GIMINGHAM, 1952 - Development of a soil microflora in relation to plant succession on sand-dunes, including the "rhizosphere" flora associated with colonizing species. *J.Ecol.,* 40, 158-178.

WENT, F.W., 1970 - Plants and the chemical environment. *In:* E. SONDHEIMER & J.B. SIMONE (Editors), *Chemical Ecology.* Acad. Press, New York, p. 71-82.

WHITTAKER, R.H., 1970 - The biochemical ecology of higher plants. *In:* E. SONDHEIMER & J.B. SIMONE (Editors), *Chemical Ecology.* Acad. Press, New York, p. 43-70.

WHITTAKER, R.H. & P.P. FEENY, 1971 - Allelochemicals. Chemical interactions between species. *Science, N.Y.,* 171, 757-770.

WOLDENDORP, J.W., 1963 - The influence of living plants on denitrification. *Meded. LandbHoogesch., Wageningen,* 63 (13), 1-100.

WOLDENDORP, J.W., 1968 - Losses of soil nitrogen. *Stikstof,* 12, 32-46.

WOLDENDORP, J.W., 1975a - Biologische systemen met lage of hoge produktiviteit. *In:* G.J. VERVELDE (Editor), *Produktiviteit in biologische systemen.* Pudoc, Wageningen, p. 128-156.

WOLDENDORP, J.W., 1975b - Nitrification and denitrification in the rhizosphere. *Soc. bot. Fr., Coll. Rhizosphère,* 89-107.

9. DISCUSSION

WENT (Nevada): You mentioned that often in grassland soils hardly any nitrification occurs. Is it also suppressed in forest soils?

WOLDENDORP: According to the literature, nitrification is often suppressed in forest soils, too. Chemolitholiophic nitrifying bacteria, which are an indication for the occurrence of the process, are usually found in low numbers. But in my opinion, root distribution, which is related to the suppression of nitrification, is much more inhomogeneous in forest soils than in grassland soils. Therefore, in forest soils the conditions for microbial activity may vary from point to point more than they do in grassland soils. Also, when animals such as beetles die in the soil there is a flush of ammonia during the decomposition of the proteinaceous compounds. This temporary presence of ammonia may give rise to a local flush in nitrification.

MINDERMAN (Arnhem): I have always found some nitrate formation in oak-forest soils, but I expect it to be absent in the acid pine forest soils of The Netherlands.

LOSSAINT (Montpellier): What is your opinion on the inhibition of nitrification in climax vegetations?

WOLDENDORP: The evidence put forward to prove the inhibition of
nitrification in climax vegetations is in my opinion not very convincing.
Most investigators have worked with extracts or leachates of plants of climax
vegetations and found inhibition of nitrification. It is well known that the
nitrifying bacteria are inhibited by many kinds of organic compounds, for
instance by amino acids. Since such compounds were probably present in the
leachates, the inhibition is not surprising. But under normal field conditions
in the temperate regions, these compounds will be released slowly and are
probably decomposed as soon as they are released by the plant. I will not go
as far to say that inhibition does not occur, but there is no proof. The
absence of nitrification in climax vegetations, which without any doubt has
often been found, can equally well be ascribed to the absence of ammonia,
because all mineralized ammonia will be re-immobilized immediately when there
is a surplus of carbon. The latter is often the case in climax vegetations.

MULDER (Wageningen): Is anything known about the quantities excreted under
tropical conditions? Are they higher than in temperate regions? In our
laboratory BESSEMS found the quantities excreted in the phyllosphere to be
higher in the tropics.

WOLDENDORP: I know of no clearcut evidence which demonstrates that the
quantities of the root excretions in the tropics are higher too. For that
purpose the same plant species should be compared, and that has not been done
under field conditions as far as I know. However, there is some indirect
evidence. In laboratory experiments the quantities of root excretions,
particularly carbohydrates, were found to be higher at high light intensities.
Also, the quantities of atmospheric nitrogen fixed in the rhizosphere, for
which excreted carbonaceous compounds are needed, are generally higher in the
tropics, as has been shown by BALANDREAU (1975). These findings suggest that
in the tropics root excretions are quantitatively more important than in
regions with lower light intensities.

MULDER: I agree that the same species should be compared.That is why
BESSEMS studied the phyllosphere of maize in Surinam as well as in The
Netherlands.

WENT: Are root excretions the sole source of carbohydrates in the soil?
In my opinion the fungi may contribute to it too.

WOLDENDORP: Because the fungi are heterotrophic organisms, they are
dependent on the plant for their carbon supply. They can only contribute to
the carbohydrate level in the soil by the transformation of other compounds,
which are derived from the plant. As could be seen in my Table 2, the numbers
of fungi are also higher in the rhizosphere than in the non-rhizosphere soil.

ANONYMOUS: Are any unambiguous cases of allelopathy known under temperate
conditions?

WOLDENDORP: Such cases are not known to me as far as excretions by viable
plant roots are concerned. Under the moist soil conditions of temperate

regions, toxic organic compounds which are gradually released by the plant, are decomposed by the microflora. Only in dry soils can they accumulate to sufficiently high levels. There are some cases of allelopathy in the upper soil levels that are due to leaching from the leaves, e.g. those of walnut trees. And of course the decomposition of dead plant materials can inhibit germination and growth of other plant species. For instance, when grass is resown in ploughed meadows, germination is sometimes very bad because of the accumulation of organic acids (e.g. butyric acid) formed during the decomposition of the plant residues.

Mycorrhiza and plant growth

1. INTRODUCTION

Mycorrhizas are mutualistic associations between soil fungi and plant roots.
The fungi inhabit living roots, causing very little tissue damage, and are
nutritionally biotrophic (LEWIS 1973). Fungal spread within the root system is
limited, and usually only the fine absorbing rootlets become mycorrhizal. In
all but one group the fungus penetrates into the cells, but the intracellular
infection is usually short lived and the host cell survives after collapse of
the invading fungus.

Mycorrhizal infections are not separate phenomena clearly demarcated from
all other fungal infections of living plants. Under particular conditions their
mutualism may not find expression and the more usual growth depressant effect
of fungal infection may become operative. There are some interesting analogies
(LEWIS 1974) in reaction at the cellular level to infection by mycorrhizal and
by more pathogenic fungi. For instance, respiration and the volume of the host
cytoplasm increase in infected cells, starch reserves are mobilized, and there
is an inflow of carbohydrate to the infection sites. MEYER (1968) stresses the
underlying parasitism in his definition of mycorrhizas as 'a living together of
partners that attack each other but, having attained an equilibrium of
struggle, can coexist for more or less prolonged periods. One prerequisite of
this partnership is that either partner can intervene in the metabolism of the
other without disturbing it too greatly'. Nevertheless, mycorrhizas are
mutualistic associations and it has now been demonstrated beyond doubt that
they can be beneficial for plant growth.

Since FRANK (1885) first described the specialized modified root structures
of the *Cupulifereae* and named them 'mycorrhiza', the term has been greatly
extended and now covers an enormous range of root-fungus associations in which
the root structure undergoes little change and there is no anatomical evidence
of tissue damage or obvious pathogenic effect on plant growth. HARLEY's (1969)
detailed account of the biology of mycorrhiza leaves the reader bewildered by
the diversity of structures, fungal species, plant families and interactions
that occur within mycorrhizal systems. HARLEY warns against the assumption that
all these associations necessarily have similar effects on plant growth. Many
workers in this field now feel that the general term mycorrhiza is of little

value and that a subdivision into clearly definable types has become essential. The division proposed by LEWIS (1973) into ecto- or sheathing mycorrhiza, vesicular-arbuscular (VA) mycorrhiza, and ericaceous and orchidaceous mycorrhiza will be followed here. Thanks to improved techniques for producing VA and ericaceous mycorrhiza experimentally, enough information has been accumulated during the last decade to permit a tentative outline of mechanisms by which the different types of mycorrhiza affect plant growth (Table 1).

TABLE 1. *Effects of different types of mycorrhiza on the host*

Mycorrhizal type	Effect on plant growth	Chief nutrients supplied
V.A.	Improved nutrient uptake	P, Zn, Sr, S, K
Sheathing	Improved nutrient uptake*	P, K
Ericaceous	Improved nutrient uptake	Organic N, P
Orchidaceous	Improved seed germination and flower induction in some saprophytic plants	sucrose, carbohydrate

* and hormonal effects on root morphology

In view of the diversity of mycorrhizal systems and the very different types of fungi involved (*Endogonaceae* (Phycomycetes) in Va mycorrhiza, Agaricales, Gasteromycetes (Basidiomycetes) and some Ascomycetes in sheathing mycorrhiza, *Pezizalae* (Discomycetes) in ericaceous, and *Rhizoctonia* spp. (Basidiomycetes and imperfect fungi) in orchid mycorrhiza), the functional similarities between the ecologically important groups viz. VA, sheathing and ericaceous mycorrhizas are perhaps surprising. Further study may reveal greater divergence. At the moment, however, the orchidaceous mycorrhizas stand out as being markedly different not only in their function - aiding plant growth during a temporary or permanent saprophytic stage - but also in the organs infected (protocorms, rhizomes, aerial roots and only sometimes terrestrial roots) and in the fungi concerned. Unlike the other mycorrhizal fungi the orchid fungi are cellulose and lignin decomposers. They can be virulent pathogens (*Armilliaria mellea* and some *Rhizoctonia* spp.) for non-orchidaceous hosts and on nutrionally rich media can even kill their own symbiotic partners (BERNARD 1909; WARCUP 1975). In contrast the endophytes of VA mycorrhiza are obligate symbionts with low pathogenic potential, and there is little evidence that the fungi of sheathing or ericaceous mycorrhizas ever have the pathogenic potential of the orchid endophytes.

In all mycorrhizal systems nutrients are transferred from the external medium, which in nature is the soil, to a higher plant by means of a continuous hyphal network based partly in the soil and partly in the root or closely adpressed to its surface. Fig. 1 shows some of the environmental factors that may influence the system. Unlike the other important symbiotic system of the leguminous nodules, which operates by fixing nitrogen from the air, mycorrhizal systems do not add to the total store of nutrients in the soil, though they may increase their availability.

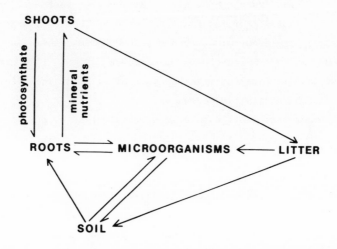

FIG. 1.   *Some pathways of nutrient transfer affecting mycorrhizal systems*

Two questions in particular interest ecologists:
1) how widespread are mycorrhizas in natural ecosystems, and 2) how important are they in nutrient uptake and cycling ? In the following an attempt is made to provide some answer or at least to show why a definitive answer to these questions is so difficult.

2. THE OCCURRENCE OF MYCORRHIZA

2.1. THE HOSTS

It has long been known that in most undisturbed ecosystems roots are normally mycorrhizal (FRANK 1885; SCHLICHT 1889; JANSE 1897; STAHL 1900). Va mycorrhizas have the widest host range and are by far the commonest type. They occur in liverworts, Pteridophytes, some Gymnosperms, and most Angiosperms, but appear to be absent in the mosses. GERDEMANN (1968) lists fourteen families that are never or rarely mycorrhizal, including *Cruciferae, Chenopodiaceae, Cyperaceae, Juncaceae* and *Caryophyllaceae*. It is widely believed that most trees have sheathing mycorrhizas but this is erroneous. *Pinaceae* and *Fagaceae* are exclusively ectomycorrhizal but many other temperate trees, most *Rosaceae*, sixty-three out of sixty-six tropical tree species in Nigeria (REDHEAD 1968) and many trees in the Amazon rain forest (T. St. JOHN personal communication) regularly have VA mycorrhizas. So do the important tropical tree crops, cocoa, coffee, tea, rubber, and citrus. A few plants, for example juniper, apple, and hazel, can have both VA and sheathing mycoorhizas, but one of these is usually the normal while the other may be site-determined. Sometimes the type of mycorrhiza may change as the host plant matures. For instance poplars, in which DANGEARD (1900) first saw the fungi he described as "vesicular-arbuscular", often have sheathing mycorrhizas when mature (FONTANA 1961). READ *et al.* (1977) recently reported that *Helianthemum chamaecistus* growing in a grassland sward was at first heavily infected with VA mycorrhizal fungi and later formed sheathing mycorrhiza with *Cenococcum* sp.

The most important plants with ericoid mycorrhizas belong to the genera *Erica, Vaccinium, Rhododendron* and *Calluna* (Read & Stribley 1975).

2.2. THE FUNGI

VA endophytes are so widely distributed that virtually no soils are reliably free from infection. However, levels of infectivity differ (MOSSE 1977a) and it is now clear that VA endophytes comprise a range of different fungi, so far uncultured. They are distinguished by their characteristic resting spores, often more than 100 μm in diameter, but non-sporing endophytes also exist and are common in forests. Spore numbers are not a reliable measure of root infection in natural situations (HAYMAN 1975, 1978) but often are in experimental systems. Although many of the known endophytes have a world wide distribution, their occurrence within small areas can be localized and is often inexplicable in terms of vegetation or soil differences. Soil pH can be unsuitable for particular species (MOSSE 1972) and species also differ in optimum temperature range (FURLAN & FORTIN 1973). Unlike some other mycorrhizal fungi VA endophytes have very little host specificity and can infect any potential host plant. Low spore mobility may account for the sporadic occurrence of different endophyte species, but this is difficult to reconcile with their world wide distribution. It is hard to believe that this could be man-made in view of the extreme age of VA infections which date back to Devonean fossil plants (BUTLER 1939).

Spores of the agarics, common fungal associates of many sheathing mycorrhizas, are small and wind dispersed. They are often absent from prairie and steppe soils. According to many, more or less well authenticated reports (summarized by REDHEAD 1974), forest soil or pure inocula must be added to

obtain satisfactory growth of introduced trees, particularly *Pinus* spp.
(BRISCOE 1959; HACSKAYLO & VOZZO 1967). BOWEN (1963) found that *Pinus radiata*
could form mycorrhizas in soils on which eucalypts had grown but such
mycorrhizas were less efficient than those formed in soils on which pines had
previously grown. Some ectomycorrhizal hosts are relatively restricted in their
fungal associates, but others form associations with a wide range of fungi.
ZAK (1973) reported over a hundred different mycorrhizal associations of one
Pseudotsuga menziesii tree. SHEMAKHANOVA (1967) concluded from a survey of
Russian work that one- and two-year-old oak seedlings became mycorrhizal, not
only in soils long denuded of forests, but also in soils far removed from them.

2.3. THE ASSOCIATION

Early investigators strongly believed that mycorrhizal development was linked
to the humus content of the soil. More systematic surveys such as those of
LIHNELL (1939), STRZEMSKA (1955), REDHEAD (1974), READ *et al*. (1976) have not
shown any correlation between levels of Va infection and soil conditions such
as pH, humus content or soil type. Mycorrhizal roots are usually most abundant
in the top 15-20 cm of the soil (MEYER 1973; READHEAD 1968; SPARLING & TINKER
1975) but this may simply reflect greater root density in the relatively
nutrient rich surface layer. Records of percentage infection alone can be
misleading and root density, itself related to nutrient levels, must also be
taken into account. MEYER (1973) reported that a beech seedling with a split
root system formed 500 root tips/100 ml soil in a mull soil and 45,000 in a
mor soil. Eighty eight percent of the roots in the mull soil were mycorrhizal
and only 51 per cent of those in the mor; the total number of mycorrhizal
roots was obviously far greater in the mor soil. Plant species also differ in
rooting habit. In a mixed forest stand containing three times as many pines
as spruces, more spruce mycorrhiza occurred in the humus layer (85 per cent
of the total) than pine (45 per cent of the total). The highest density of
mycorrhizal tips was up to $10/cm^3$ (MIKOLA *et al*. 1966). Root density may
itself affect rates of infection at least during seedling establishment.
Onions, which have a sparse root system, dit not become mycorrhizal during 10
weeks growth, whereas subterranean clover in the same field soil became
heavily infected, as were two weed and two grass species in the field (BARROW
et al. 1977). Many surveys of natural ecosystems reported 50-90 per cent of
roots as being mycorrhizal. Sometimes certain plant species appear to be more
heavily infected. READ *et al*. (1976) pointed out *Festuca ovina* as such a
species. In a survey of mycorrhizal infection in plant species of a mixed
deciduous forest *Fraxinus excelsior* and *Lonicera periclymenum* ranged from
90-100 per cent and 50-100 per cent, respectively but *Rubus vestitus* only had
28-53 per cent mycorrhizal roots. Most grasses in the tussock grasslands of
New Zealand only contained about 30 per cent infection except in the subalpine
scrub where it was higher (CRUSH 1973a). In pot experiments with *Lolium perenne*

273

I rarely found more than 30 per cent infection in a range of soils whereas tropical grasses such as *Brachiaria* sp. and *Paspalum notatum* commonly have 80-95 per cent mycorrhizal roots. Species differences have also been reported for crop plants. STRZEMSKA (1955) considered mycorrhiza to occur more rarely in rye than in wheat, barley, or oats, and also found infection differences in a range of legumes (STRZEMSKA 1975a). KRUCKELMANN (1975) found legumes and maize to be more strongly infected than potatoes. After prolonged monoculture potatoes also markedly reduced spore numbers in the soil and had some depressant effects on spore numbers in cropping sequences.

In general, mycorrhizal infection decreases in fertile soils and tends to be absent from garden soils, but field responses to added fertilizer have often been unpredictable (SHEMAKHANOVA 1967; MOSSE 1973a; STRZEMSKA 1975b; KRUCKELMANN 1975). A pot experiment cited by SHEMAKHANOVA (1967) illustrates the marked effects added nitrogen and phosphorus can have on mycorrhiza formation (Table 2).

TABLE 2. *Effects of P and N on mycorrhiza formation in* Pinus sylvestris *(after* SHEMAKHANOVA 1960*)*

N mg/flask*	7.9				4.0		2.9		1.3	
P mg/flask	22.1	15.5	12.1	11.1	17.8	11.1	15.5	5.5	14.4	3.3
Mycorrhizal tips/sdg	2	3	0	0	12	23	3	7	0	0

* each flask contained 150 ml nutrient solution

In these experiments, as in some field observations on VA mycorrhizas (HAYMAN 1975), added nitrogen determined mycorrhizal frequency even more than added phosphate. THEODOROU & BOWEN (1969) studied the effects of nitrate and pH on colonization of pine roots and glass fibres. At pH values suitable for fungal growth, added nitrate reduced fungal growth along the roots but not along the fibres. Circumstantial evidence and studies on transplantation of seedlings raised in high and low phosphorus media suggested that phosphorus levels in the plant rather than those in the medium affect the establishment and disappearance of VA infection (MOSSE 1973b and unpublished data). SANDERS (1975) took this further by applying phosphorus via the leaves of mycorrhizal and non-mycorrhizal onions. Mean phosphorus inflow rates over the five-week period of foliar feeding were similar for both sets of plants, thus indicating that the fungus in the mycorrhizal plants had ceased to function. External mycelium of the fertilized plants did not increase further after three weeks and at the end of the experiment weighed only half as much as that attached to unfertilized plants. This suggests that the absence of infection in fertile conditions is plant operated and depends on internal nutrient levels.

Taking into account results from many field and pot experiments as well as synthesis experiments with both VA and sheathing mycorrhizas (MOSSE & PHILLIPS 1971; MULLETTE 1976) one might construct a hypothetical response curve (Fig. 2)

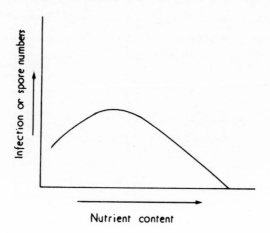

FIG. 2.   *Effect of nutrient level on mycorrhizal infection*

relating mycorrhizal infection to nutrient addition. The turning point would depend on plant species. Such a curve might help to explain the divergent results reported in the literature because the starting point, i.e., the level of available soil P as well as amounts of nutrients added, would determine whether mycorrhizal frequency increased or decreased. Infection, then, is controlled by

    a) Root density x soil infectivity,

    b) Plant nutrient deficit = plant species x nutrient availability
                            in the soil.

It is interesting that any pathogenic effects of VA mycorrhiza generally occur just before or at the time of decreasing infection (COOPER 1975; JOHNSON 1976; CRUSH 1976; MOSSE 1973b), when the plant is throwing off the invasion. The anatomical appearance of the infection also changes at that time.

Other factors determining mycorrhizal infection have been suggested, chiefly C:N ratio (HATCH 1937), light (PEYRONEL 1940; BOULLARD 1974; SHEMAKHANOVA 1967; JOHNSON 1976) and soluble sugar levels in the root (BJÖRKMAN 1942, 1970; HAYMAN 1974). Reactions to BJÖRKMAN's hypothesis have been reviewed by HACSKAYLO (1973) and put into perspective by LEWIS (1975). Infection in oaks was markedly reduced at 771 lux (SHEMAKHANOVA 1967) but with so little light plant growth must have been affected in a variety of ways. In experiments with VA mycorrhiza, 13,000 lux reduced the dry weight of onions to one fifth of that after full light but infection only from 80 to 65 per cent (HAYMAN 1974). JOHNSON (1976) found rather complicated interactions between soil phosphate levels and shade in a range of plants. To some extent the phosphorus, nitrogen and sugar levels in plants are interrelated and depend not only on photosynthesis and external nutrient levels but also on plant growth rates and dry matter production. In practice, plant nutrient levels high enough to reduce or exclude mycorrhizal infection are only likely to occur in fertile agricultural soils, and plants from most natural ecosystems are likely to be on the upward slope of the response curve.

3. THE MYCORRHIZAL POTENTIAL

3.1. THE MECHANISM OF IMPROVED NUTRIENT UPTAKE

In the early literature growth effects of mycorrhiza are often attributed to a better utilisation of organic nitrogen compounds in humus. This is generally discounted at present (LUNDEBERG 1970). However, using ^{15}N labelling STRIBLEY & READ (1974) recently showed that the ericoid mycorrhiza of *Vaccinium* took up from the organic fraction nitrogen that was unavailable to non-mycorrhizal plants and to other soil fungi. After six months the mycorrhizal *Vaccinium* weighed 50 per cent more and contained 20 per cent more nitrogen per unit dry matter than the non-mycorrhizal. Mycorrhiza also increased nitrogen uptake from added ammonium sulphate at intermediate but not at high or low concentrations (STRIBLEY & READ 1976).

The effects of VA and sheathing mycorrhizas have mainly been explained on the basis of a better phosphorus supply. Recent studies on VA mycorrhiza have concentrated on the source of the extra phosphate and the role of the fungus in the soil, which has widened our understanding of how the system functions in its natural environment. The rapidly increasing mass of data concerned with plant growth responses to inoculation with VA endophytes has been well covered in recent reviews and the proceedings of a symposium. Their titles give an indication of the particular aspects covered. They are *Vescicular-arbuscular mycorrhiza* (GERDEMANN 1975), *Effects of vesicular-arbuscular mycorrhiza on higher plants* (TINKER 1975), *Microorganisms, the third dimension in phosphate nutrition and ecology* (BOWEN & BEVEGE 1977) and *The role of mycorrhiza in legume nutrition on marginal soils* (MOSSE 1977b). The proceedings of a symposium on *Endomycorrhizas* have also been published (SANDERS *et al.* 1975). Only the main points need be summarised here.

Most phosphate in soils is insoluble or strongly absorbed to soil particles. Up to 50 per cent may be present in organic form, and a third of that may be present as phytate. Only a small proportion, often less than 5 per cent of the total soil P, is available to plants. In phosphate deficient soils the uptake of phosphate ions by plant roots exceeds the rate at which new ions move to the root surface and this leads to the development of a depletion zone, measurable in mm, around the roots. The depletion zone widens with root age as long as the root remains functional. Phosphorus uptake therefore depends primarily on the number of phosphate ions reaching the root surface, which in turn depends on their diffusion rate in the particular soil. Changes in root activity can only influence uptake if they result in the mobilization of otherwise insoluble phosphate.

Two mechanisms could account for the greatly increased phosphate uptake and growth of VA mycorrhizal plants that has been demonstrated in a range of experiments (Fig. 3).

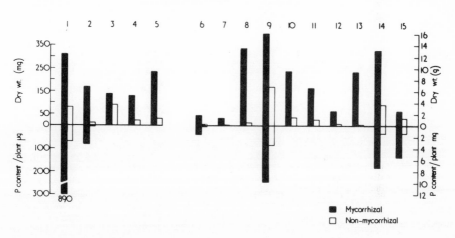

FIG. 3. *Effects of inoculation with VA endophytes on weight and phosphorus content of plants grown in sterilized soils.* 1. Allium cepa (HAYMAN & MOSSE 1971); 2. Lolium perenne (CRUSH 1973b); 3. Melinis minutifolia (*unpubl.*); 4. Paspalum notatum (*unpubl.*); 5. Stylosanthes guyanensis (*unpubl.*); 6. Brachypodium sylvaticum (*unpubl.*); 7. Centrosema pubescens (*unpubl.*); 8. Citrus (KLEINSCHMIDT & GERDEMANN 1972); 9. Fragaria *sp.* (HOLEVAS 1966); 10. Liquidambar styraciflua (*unpubl.*); 11. Liriodendron tulipifera (CLARK 1969); 12. Podocarpus totara (BAYLIS *et al.* 1963); 13. Vitis vinifera (POSSINGHAM & OBBINK 1971); 14. Zea mays (GERDEMANN 1964); 15. Glycine max (ROSS 1971)

Mycorrhiza could increase the effective absorbing surface of the root by means of the strategically situated network of hyphae that grow from the root surface and explore the soil beyond the depletion zone, or the fungus in the soil could mobilize -or induce the infected root to mobilize- insoluble phosphate unavailable to the non-mycorrhizal root. Soil labelling experiments with ^{32}P indicate that similar kinds of phosphate are taken up by VA mycorrhizal and non-mycorrhizal plants but that the former take up more, i.e., mycorrhizal plants do not mobilize insoluble soil phosphate. Nevertheless, many experiments have shown that mycorrhizal plants grow much better than non-mycorrhizal in soils enriched with such relatively insoluble phosphorus sources as apatite, bone meal, tricalcium phosphate, iron and aluminium phosphate and phytates. A more effective uptake of small amounts of soluble P associated with such phosphates is thought to explain this. It is not yet known whether sheathing mycorrhizas function in the same way.

Once taken up into the mycelium, phosphate is protected against further adsorption in the soil. Polyphosphate granules have been demonstrated in the fungi of both VA and sheathing mycorrhizas and are thought to be the form in which phosphate moves within the fungal system. Phosphate transfer from fungus to host is probably an active metabolic process that is not, or at least only partially dependent on breakdown of the arbuscule. The process of mycorrhizal uptake thus proceeds in three steps 1) uptake from the soil by the soil mycelium; 2) translocation from the mycelium in the soil to the mycelium in the root; and 3) transfer from the fungus to the plant. Any of these stages can be rate limiting but uptake from the soil is probably the most important one. Any factor affecting fungal spread in the soil such as pH, moisture content, aeration, temperature and interactions with other soil microorganisms will affect the uptake potential of the mycorrhizal system. The most important factor is probably the amount of phosphorus available to the fungus in the soil.

Some plants species only took up ^{32}P from labelled soils containing extremely little available phosphate when they were mycorrhizal.

In sheathing mycorrhiza the sheath may fulfill a storage function not only for nutrients taken up from the soil but also for those lost by efflux from the root.

Plant species differ markedly in their ability to grow in phosphorus deficient soils (Plate 1). Those with restricted root systems, short stubby roots and few root hairs (e.g. *Liriodendron,* onion and citrus) are generally less able to extract phosphorus and are therefore more dependent on mycorrhizal association than species with many root hairs, long fine roots and many root tips (e.g. *Gramineae*). Nevertheless even such species can benefit from mycorrhizas if soil phosphate is sufficiently low. Even in relatively fertile soils plants can benefit from mycorrhizas if growing conditions are good and rapid growth requires a high rate of phosphorus inflow. On the other hand,

PLATE 1.   *Response of different plant species, grown in the same soil,*
           to calcium dihydrogen phosphate (PO$_4$) (100 mg P/kg soil),
           to inoculation with VA endophytes (inoc.), and
           to filtered washings (leachings) from the inoculum

mycorrhizal plants can in some circumstances take up extra phosphate that is not reflected in extra growth.

Finally, mycorrhizal fungi differ in their capacity to aid nutrient uptake. This difference may depend on their adaptation to particular soils, perhaps on a tolerance of high levels of certain substances such as Al or NaCl, and reaction to physical soil conditions such as aeration, moisture content, and temperature. *In vitro,* in pots, and in the field, both VA and ecto-mycorrhizal fungi have been shown to differ in ability to utilize rock phosphate. However, in the same soil different VA endophytes used similarly labelled sources of soil phosphate. Sheathing mycorrhizas also differ in surface phosphatases and phytases.

The effect of mycorrhiza is therefore determined by plant species, fungal species and nutrient level in the soil.

The uptake of other ions can also be affected by mycorrhiza. Increased uptake of zinc (GILMORE 1971; BOWEN *et al.* 1974), potassium (POWELL 1975), strontium (JACKSON *et al.* 1973), and copper (MOSSE 1973a) have been reported.

3.2. QUANTITATIVE ASPECTS

Many interesting calculations have been made on fungal mass, the storage and translocating capacity of mycorrhizal systems, the spread of mycelium into soil, inter-root distances, and other factors of importance for the elucidation of mycorrhizal function and potential. It will be useful to bring some of them together here.

Mean inter-root distances in the top 10 cm of a 20 year old stand of *Pinus radiata* were 12.8 mm for long roots and 9 mm for short roots. In a 27 year old stand on another soil the values were 14.2 mm and 5.9 mm, respectively (BOWEN & THEODOROU 1967). Root length of *Gramineae* may be up to 50 cm/cm^3 of soil, which is twice that of legumes and other non-graminaceous herbs and 10-15 times that of woody plants (BARLEY 1970). PAVLYCHENKO (1942) recorded 53 cm root length/cm^3 soil in the top 10 cm of a wheat field in June. At 30 cm/cm^3 and a calculated inter-root distance of 1.8 mm there was no competition for phosphorus between wheat roots (NEWMAN & ANDREWS 1973). The mean distance between roots in a brown earth grassland at depths of 5, 10, and 15 cm was 0.39, 0.89, and 1.23 cm, respectively (SPARLING & TINKER 1975). Per unit length, clover roots including root hairs occupied one-fourth of the volume occupied by grass roots. Since inflow rates were one order of magnitude greater in the clover, phosphate depletion zones would also be greater (SPARLING 1976).

There was a constant relationship between VA fungus in roots and external media when $CaHPO_4$ was used as a phosphate source. This relationship no longer held, and relatively more external mycelium grew when Ca phytate was used as the phosphate source (MOSSE & PHILLIPS 1971). Furthermore, SANDERS & TINKER (1973) found a constant relationship between external mycelium and the length

of infected onion roots. Assuming a specific gravity of 1 for the hyphae, they
calculated a length of 80 cm external hyphae/cm infected root. BEVEGE et al.
(1975) gave the fresh weight of external hyphae as approximately 1 per cent of
the total root weight in mycorrhizal clover, and MOSSE (1956) harvested 5 per
cent from heavily infected apple rootlets. BURGES & NICHOLAS (1961) reported
hyphal lengths of up to 6 m/cm^3 in the humus layer and up to 4 m/cm^3 in the
A_1 horizon of a 60 year old pine stand in July. These amounts fell to 2.4 m/cm^3
and 1.3 m/cm^3, respectively, in September. These figures include live and dead
hyphae and mycorrhiza as well as other fungi present. Mycelial strand growth
of *Rhizopogon* varied widely with soil type and compaction reduced it greatly
(SKINNER & BOWEN 1974a).

Translocation rates and distances covered by the external mycelium have
been measured. Strands of *Rhizopogon luteus* conducted ^{32}P (applied at
concentrations comparable to those in soil) towards mycorrhizal roots over
distances of 1.2 cm in a pot experiment and 12 cm in the field and 30–80 per
cent of the absorbed phosphorus was translocated (SKINNER & BOWEN 1974b).
Within four days the external mycelium of mycorrhizal onion roots translocated
into the plant appreciable amounts of ^{32}P applied 8 cm from the root surface.
In the absence of mycorrhizas no ^{32}P reached the plant and radioactivity
diffused in the soil over 1 cm (RHODES & GERDEMANN 1975). Very much lower
diffusion rates to a root surface were measured by LEWIS & QUIRK (1967). In a
soil given 100, 300, or 1,000 µg P/g soil, the velocity of phosphate ions was
0.04, 0.08, and 0.26 mm/day, respectively. BIELESKI (1973) calculated that in
a VA mycorrhizal system having per mm root length four hyphae each 2 cm long
and 25 µm in diamter, phosphate uptake could increase 60 fold if diffusion was
limiting and 10 fold if uptake was proportional to surface area.

The number of entry points of the fungus into the root may have some
importance for translocation since the movement of material from soil-based to
root-based mycelium has to occur via these connecting points. The numbers
reported in the literature range from 4–16/mm root length in juniper (LIHNELL
1939), 2.1–16.9/mm^2 root surface in strawberries and 4.6–10.7/mm root length
in apples (MOSSE 1959), 0.6/mm root length in onion (SANDERS & TINKER 1973),
1.5/mm in *Festuca*, and 25–200/mm in *Calluna* (READ & STRIBLEY 1975). The
number of entry points in *Calluna* is of an altogether different order of
magnitude. This difference may be related to the different function of
ericoid mycorrhizas; to make an impact on growth far greater quantities of
nitrogenous substances than of phosphate would need to reach the plant.

The total mass of fungus associated with mycorrhizal roots is also of
interest. Dissection of beech mycorrhizas indicated that 40 per cent of the
total dry weight consisted of mantle. On the assumption that the feeding roots
represent 10 per cent of total root mass something like 4 per cent of this
would consist of fungus (HARLEY 1971). On the basis of a glucosamine assay
HEPPER (1977) estimated that 4, 9 and 17 per cent of the dry weight of a

mycorrhizal root might consist of fungus in light, medium and heavily infected roots, respectively. Per gram weight, hyphae are several hundred times longer than roots and have the same total length as root hairs or a few times more. Therefore hyphae produce absorbing surface at much lower energy cost than roots (BOWEN & BEVEGE 1977).

3.3. OTHER EFFECTS OF MYCORRHIZA

3.3.1. Water uptake and longevity

It has repeatedly been suggested that mycorrhiza may help in the uptake of water and that mycorrhizal roots live longer. It is well known that phosphate deficient plants are more susceptible to drought (ATKINSON & DAVISON 1972) and it has been shown (SAFIR et al. 1971,1972) that the enhancement of water transport in mycorrhizal soya bean can be ascribed to a better phosphate nutrition. It is also well documented that soil fungi are able to survive much higher moisture tensions than plants and that they may therefore continue to take up nutrients when root hairs have collapsed. But because most mycorrhizas occur in the surface layer of soils and the water supply is obtained from the lower layers in times of drought, the more likely significance of mycorrhiza in such a situation is their continued ability to take up nutrients from relatively dry soil. Sheathing mycorrhizas in particular may act as a reservoir and as a protective layer against desiccation.
STRZEMSKA (1975a) suggests that in the *Gramineae* many non-mycorrhizal roots are deformed and presumably functional for shorter periods, but there are no confirmatory observations. Longevity is often stressed as an advantage of sheathing mycorrhizas. Although this may be advantageous for the uptake of nutrients other than phosphate, the chief advantage may be that such roots continue to support a network of easily replaceable soil hyphae that are not subject to the same ageing process as roots in terms of suberization.

3.3.2. Antagonism to pathogens

Field observations and laboratory results relating mycorrhizal infection to feeder root diseases have been well reviewed by MARX (1973). Although mycorrhizas never confer complete immunity, they often appear to reduce the severity of infection by other fungi or at least its symptom expression. MARX (1973) put forward several possibilities that might explain this: mechanical protection by the sheath, better nutrition of the plant, production of antibiotics by the mycorrhizal fungus, competition for infection sites, formation of phytoalexins by the plant (as in orchids) and changes in root exudates leading to a build up of protective rhizosphere organisms. To these may be added catabolite repression of degradative enzymes, proposed by LEWIS (1974) as a step towards mutualism. Evidence is accumulating that VA endophytes may exert similar effects. There is a competition for sites between endophytes and *Pyrenochaeta terrestris* in onions (BECKER 1976) and a reduction in

pathogenicity and in sporulation of *Thielaviopsis basicola* in tobacco
(BALTRUSCHAT & SCHÖNBECK 1972). Symptom severity of *Fusarium* wilt is less in
mycorrhizal tomatoes (DEHNE & SCHÖNBECK 1975), and there are instances of
interactions reducing the population of pathogenic nematodes (SIKORA &
SCHÖNBECK 1975). JANOS (1975) observed that mycorrhizal seedlings were more
resistant to herbivory in a tropical forest. Nevertheless, it is by no means
rare to find mycorrhizal infection in root lesions caused by other pathogenic
fungi and some root rots have been ascribed, albeit erroneously, to mycorrhizal
infection.

4. MYCORRHIZA IN THE NATURAL ECOSYSTEM

4.1. AGRICULTURE, PASTURE AND FOREST

Even though we have some knowledge and appreciation of mycorrhizal potential,
a quantitative assessment of the mycorrhizal contribution to nutrient uptake
in an undisturbed ecosystem presents many difficulties,chiefly because these
are dynamic systems. Nutrient requirements vary both during the annual growth
cycle and, for perennials, during their life span. The possible significance
of mycorrhizas is linked to these fluctuating needs. In annual crops an adequate
P supply is particularly critical during early growth when root density is low
and depletion zones do not yet overlap. Because phosphate is removed by the
crop and high yields are usually desired, most annual crops are grown with
some fertilizer input. This addition would have to be larger if there were no
mycorrhizas. COOKE (1965) stated that in general 25 per cent of applied
phosphate is taken up by the crop in the first year and the rest reverts to
unavailable forms unless a reserve of potentially available P has been built
up over the years by repeated fertilizer application. Many wheat growing soils
in Australia are so P deficient and have such high fixation rates that the
yield is directly proportional to fertilizer P input. Even in Danish
agricultural soils with a long history of generous fertilizer application,
further phosphate addition became necessary to maintain high productivity when
no phosphate had been given for two years (LARSEN 1976). The only question is
the extent to which mycorrhizal infection is suppressed by heavy fertilization
under agricultural conditions. Plate 2 shows the effect of removing VA
endophytes from an agricultural soil at Rothamsted. Where poor growth occurs
after soil sterilization, particularly of a relatively mild kind such as
steaming, low dosage irradiation (below 1 Mrad), or fumigation, one suspects
that this is due to lack of mycorrhizas. It sometimes occurs in nurseries and
has been overcome by re-inoculation with VA endophytes (KLEINSCHMIDT &
GERDEMANN 1972; HATTINGH & GERDEMANN 1975). In the past, poor growth after
soil sterilization was often attributed to soil toxicity. The typical response
to toxicity is initially poor growth that improves as a microbial population

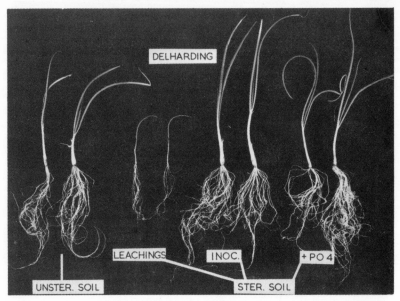

By courtesy of D.S. Hayman

PLATE 2.   *Growth of onions in a Rothamsted soil. Plants in the*
unsterile soil and those labelled 'inoc.' are
mycorrhizal; 'leachings' and '+PO$_4$' plants are without
mycorrhiza.
PO$_4$ = calcium dihydrogen phosphate, 100 mg P/kg soil

re-established itself and detoxifies the soil. Poor growth due to lack of
mycorrhiza proceeds in the opposite way, good initial growth while the seed
reserves last, followed by a gradual decline as nutrients run out.

    In forests, only a small proportion of the annual P requirement (less than
2 per cent in a 20 year old pine stand (SWITZER & NELSON 1972) comes from the
soil. The extreme phosphorus deficiency in soils under luxurious tropical
forest in the Amazon basin shows that in such situations nutrient cycling is
very tight. Of the annual phosphorus requirement sixty to eighty per cent may
be recycled within the plant and the balance taken up again directly from the
decaying litter without ever reaching the soil. The function of mycorrhiza in
such a system could be twofold. The fungi could help to accelerate the cycle
by assisting in litter breakdown and by acting as a direct link for
transferring nutrients from the litter into the plant. They might also act,
as in the soil, as an extension of root absorbing surface and, in the case of
sheathing mycorrhiza, as a temporary storage tissue during flushes of nutrient
release. Fresh pine litter contains 1 per cent total P of which approximately
20 per cent is inorganic and 20 per cent is phytate (BOWEN & THEODOROU 1967).
Mycorrhizal fungi in culture can utilize phytate (THEODOROU 1968; SHEMAKHANOVA
1967) but so can sterile plants. The mycelium and rhizomorphs often seen in

litter usually belong to saprophytes, fungi such as *Mucor, Mortierella* and *Penicillium* (WENT 1971) which are not mycorrhiza formers. GADGIL & GADGIL (1971) reported that litter decomposition actually appeared to be retarded by mycorrhizal fungi and MACAULEY (1975) concluded that while fungi released P from litter they retarded the mineralization of N.

The important effects of mycorrhiza on the establishment and early growth of seedling trees planted in the field is well illustrated by Russian work in which pure cultures of ectomycorrhizal fungi were used as inoculum. SHEMAKHANOVA (1967) cited many instances of better survival (30-80 per cent in 2-year-old pines, 2-36 per cent in 3-year-old oaks) and of increased dry weight (0-131 per cent and 25-130 per cent in pine seedlings and up to 140 per cent in oak) according to soil and site. Inoculation sometimes caused small growth depressions. The average diameter of oaks was increased by up to 25 per cent in a 9-year-old oak stand. General experience in forestry has shown that initial advantages in early growth persist in the mature stand.

Grasslands resemble forests in that phosphorus moves to the roots during winter, but they resemble agricultural systems which are subject to a net loss of nutrients due to grazing followed by removal of the animals. CRUSH (1973a), SPARLING (1976) and KOUCHEKI & READ (1976) who studied inoculation responses of grasses in upland and tussock grassland soils, concluded that despite abundant infection at these sites the precise role of mycorrhiza was difficult to interpret and of doubtful significance. One difficulty encountered was that considerable release of nutrients occurred when these high organic soils were sterilized. Therefore, no growth responses were observed until some time after inoculation, i.e., four months in the investigations of KOUCHEKI & READ (1976) and twelve months in one of the three soils studied by SPARLING (1976). By contrast clover in all three soils showed very large responses to inoculation. CRUSH (1973a) found that out of five grass species tested only two benefitted from inoculation and then only in soils at high altitudes which contained less than 8 ppm-Truog soluble P. Growth of the other species was slightly depressed by infection. CRUSH thought that the above effects, also observed by MAGROU (1938) and MOSER (1963), might be explained by the slower mineralization of phosphate at high altitudes.

The need for soil sterilization, arising from the universality of VA endophytes, is not the only factor that makes interpolation from pot experiments to field situations difficult. A short account of an attempt to assess the involvement of V˙ corrhiza in phosphate uptake in a mixed deciduous forest in England w. .l illustrate some of the other problems.

4.2. THE ROLE OF MYCORRHIZA IN PHOSPHATE UPTAKE AT MEATHOP WOOD IBP SITE

Species included in this study were *Fragaria vesca, Viola riviniana, Brachypodium sylvestris, Rubus vestitus, Fraxinus excelsior* and *Betula pubescens*. Preliminary experiments (Plate 3) with the grass *Melinis*

PLATE 3.   Melinis minutifolia *in irradiated Meathop Wood soil*
P = KH$_2$PO$_4$, 100 mg P/pot; N = NH$_4$NO$_3$, 200 mg N/pot

minutifolia, a species very efficient in phosphorus uptake, showed that even for this species phosphorus was growth limiting in the soil but when phosphate had been added there was also a response to nitrogen. Inoculation experiments in the irradiated soil (Table 3) showed that growth of five locally common species was greatly improved by inoculation with the indigenous mycorrhizal fungi.

TABLE 3.   *Dry weight (mg) of plants with and without mycorrhiza*

Plant	Inoculated with indigenous fungi	Non-inoculated
Brachypodium	453	15
Viola	349	146
*Rubus**	2756	1122
*Fraxinus***	174	125
Betula	1718	176

*   including weight of original cutting

**   excluding weight of original cutting

The addition of monocalcium phosphate at the rate of 100 mg P/kg soil produced even better growth responses except in *Brachypodium*. Inoculation responses in irradiated soil are frequently better than in unsterilized soil and this was also the case here (Table 4; Plate 4).

TABLE 4.   *Dry weight (DW) and P uptake (P) of inoculated plants in irradiated (I) and unsterlized (U) soil*

Plant	Soil	DW (mg)	P (µg)
Viola	I	184	403
	U	91	156
Fragaria	I	1124	1253
	U	549	410
Fraxinus*	I	160	470
	U	44	241
Betula	I	160	319
	U	88	190

* weight and P gain of cuttings from one year old seedlings

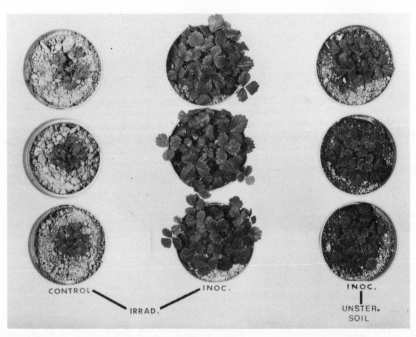

PLATE 4.   *Inoculated and non-inoculated* Fragaria vesca *in unsterile and irradiated Meathop Wood soil*

That soil-available phosphorus was little changed by the irradiation treatment was shown by chemical analysis and also by the early growth of strawberry seedlings (Table 5).

TABLE 5.   *Dry weight (DW) and mycorrhizal infection of* Fragaria *seedlings in irradiated and unsterlized soil*

Treatment	3 weeks		5 weeks	
	DW(mg)	Infection (%)	DW(mg)	Infection (%)
Irradiated soil	22	0	28	0
Unsterilized soil	22	0	35	30

Their growth was equally good in irradiated and unsterlized soils until the seedlings in the unsterilized soil became mycorrhizal, after which they grew better. It was therefore decided to compare non-mycorrhizal plants (NM) in irradiated soil with inoculated plants (M) in unsterilized soil. This would, if anything, give an advantage to the non-mycorrhizal plants. Indigenous mycorrhizal fungi were used as inoculum. Results are shown in Table 6.

TABLE 6.   *Dry weight (DW) and P content (P) of inoculated plants in unsterilized soil (M) and non-inoculated plants in irradiated soil (NM)*

Plant	DW(mg)		P(μg)	
	M	NM	M	NM
Brachypodium	355	38	182	18
Viola	142	22	157	22
Fragaria	214	47	165	37
Fraxinus*	173	125	304	95
Rubus**	1115	1122	800	394

*   gains made by cuttings

**   total weight and P content of leaf-bud cuttings

The difference between M and NM plants, although originally induced by the inoculation treatment, cannot be attributed solely to fungal uptake. As the mycorrhiza improved plant growth, the plants themselves developed larger root systems which would also contribute to greater nutrient uptake. Small phosphate starved plants sometime have abnormally high phosphate concentrations. When therefore adjustments were made to allow for the larger size of M roots on the basis of P content and weight of the NM roots, the calculations often showed a very small or even negative involvement of mycorrhiza in P uptake. This was obviously absurd because inoculation had induced much better growth. Perhaps a better measure of mycorrhizal involvement would be to determine the size of phosphorus application necessary to produce plants as large as the inoculated ones. The best solution would be to find a systematic fungicide that would inhibit fungal uptake *in situ*, but this is likely to be difficult because few systemics act against Phycomycetes and most are translocated upwards and often are relatively insoluble. In a preliminary study of this problem Benomyl added to soil inhibited mycorrhizal infection (BOATMAN *et al*. 1978). This was also reported by BERTOLDI *et al*. (1977).

Another aspect that emerged during our investigations indicates even more strongly that *in situ* inhibition of fungal uptake is the best approach. When soil was collected for the experiments the litter layer was removed. This contained four times as much $NaHCO_3$-soluble P as the soil below it. If this layer had not been removed but mixed with the soil it would have added only 1.5 per cent extra available P, but *in situ*, as a top layer, it had very marked effects on seedling growth. The soil profile and the thickness of the litter layer were found to be highly variable in the wood. We therefore compared results of inoculation in three types of reconstructed profiles similar to those found in the wood. Table 7 shows the results obtained.

TABLE 7. *Dry weight (mg) of mycorrhizal (M) and non-mycorrhizal (NM) plants in soil reconstructed in its natural profile. Litter depth decreased from profile 1 to profile 3*

Plant	Profiles							
	Subsoil		1		2		3	
	NM	M	NM	M	NM	M	NM	M
Fragaria	29	143	1141	385	13	263		
Brachypodium	32	28	1133	604	174	469	43	49
Viola							22	138
Rubus seedlings					144	509		
stem cuttings					819	735		

M = inoculated plants grown in unsterilized soil
NM = non-inoculated plants grown in irridiated soil

Profile 2 was considered to represent the most common type in the wood. In profile 1 litter occupied more than half the pot. It is likely that the exceptionally good growth of the NM plants in this profile was due to nutrient release following sterilization. The comparable M plants were grown in unsterilized soil probably containing less available P. The potential pathogenicity of mycorrhizal infection may also have found expression in this relatively fertile soil where plants generally grew much better. It is evident that considerable practical difficulties will be encountered in any attempt to make a quantitative assessment of mycorrhizal significance in such a variable environment.

The investigation suggested one further conclusion, namely that mycorrhiza might be particularly important in early seedling establishment. In profile 2 (Table 7) *Rubus* seedlings benefitted from inoculation but stem cuttings, which had an initially higher P reserve, dit not. Practical experience with forest trees has already established the importance of mycorrhiza for seedling development and the extremely small size of orchid seed and lack of seed reserves has always been regarded as a probable explanation for their mycorrhiza dependence. The mycorrhizal *Rubus* cuttings were marginally larger and contained 20 per cent more P after 6 weeks but after 12 weeks (Table 7) infection had reduced growth. *Rubus* has a very extensive root system with many long root hairs. It grew vigorously, and in small pots this resulted in great root density. It is noteworthy that in the survey of Meathop Wood plants *Rubus* had the lowest infection level, which is probably an indication of the self-adjusting character of VA infection.

4.3. SPECIAL SITUATIONS

In ecological studies the possibility of species interactions always attracts attention. On the basis of results obtained by WOODS & BROCK (1964) and BJÖRKMAN (1960) it is occasionally suggested that movement of nutrients could occur both between individuals and also between plant species via common mycorrhizal fungi. Although one frequently sees connections between roots of the same species one rarely finds interspecies connections. Since infection can spread from one species to another some links must occur. The orchid studies of BJÖRKMAN (1964) are a special case, and the interspecies transfer of ^{32}P and ^{45}Ca reported by WOODS & BROCK (1964) could have occurred in many ways. Much more likely and interesting for mycorrhizal studies is the question of interspecies competition. STONE (1949) showed that ectomycorrhizal Monterey pine competed successfully with Sudan grass for poorly soluble P and obtained a relatively large share of a limited supply when it was mycorrhizal. A similar situation probably exists between grasses and clovers or other pasture legumes. SPARLING (1976) showed that clovers had about half as much root length per unit weight as grasses and without mycorrhiza grew very badly in soils where grasses grew quite well. In mixed swards clover is much more likely to become established and to persists if it is mycorrhizal.

In phosphate deficient soils mycorrhizas are potentially important for the proper functioning of symbiotic nitrogen fixation by rhizobia. This has been demonstrated in sterilized (MOSSE *et al*. 1976; SMITH & DAFT 1977) and unsterilized soils (MOSSE 1977a); the subject has been reviewed by MOSSE (1977b). Mycorrhizas enable the plant to take up the phosphorus necessary for successful nodulation. They may also interact in other ways with rhizobia, but this has not yet been proved. Mycorrhizas also occur in non-leguminous plants with nitrogen fixing nodules. A combination of the two symbiotic systems offers great advantage to pioneer plants on reclaimed waste land and coal spoils (DAFT & NICOLSON 1974; DAFT *et al*. 1975; DAFT & HACSKAYLO 1976). Trees with sheathing mycorrhizas formed by *Pisolithus tinctora* are also good colonizers of such sites since the fungus develops prolific mycorrhizas with pine at soil temperatures ranging from 19 to 47°C (MARX & BRYAN 1975). High soil temperatures, low pH and sometimes heavy metal toxicites, are particular obstacles to the colonization of such sites.

VA endophytes are also very effective in the aggregation of dune sands (KOSKE *et al*. 1975; SUTTON & SHEPPARD 1976). They increased aggregate weight from 0.9 to 2.5 g per kg dry soil and to 127 g after a second crop. Although other soil fungi were present, mycelium of VA endophytes predominated in the aggregates. NICOLSON (1960) considered heavy mycorrhizal infection as a developmental stage in the colonization of dune sands. If the fungi have special binding power they may also have a function in erosion control.

Many possibilities and many problems remain in mycorrhiza research. Experimental techniques are available to study particular problems but results must be treated with caution when one tries to relate them to natural situations. The potential value of myccorhiza has been clearly demonstrated but how far this can be utilized in crop production and to what extent it may be endangered by new agricultural and forestry practices remains to be determined.

5. ACKNOWLEDGEMENTS

The work at Meathop Wood was done in conjunction with Dr. D.S. Hayman. I would like to thank the International Biological Programme for a grant towards the Meathop Wood investigation and I am indebted to Mr. D. Arnold and Mr. D. Paget who did much of the work.

6. REFERENCES

ATKINSON, D. & A.W. DAVISON, 1972 - The effects of phosphorus deficiency on water content and response to drought. *New Phytol.*, 72, 307-313.

BALTRUSCHAT, H. & F. SCHÖNBECK, 1972 - Untersuchungen über den Einfluss der endotrophen Mycorrhiza auf die Chlamydosporenbildung von *Thielaviopsis basicola* in Tabakwurzeln. *Phytopath. Z.*, 74, 358-361.

BARLEY, K.P., 1970 - The configuration of the root system in relation to nutrient uptake. *Adv. Agron.*, 22, 159-201.

BARROW, N.J., MALAJCZUK, N. & T.C. SHAW, 1977 - A direct test of the ability of vesicular-arbuscular mycorrhiza to help plants take up fixed soil phosphate. *New Phytol.*, 78, 269-276.

BAYLIS, G.T.S., McNABB, R.F.R. & T.M. MORRISON, 1963 - The mycorrhizal nodules of podocarps. *Trans. Br. mycol. Soc.*, 46, 378-384.

BECKER, W.N., 1976 - Quantification of onion vesicular-arbuscular mycorrhizae and their resistance to *Pyrenochaeta terrestris*. *Ph. D. thesis,* Univ. of Illinois at Champaign-Urbana.

BERNARD, N., 1909 - L'évolution dans la symbiose. *Ann. Sci. Nat. Bot.*, 9, 1-96.

BERTOLDI, M., GIOVANETTI, M., GRISELLI, M. & A. RAMBELLI, 1977 - Effects of soil applications of benomyl and captan on growth of onions and the occurrence of endophytic mycorrhizas and rhizosphere microbes. *Ann. Appl. biol.*, 86, 1111-1115.

BEVEGE, D.I., BOWEN, G.D. & M.F. SKINNER, 1975 - Comparative carbohydrate physiology of ecto- and endomycorrhizas. *In:* F.E. SANDERS, B. MOSSE & P.B. TINKER (Editors), *Endomycorrhizas,* Acad. Press, New York, p.149-174.

BIELESKI, R.L., 1973 - Phosphate pools, phosphate transport, and phosphate availability. *A. Rev. Pl. Physiol.*, 24, 225-252.

BJÖRKMAN, E., 1942 - Über die Bedingungen der Mykorrhizabildung bei Kiefer und Fichte. *Symb. botan. upsal.*, 6, 1-191.

BJÖRKMAN, E., 1960 - *Monotropa hypopitys* L. - an epiparasite on tree roots. *Physiol. Pl.*, 13, 308-327.

BJÖRKMAN, E., 1970 - Forest tree mycorrhiza. The conditions for its formation and the significance for tree growth and afforestation. *Pl. Soil*, 32, 589-610.

BOATMAN, N., PAGET, D., HAYMAN, D.S. & B. MOSSE, 1978 - The effects of systemic fungicides on vesicular-arbuscular mycorrhizal infection and on plant phosphate uptake. *Trans. Brit. myc. Soc.*, (in press).

BOULLARD, B., 1974 - A propos de mycothalles, mycorrhizomes et mycorrhizes. *Ann. Microbiol.*, 24, 105-123.

BOWEN, G.D., 1963 - The natural occurrence of mycorrhizal fungi for *Pinus radiata* in South Australian soils. *C.S.I.R.O. Div. Soils, Adelaide, Austr., Div. Rep. 6/63,* 11.

BOWEN, G.D. & D.I. BEVEGE, 1977 - Micro-organisms, the third dimension in phosphate nutrition and ecology. *In:* G.J. BLAIR (Editor), *Improving the efficiency of phosphorus utilization.* Reviews in Rural Science, 3, Univ. New England Press, Armidale N.S.W. (in press).

BOWEN, G.D., SKINNER, M.F. & D.I. BEVEGE, 1974 - Zinc uptake by mycorrhizal and uninfected roots of *Pinus radiata* and *Araucaria cunninghamii*. *Soil Biol. Biochem.*, 6, 141-144.

BOWEN, G.D. & C. THEODOROU, 1967 - Studies on phosphate uptake by mycorrhizas. *Proc. 14th. IUFRO Congress*, 24, 116-138.

BRISCOE, C.B., 1959 - Early results of mycorrhizal inoculation of pine in Puerto Rico. *Caribb. Forester*, 20, 73-77.

BURGES, A. & D.P. NICHOLAS, 1961 - Use of soil sections in studying amounts of fungal hyphae in soil. *Soil Sci.*, 92, 25-29.

BUTLER, E.J., 1939 - The occurrences and systematic position of the vesicular-arbuscular type of mycorrhizal fungi. *Trans. Br. mycol. Soc.*, 22, 274-301.

CLARK, F.B., 1969 - Endotrophic mycorrhizal infection of tree seedlings with *Endogone* spores. *Forest Sci.*, 15, 134-137.

COOKE, G.W., 1965 - The responses of crops to phosphate fertilizers in relation to soluble phosphorus in soils. *Soil Phosph. Techn. Bull.*, 13, 64-74.

COOPER, K.M., 1975 - Growth responses to the formation of endotrophic mycorrhizas in *Solanum, Leptospermum* and New Zealand ferns. *In:* F.E. SANDERS, B. MOSSE & P.B. TINKER (Editors), *Endomycorrhizas,* Acad. Press, New York, p. 391-407.

CRUSH, J.R., 1973 a - Significance of endomycorrhizas in tussock grassland in Otago, New Zealand. *N.Z.Jl Bot.*, 11, 645-660.

CRUSH, J.R., 1973b - The effect of *Rhizophagus tenuis* mycorrhiza on ryegrass, cocksfoot and sweet vernal. *New Phytol.*, 72, 965-973.

CRUSH, J.R., 1976 - Endomycorrhizas and legume growth in some soils of the Mackenzie Basin, Canterbury, New Zealand. *N.Z.Jl agric. Res.*, 19, 473-476.

DAFT, M.J. & E. HACSKAYLO, 1976 - Arbuscular mycorrhizas in the anthracite and bituminous coal wastes of Pennsylvania. *J. appl. Ecol.*, 13, 523-531.

DAFT, M.J., HASCKAYLO, E. & T.H. NICOLSON, 1975 - Arbuscular mycorrhizas in plants colonising coal spoils in Scotland and Pennsylvania. *In:* F.E. SANDERS, B. MOSSE & P.B. TINKER (Editors), *Endomycorrhizas,* Acad. Press, New York, p. 561-580.

DAFT, M.J. & T.H. NICOLSON, 1974 - Arbuscular mycorrhizas in plants colonising coal wastes in Scotland. *New Phytol.*, 73, 1129-1138.

DANGEARD, P.A., 1900 - Le *Rhizophagus populinus* Dangeard. *Botaniste,* 7, 285-291.

DEHNE, H.W. & F. SCHÖNBECK, 1975 - Untersuchungen über den Einfluss der endotrophen Mykorrhiza auf die Fusarium-Welke der Tomate. *Z. PflKrankh. PflSchutz,* 82, 630-632.

FONTANA, A., 1961 - Primo contributo allo studio delle micorrhiza dei Pioppi in Piemonte. *Allionia,* 7, 87-129.

FRANK, A.B., 1885 - Über die auf Wurzelsymbiose beruhende Ernährung gewisser Bäume durch unterirdische Pilze. *Ber. dt. bot. Ges.*, 3, 128-145.

FURLAN, V. & J.A. FORTIN, 1973 - Formation of endomycorrhizae by *Endogone calospora* on *Allium cepa* under three temperature regimes. *Naturaliste can.*, 100, 467-477.

GADGIL, R.L. & P.D. GADGIL, 1971 - Mycorrhiza and litter decomposition. *Nature, Lond.*, 233, 133.

GERDEMANN, J.W., 1964 - The effect of mycorrhiza on the growth of maize. *Mycologia,* 56, 342-349.

GERDEMANN, J.W., 1968 - Vesicular-arbuscular mycorrhiza and plant growth. *A. Rev. Phytopath.*, 6, 397-418.

GERDEMANN, J.W., 1975 - Vesicular-arbuscular mycorrhizae. *In:* J.G. TORREY & D.T. CLARKSON (Editors), *The development and function of roots,* Acad. Press, New York, p. 575-591.

GILMORE, A.E., 1971 - The influence of endotrophic mycorrhizae on the growth of peach seedlings. *J. Am. Soc. Hort. Sci.*, 96, 35-38.

HACSKAYLO, E., 1973 - Carbohydrate physiology of ectomycorrhizae. *In:* G.C. MARKS & T.T. KOZLOWSKI (Editors), *Ectomycorrhizae,* Acad. Press, New York, p. 207-230.

HACSKAYLO, E. & J.A. VOZZO, 1967 - Inoculation of *Pinus caribea* with pure cultures of mycorrhizal fungi in Puerto Rico. *Proc. 14th. IUFRO Congress,* 5, 139-148.

HARLEY, J.L., 1969 - *The biology of mycorrhiza.* Leonard Hill, London, 334 p.

HARLEY, J.L., 1971 - Fungi in ecosystems. *J. Ecol.*, 59, 635-668.

HATCH, A.B., 1937 - The physical basis of mycotrophy in the genus *Pinus. Black Rock Forest Bull.*, 6, 1-168.

HATTINGH, M.J. & J.W. GERDEMANN, 1975 - Inoculation of Brazilian sour orange seed with an endomycorrhizal fungus. *Phytophathology,* 65, 1013-1016.

HAYMAN, D.S., 1974 - Plant growth responses to vesicular-arbuscular mycorrhiza. VI. Effect of light and temperature. *New Phytol.*, 73, 71-80.

HAYMAN, D.S., 1975 - The occurrence of mycorrhiza in crops as affected by soil fertility. *In:* F.E. SANDERS, B. MOSSE & P.B. TINKER (Editors), *Endomycorrhizas,* Acad. Press, New York, p. 495-509.

HAYMAN, D.S., 1978 - Mycorrhizal populations of sown pastures and native vegetation in Otago, New Zealand. *N.Z.Jl agric. Res.* (in press).

HAYMAN, D.S. & B. MOSSE, 1971 - Plant growth responses to vesicular-arbuscular mycorrhiza. I. Growth of *Endogone*-inoculated plants in phosphate-deficient soils. *New Phytol.*, 70, 19-27.

HEPPER, C.M., 1977 - A colorimetric method for estimating vesicular-arbuscular mycorrhizal infection in roots. *Soil Biol. Biochem.*, 9, 15-18.

HOLEVAS, C.D., 1966 - The effect of a vesicular-arbuscular mycorrhiza on the uptake of soil phosphorus by strawberry (*Fragaria* sp. var. *Cambridge Favourite*). *J. hort. Sci.*, 41, 57-64.

JACKSON, H.E., MILLER, R.H. & R.E. FRANKLIN, 1973 - The influence of vesicular-arbuscular mycorrhiza on uptake of ^{90}Sr from soil by soybeans. *Soil Biol. Biochem.*, 5, 205-212.

JANOS, D.P., 1975 - Effects of vesicular-arbuscular mycorrhizae on lowland tropical rainforest trees. *In:* F.E. SANDERS, B. MOSSE & P.B. TINKER (Editors), *Endomycorrhizas,* Acad. Press, New York, p. 437-446.

JANSE, J.M., 1897 - Les endophytes radicaux de quelques plantes Javanaises. *Ann. Jard. Botan. Buitenz.*, 14, 53-212.

JOHNSON, P.N., 1976 - Effects of soil phosphate level and shade on plant growth and mycorrhizas. *N.Z.Jl Bot.*, 14, 333-340.

KLEINSCHMIDT, G.D. & J.W. GERDEMANN, 1972 - Stunting of citrus seedlings in fumigated nursery soils related to the absence of endomycorrhizae. *Phytopathology,* 62, 1447-1453.

KOSKE, R.E., SUTTON, J.C. & B.R. SHEPPARD, 1975 - Ecology of *Endogone* in Lake Huron sand dunes. *Can. J. Bot.*, 53, 87-93.

KOUCHEKI, H.K. & D.J. READ, 1976 - Vesicular-arbuscular mycorrhiza in natural vegetation systems. II. The relationship between infection and growth in *Festuca ovina* L. *New Phytol.*, 77, 655-666.

KRUCKELMANN, H.W., 1975 - Effects of fertilizers, soils, soil tillage and plant species on the frequency of *Endogone* chlamydospores and mycorrhizal infection in arable soils. *In:* F.E. SANDERS, B. MOSSE & P.B. TINKER (Editors), *Endomycorrhizas,* Acad. Press., New York, p. 511-525.

LARSEN, S., 1976 - Phosphorus in past, present and future agriculture. *Phosphorus in Agriculture,* 68, 1-9.

LEWIS, D.H., 1973 - Concepts of fungal nutrition and the origin of biotrophy. *Biol. Rev.*, 48, 261-278.

LEWIS, D.H., 1974 - Micro-organisms and plants: the evolution of parasitism and mutualism. *Symp. Soc. gen. Microbiol.*, 24, 367-392.

LEWIS, D.H., 1975 - Comparative aspects of the carbon nutrition of mycorrhizas. *In:* F.E. SANDERS, B. MOSSE & P.B. TINKER (Editors), *Endomycorrhizas,* Acad. Press, New York, p. 119-148.

LEWIS, D.G. & J.P. QUIRK, 1967 - Phosphate diffusion in soil and uptake by plants. III. P^{31} movement and uptake by plants as indicated by P^{32} autoradiography. *Pl. Soil,* 26, 445-453.

LIHNELL, D., 1939 - Untersuchungen über die Mykorrhiza und die Wurzelpilze von *Juniperus communis. Symb. bot. upsal.*, 3, 1-143.

LUNDEBERG, G., 1970 - Utilisation of various nitrogen sources, in particular bound soil nitrogen, by mycorrhizal fungi. *Studia Forestalia Suecica,* 79, 1-75.

MACAULEY, J., 1975 - Biodegradation of litter in *Eucalyptus pauciflora* communities. I. Techniques for comparing the effects of fungi and insects. *Soil Biol. Biochem.*, 7, 341-344.

MAGROU, J., 1938 - Sur la tubérisation de la pomme de terre. *C. r. Séanc. Soc. Biol.*, 127, 793-796.

MARX, D.H., 1973 - Mycorrhizae and feeder root diseases. *In:* G.C. MARKS & W.C. BRYAN (Editors), *Endomycorrhizae,* Acad. Press, New York, p. 351-382.

MARX, D.H. & W.C. BRYAN, 1975 - Growth and ectomycorrhizal development of Loblolly pine seedlings in fumigated soil infested with the fungal symbiont *Pisolithus tinctorius. Forest Sci.*, 21, 245-254.

McLUCKIE, J. & A. BURGES, 1932 - Mycotrophism in the *Rutaceae.* I. The mycorrhiza of *Eriostemon crowei. Proc. Linn. Soc. N.S.W.,* 57, 291-312.

MEYER, F.H., 1968 - Auxin relationships in symbiosis. *In: Transport of plant hormones.* Nato/EGE university summer Institute, p. 320-328.

MEYER, F.H., 1973 - Distribution of ectomycorrhizae in native and man-made forests. *In:* G.C. MARKS & T.T. KOZLOWSKI (Editors), *Ectomycorrhizae,* Acad. Press, New York, p. 79-106.

MIKOLA, P., HAHL, J. & E. TORNIAINEN, 1966 - Vertical distribution of mycorrhizae in pine forests with spruce undergrowth. *An.bot.Fen.,* 3, 406-409.

MOSER, M., 1963 - Die Bedeutung der Mykorrhiza bei Aufforstungen unter besonderer Berücksichtigung von Hochlagen. *In: Mycorrhiza,* Int. Mykorrhiza Symp. Weimar, 1960, p. 407-422.

MOSSE, B., 1956 - Studies on the endotrophic mycorrhiza of some fruit plants. *Ph. D. thesis,* Univ. London.

MOSSE, B., 1959 - Observations on the extra-matrical mycelium of a vesicular-arbuscular endophyte. *Trans. Br. mycol. Soc.*, 42, 439-448.

MOSSE, B., 1972 - The influence of soil type and *Endogone* strain on the growth of mycorrhizal plants in phosphate-deficient soils. *Rev. Ecol. Biol. Sol.*, 9, 529-537.

MOSSE, B., 1973a - Advances in the study of vesicular-arbuscular mycorrhiza. *A. Rev. Phytopath.*, 11, 171-196.

MOSSE, B., 1973b - Plant growth responses to vesicular-arbuscular mycorrhiza. IV. In soil given additional phosphate. *New Phytol.*, 72, 127-136.

MOSSE, B., 1977a - Plant growth responses to vesicular-arbuscular mycorrhiza. X. Responses of *Stylosanthes* and maize to inoculation in unsterile soils. *New Phytol.*, 78, 277-288.

MOSSE, B., 1977b - The role of mycorrhiza in legume nutrition on marginal soils. Exploiting the legume - *Rhizobium* symbiosis in tropical agriculture. *In:* J.M. VINCENT, A.S. WHITNEY & J. BOSE (Editors), *Coll. Trop. Agric. Univ. Hawaii, Misc. Publ.*, 145, 275-292.

MOSSE, B. & J.M. PHILLIPS, 1971 - The influence of phosphate and other nutrients on the development of vesicular-arbuscular mycorrhiza in culture. *J. gen. Microbiol.*, 69, 157-166.

MOSSE, B., POWELL, C.Ll. & D.S. HAYMAN, 1976 - Plant growth responses to vesicular-arbuscular mycorrhiza. IX. Interactions between VA mycorrhiza, rock phosphate and symbiotic nitrogen fixation. *New Phytol.*, 76, 331-342.

MULLETTE, K.J., 1976 - Studies of *Eucalypt* mycorrhizas. I. A method of mycorrhiza induction in *Eucalyptus gummifera* (Gaestn. + Hochr.) by *Pisolithus tinctorius* (Pers.) Coker + Couch. *Aust.J.Bot.*, 24, 193-200.

NEWMAN, E.I. & R.E. ANDREWS, 1973 - Uptake of phosphorus and potassium in relation to root growth and root density. *Pl. Soil*, 38, 49-69.

NICOLSON, T.H., 1960 - Mycorrhiza in the *Gramineae*. II. Development in different habitats, particularly sand dunes. *Trans. Br. mycol. Soc.*, 43, 132-145.

PAVLYCHENKO, T.K., 1942 - Root systems of certain forage crops in relation to management of agricultural soils. *Publ. natl. Res. Counc. Canada*, 1088.

PEYRONEL, B., 1940 - Prime osservazioni sui rapporti tra luce e simbiosi micorrizica. *Estratto dell' Annuario N.4 del Laboratorio della Chanousia, Giardino Botanico dell' Ordine Mauriziano al Piccolo San Bernardo*, 4,1-19.

POSSINGHAM, J.V. & J.G. OBBINK, 1971 - Endotrophic mycorrhiza and the nutrition of grape vines. *Vitis*, 10, 120-130.

POWELL, C. Ll., 1975 - Potassium uptake by endotrophic mycorrhizas. *In:* F.E. SANDERS, B. MOSSE & P.B. TINKER (Editors), *Endomycorrhizas*, Acad. Press, New York, p. 461-468.

READ, D.J., KIANMEHR, H. & A. MALIBARI, 1977 - The biology of mycorrhiza in *Helianthemum* Mill. *New Phytol.*, 78, 305-312.

READ, D.J., KOUCHEKI, H.K. & J. HODGSON, 1976 - Vesicular-arbuscular mycorrhiza in natural vegetation systems. I. The occurrence of infection. *New Phytol.*, 77, 641-654.

READ, D.J. & D.P. STRIBLEY, 1975 - Some mycological aspects of the biology of mycorrhiza in the *Ericaceae*. *In:* F.E. SANDERS, B. MOSSE & P.B. TINKER (Editors), *Endomycorrhizas*, Acad. Press, New York, p. 105-117.

REDHEAD, J.F., 1968 - Mycorrhizal associations in some Nigerian forest trees. *Trans. Br. mycol. Soc.*, 51, 377-387.

REDHEAD, J.F., 1974 - Aspects of the biology of mycorrhizal associations occurring on tree species in Nigeria. *Ph. D. thesis*, Univ. Ibadan.

RHODES, L.H. & J.W. GERDEMANN, 1975 - Phosphate uptake zones of mycorrhizal and non-mycorrhizal onions. *New Phytol.*, 75, 555-561.

ROSS, J.P., 1971 - Effect of phosphate fertilization on yield of mycorrhizal and non-mycorrhizal soybeans. *Phytopathology*, 61, 1400-1403.

SAFIR, G.R., BOYER, J.S. & J.W. GERDEMANN, 1971 - Mycorrhizal enhancement of water transport in soybean. *Science N.Y.*, 172, 581-583.

SAFIR, G.R., BOYER, J.S. & J.W. GERDEMANN, 1972 - Nutrient status and mycorrhizal enhancement of water transport in soybean. *Pl. Physiol. Wash.*, 49, 700-703.

SANDERS, F.E., 1975 - The effect of foliar-applied phosphate on the mycorrhizal infections of onion roots. *In:* F.E. SANDERS, B. MOSSE & P.B. TINKER (Editors), *Endomycorrhizas*, Acad. Press, New York, p. 261-276.

SANDERS, F.E., MOSSE, B. & P.B. TINKER, 1975 - *Endomycorrhizas*. Acad. Press, New York, 626 p.

SANDERS, F.E. & P.B. TINKER, 1973 - Phosphate flow into mycorrhizal roots. *Pestic. Sci.*, 4, 385-395.

SCHLICHT, A., 1889 - Beitrag zur Kenntniss der Verbreitung und der Bedeutung der Mykorrhizen. *Ph. D. thesis*, Univ. Erlangen.

SHEMAKHANOVA, N.M., 1960 - The conditions of pine mycorrhiza formation with *Boletus luteus* (Linn.) Fr. in pure culture. *Izvestiya, Akademii Nauk. SSSR., Seriya biologicheskaya*, 2, 240-255.

SHEMAKHANOVA, N.M., 1972 - *Mycotrophy of woody plants*. (Translation of Mikotrofiya drevesnykh porod publ. Moscow, 1967), U.S. Depart. of Agric. and Nat. Sci. Foundation, Washington D.C., 329 p.

SIKORA, R.A. & F. SCHÖNBECK, 1975 - Effect of vesicular-arbuscular mycorrhiza (*Endogone mosseae*) on population dynamics of the root-knot nematodes *Meloidogyne incognita* and *Meloidogyne hapla*. VIII. Intern. Plant. Protection Congress Sect. V, 158-164.

SKINNER, M.F. & G.D. BOWEN, 1974a - The penetration of soil by mycelial strands of ectomycorrhizal fungi. *Soil Biol. Biochem.*, 6, 57-61.

SKINNER, M.F. & G.D. BOWEN, 1974b - The uptake and translocation of phosphate by mycelial strands of pine mycorrhizas. *Soil Biol. Biochem.*, 6, 53-56.

SMITH, S.E. & M.J. DAFT, 1977 - Interactions between growth, phosphate content and N_2 fixation in mycorrhizal and non-mycorrhizal *Medicago sativa* (alfalfa). *Austr. J. Pl. Physiol.* (in press).

SPARLING, G.P., 1976 - Effects of vesicular-arbuscular mycorrhizas on Pennine grassland vegetation. *Ph. D. thesis*, Leeds Univ.

SPARLING, G.P. & P.B. TINKER, 1975 - Mycorrhizas in Pennine grassland. *In: F.E. SANDERS, B. MOSSE & P.B. TINKER (Editors), Endomycorrhizas*, Acad. Press, New York, p. 545-560.

STAHL, E., 1900 - Der Sinn der Mycorrhizenbildung. *Jb. wiss. Bot.*, 34, 539-668.

STONE, E.L., 1949 - Some effects of mycorrhizae on the phosphorus nutrition of Monterey pine seedlings. *Proc. Soil Sci. Soc. Am.*, 14, 340-345.

STRIBLEY, D.P. & D.J. READ, 1974 - The biology of mycorrhiza in the *Ericaceae*. IV. The effect of mycorrhizal infection on uptake of ^{15}N from labelled soil by *Vaccinium macrocarpon* Ait. *New Phytol.*, 73, 1149-1155.

STRIBLEY, D.P. & D.J. READ, 1976 - The biology of mycorrhiza in the *Ericaceae*. VI. The effects of mycorrhizal infection and concentration of ammonium nitrogen on growth of Cranberry (*Vaccinium macrocarpon* Ait.) in sand culture. *New Phytol.*, 77, 63-72.

STRZEMSKA, J., 1955 - Investigations on the mycorrhiza in corn plants. (Polish) *Acta microbiol. pol.*, 4, 191-204.

STRZEMSKA, J., 1975a - Occurrence and intensity of mycorrhiza and deformation of roots without mycorrhiza in cultivated plants. *In: F.E. SANDERS, B. MOSSE & P.B. TINKER (Editors), Endomycorrhizas*, Acad. Press, New York, p. 537-544.

STRZEMSKA, J., 1975b - Mycorrhiza in farm crops grown in monoculture. *In: F.E. SANDERS, B. MOSSE & P.B. TINKER (Editors), Endomycorrhizas*, Acad. Press, New York, p. 527-536.

SUTTON, J.C. & B.R. SHEPPARD, 1976 - Aggregation of sand-dune soil by endomycorrhizal fungi. *Can. J. Bot.*, 54, 326-333.

SWITZER, G.L. & L.E. NELSON, 1972 - Nutrient accumulation and cycling in Loblolly Pine (*Pinus taede* L.) plantation ecosystems: The first 20 years. *Soil Sci. Soc. Amer. Proc.*, 36, 143-147.

THEODOROU, C., 1968 - Inositol phosphates in needles of *Pinus radiata* D. Don and the phytase activity of mycorrhizal fungi. *Trans. 9th. Int. Congr. Soil Sci.*, 483-490.

THEODOROU, C. & G.D. BOWEN, 1969 - The influence of pH and nitrate on mycorrhizal associations of *Pinus radiata* D. Don. *Austr. J. Bot.*, 17, 59-67.

TINKER, P.B., 1975 - Effects of vesicular-arbuscular mycorrhizas on higher plants. *Symp. Soc. exp. Biol.*, 29, 325-350.

WARCUP, J.H., 1975 - Factors affecting symbiotic germination of orchid seed. *In: F.E. SANDERS, B. MOSSE & P.B. TINKER (Editors), Endomycorrhizas*, Acad. Press, New York, p. 87-104.

WENT, F.W., 1971 - Mycorrhizae in a Montane Pine Forest. *In:* E. HACSKAYLO
    (Editor), *Mycorrhizae,* Proc. 1st. North Amer. Conf. on mycorrhizae.
    U.S.D.A. publ., 1189, p. 230-232.
WOODS, F.W. & K. BROCK, 1964 - Interspecific transfer of Ca-45 and P-32 by
    root systems. *Ecology,* 45, 886-889.
ZAK, B., 1973 - Classification of ectomycorrhizae. *In:* G.C. MARKS & T.T.
    KOZLOWSKI (Editors), *Ectomycorrhizae,* Acad. Press, New York, p. 43-78.

7. DISCUSSION

LEVIN (Texas): Is there a relationship between the likelihood of the
occurrence of infection on the one hand and the ability of species to obtain
nutrients on the other hand? In other words, are plants which are poor at
obtaining nutrients more likely to have mycorrhiza than plants that are
better adapted for obtaining nutrients?

MOSSE: I don't think there is any evidence that this is so. You cannot
make any categorical statement. It all depends on the type of soil, on the
soil fertility, and above all on the phosphorus levels in the plant.

LEVIN: I was curious as to whether there are any biochemical preadaptations
in different species which might cause some to be more likely than others to
form mycorrhizal associations.

MOSSE: I think that all these things are possible, and we will know as
soon as people go into it much further. But at the moment we are still at a
very early stage. We are just learning to manage our experimental system and
we are just beginning to formulate the problems. There are so many snags.
I don't doubt there will be interactions of the kind you suggested, but they
will be minor.

VAN DEN ANDEL (Amsterdam): Is there an effect of mycorrhiza on the
reproductive behaviour of plants? You only talked about the effects on growth.

MOSSE: There is some work which suggests that in pot experiments mycorrhizal
plants flower earlier. The suggestion has been made that this might be an
hormonal effect. But I think it is probably also a nutritional effect.

VAN DER AART (Oostvoorne): You mentioned that there are some plant families
without mycorrhiza, for instance *Carex*. How can these species compete with
species with mycorrhiza in vegetations on soils which are low in phosphates?
Is anything known about how they get their phosphates?

MOSSE: This question has been investigated, but no definite conclusion was
possible. It has something to do with growth rates. Some plants have higher
phosphorus concentrations, and some can grow with lower phosphorus inputs
than others can. It also depends on the species they are competing with.
Another thing is that *Carex* species often occur on waterlogged soils. VA
mycorrhizas do not flourish under waterlogged conditions. There also are some
reports that *Carex* species sometimes possess VA mycorrhiza. It is unusual,
but seemingly it can happen.

ANONYMOUS : You said tropical grasses show a higher infection rate of mycorrhiza than temperate grasses. Tropical grasses are most common in savannahs and woodlands were there is a strong seasonality of growth. Could it be that the rate of mycorrhiza infection is higher in all types of plants which have strong seasonality? And my second question is: can tropical grasses stand a higher infection rate of mycorrhiza than temperate grasses before it become pathogenic?

MOSSE: As to your last question, it does not become pathogenic because the infection rate is too high. It is pathogenic under circumstances where there is no growth response to extra phosphate. Then the mycorrhiza still brings in extra phosphorus, but it does not produce extra growth. The advantage of the plant-microorganism system is no longer obvious in terms of growth. At such supraoptimal phosphorus levels, the disadvantages to the plants from any fungal attack prevail. As to your question on seasonality, no one seems to have worked on it.

GIGON (Zürich): Do you have any idea of the importance of mycorrhiza in heavy-metal soils, which are rich in zinc, chromium, and nickel?

MOSSE: It is known that mycorrhiza may help in zinc uptake in zinc-deficient soils. But what happens in these abnormal soils is anybody's guess.

GRIME (Sheffield): Does the mycorrhiza system, particularly in ectotrophic evergreen species, make it possible to exploit soil in which nutrients, such as phosphorus, are only available for brief periods during the year at low concentrations.

MOSSE: This is a question of storage capacity, i.e., the volume of mycorrhizal tissue. In ectotrophic mycorrhiza the total volume is about 4 per cent of that of the total root system. Although it may contain poly-phosphate granules, these are not large amounts. In VA mycorrhiza the volume is between 4 and 17 per cent of the fine roots only. But in terms of energy the mycorrhiza is very much more efficient in making absorbing surface than the roots themselves. And it moves its cytoplasm around. So it is a much more mobile system.

GRIME: Do I interpret you correctly, is the mycorrhizal plant getting a root system at minimal synthetic cost?

MOSSE: Absolutely. There are figures that are fantastic. It does not lose anything when moving around in the soil, and the whole system is interconnected. So it can explore very large areas of soil, where the roots would never come. It is a very mobile system. The mycorrhiza has evolved over long periods of time as a beautiful system.

The biotic environment of plants

1. INTRODUCTION

The struggle for live is a romantic, anthropomorphic, and antiquated metaphor
for the behaviour of living organisms. Nonetheless, the responses of plant species
in time and the instantaneous responses of individual plants to their environment
can be understood most easily in terms of struggle and fight. Certainly, fighting,
virulence, and resistance are prominent terms in the vocabulary of that peaceful
gang engaged in crop protection.

   Crop protection, a specialized branch of ecology geared to practical
applications, can contribute to ecology at large some views based on extensive
experimentation and thorough theoretical elaboration.

2. TYPOLOGY OF THE BIOTIC ENVIRONMENT OF PLANTS

2.1. MAN

The biotic environment of plants consists of man, plants, animals, and microorganisms. Man, originally a modest consumer, became a modifyer, a competitor,
and a destroyer of plants and vegetations. To phrase it in crop protection
terminology: man became a pest. This topic will be left aside here.

2.2. PLANTS

Plants form the most conspicuous part of the biotic environment of plants. Plants
affect the life of other plants in four different roles: competitors, epiphytes,
parasites, and alternate hosts.

   Plants interact with plants, within and between species, within and between
seasons. The major within-season interaction is competition; the major between-
season interaction is succession. Both phenomena are well known to ecologists;
they need only few words here. In agriculture, competition with and succession
of crop plants is undesirable. The Dutch shephard of older times periodically
set the heaths to fire. One effect was the death of self-sown *Betula* and *Pinus*
trees; succession was stopped to preserve food for the sheep. The heather itself
formed fresh sprouts more palatable to the sheep than the old woody ones. Most
of todays pesticides are used to stop competition by weeds. Weeds are simply

plants in places where man does not want them. They compete with crop plants for space, light, water, or nutrients. If there are too many plants of any species of weed, this becomes a pest. A heavy weed infestation can be quite satisfactory from the aesthetic point of view (e.g. *Centaurea cyanus* and *Papaver rhoeas* in corn fields) but, unfortunately, economic necessities do not comply with aesthetics.

Competition is thoroughly studied in agriculture (DE WIT 1960; see also GRIME in this volume). An analysis of the classical barley-oats mixture formerly grown on the sandy soils of the eastern Netherlands revealed that there is not only competition between the two species but also complementation, because they differ in growth rhythm and water requirements. What species outyields the other depends on the season, but the total biomass of the seed is on the average higher in the mixture than in any of the two species grown in mono-culture. Other advantages of species mixtures, not studied in competition experiments, will be discussed later.

Some plants are epiphytes; they find a suitable niche upon other plants without damaging their supports. *Orchidaceae* and *Bromeliaceae* are well known examples from the tropics; in The Netherlands we have to be content with algae, lichens, and occasional mosses on tree trunks. The *horror vacui* displayed by nature is ecologically interesting and aesthetically pleasing. The supporter plants have no visible advantage from their epiphytes.

Several plants take nutrients and water from their supports, plants becoming consumers of plants. Examples are the mistletoe (*Viscum album*), dwarf mistletoe (*Arceuthobium pusillum*), *Cuscuta, Orobanche,* and *Striga* spp. The mistletoe is a minor commercial plant, sold in the U.K. at Chris⊤ˍas time; the dwarf mistletoe can become a serious pest in forestry, especially in the NW Pacific area of North America. *Orobanche* and *Striga* can become pests to agricultural crops.

Plants can become alternate hosts to pests and diseases of other plants. Sometimes, the primary host suffers little or nothing in this role. *Euonymus europaeus* is the winter host of the black bean aphid, *Aphis fabae,* that does no harm to *Euonymus* but is a serious pest to agricultural crops. In other aphid species both the primary and the secondary host may be damaged. *Berberis* species have a similar function for the wheat rust fungus *Puccinia graminis*. A difference is that the barberry is parasitized but *Euonymus* is inhabited only. *P. graminis* develops profusely on several *Gramineae*, and can do great damage to wheat and oat crops. When the host plants mature, the rust forms resting spores. After the winter these germinate, produce tiny wind-borne spores which only infect recently emerged barberry leaves. On the barberry the rust forms another type of spore which is wind-blown towards surrounding grasses and cereal crops: the annual cycle is closed (HOGG *et al.* 1969). Host alternations, sometimes with complicated cycles, are frequent in nature e.g. among insects which cause plant galls (ALTA & DOCTERS VAN LEEUWEN 1946).

2.3. ANIMALS

Animals as part of the biotic environment of plants are given three roles in this paper: consumers, symbionts, and vectors.

Animals as plant consumers have changed plants, species, and vegetations in the evolutionary and in the historical time scale. This is too well known to be discussed here. The smaller insects and the nematodes are voracious consumers, which can develop into serious agricultural pests.

Animals as symbionts of plants appear in many ways. They carry seeds around in their intestines after having eaten the berries. Some seeds only germinate well after intestinal passage. Mistletoe seeds must be deposited *in situ* by birds. Fertilization of plants is performed by birds and all kinds of insects in exchange of food and building material. A complex situation, demonstrated at the poster-session of this conference by Valdeyron and Dommee (Poster 22), is encountered by figs (*Ficus carica*) where a wasp (*Blastophaga psenes*) produces galls in the inflorescences of the female trees in the service of reproduction of both the fig and the wasp.

Animals can be vectors of plant diseases, whether caused by other animals, fungi, bacteria, or viruses. Man is the vector species most dangerous to agricultural crops (ZADOKS 1966), but his theme cannot be elaborated here. The palm weevil *Rhynchophorus palmarum* is the vector of the nematode *Rhadinaphelenchus cocophilus* causing the much feared red-ring disease of coconut trees in the Caribbean area. Similarly, insects can disperse fungal diseases like the "Dutch" elm disease on *Ulmus* ssp., caused by the fungus *Ceratocystis ulmi*, dispersed by *Scolytus* beetles. Migrating birds carried the fungus *Endothia parasitica*, which caused the devastating chestnut (*Castanea dentata*) blight in the eastern U.S.A., over large distances. Recently migrating sparrows have been inculcated with the spread of the bacterium *Erwinia amylovora*, causing the destructive fire blight disease of pears, from the U.K. far into the baltic area (SCHROTH *et al*. 1974). Aphids and many other types of insects are well-known carriers of viruses; control of insects is a much-used procedure in crop protection to prevent insect-borne spread of viruses through crops. Nematodes, of which there go thousands to the thimble, do not only cause plant diseases themselves but can also transmit viruses from plant to plant. In addition, plant sucking nematodes may cause wounds that function as point-of-entry for soil-inhabiting fungi which, otherwise relatively innocuous, become serious pathogens once they have found a way to enter the plant. The reverse may also be true: the pathogenic fungi promote attack by nematodes. In this fungus-nematode interplay the attack by either type of organism may be enhanced by the other type (EDMUNDS & MAY 1966).

In nature the vector function of animals is real but relatively inconspicuous. In agriculture, however, it can become a major problem. Here we refrain from entering the area of the crop protection sciences.

In this paper the various roles of microorganisms as elements in the biotic environment will be treated in some detail. The following roles are distinguished: pathogens, symbionts, epiphytes, and reducers.

Diseases of plants caused by protozoa are rare; a disease of coffee trees (*Coffea liberica*) is caused by flagellates (VERMEULEN 1963). Fungi cause many diseases, as do bacteria. Mycoplasmas form a class apart, in their etiology not unlike viruses (DAVIS & WHITCOMB 1971). Viruses are common again.

Viruses and plant mycoplasmas cannot be grown outside the living plant cell; the same is true for a number of fungi among which are those causing "rusts" (*Uredinales*) and "powdery mildews" (*Erysiphales*). The biotrophic fungi, those which cannot grow without a living host, can be real parasites which consume the host plant without giving anything in return. But another group of biotrophes, parasitic in behaviour, is highly beneficial to the host plant; among them are the endomycorrhizal fungi of the *Endogone* group (see MOSSE in this volume). Plants can do without them but grow better and use phosphate more efficiently with the assistance of *Endogone* spp.

Among the fungal species there is a gradual transition from real parasites, that cannot multiply without a suitable host, through parasites that can live saprobically for some time at least, and fungi that live saprobically but attack weekened plants occasionally, to the real saprobic fungi that are decomposers of dead plant material. There is, again, a gradual transition from micro-organisms, that are consumers of plants only, unto those, that live in a balanced give-and-take with their host plants. The latter microorganisms feed of photosynthates from the plant and assist the plant in the uptake of nutrients from the soil (the mycorrhizal fungi) or nitrogen from the air (the N-fixating microorganisms, bacteria, actinomycetes, and blue algae) (RUINEN 1965).

In the extreme forms of symbiosis the process of penetration and colonization by the symbiont in the plant cannot be distinguished from pathogenesis. The malformation caused, as root nodules of *Leguminosae* by *Rhizobium radicicola* are physiologically akin to the malformations (crown galls) caused by real pathogens like *Agrobacterium tumefaciens* on raspberry (*Rubus idaeus*).

A typical form of symbiosis of a plant community with N-fixating micro-organisms occurs in the tropics, where blue algae and bacteria live epiphytically on the leaves; they utilize plant exudates and decomposition products of the cuticle and they fixate nitrogen from the air. They are said to form the major nitrogen source of the tropical rain forest biocoenosis (RUINEN 1965).

From epiphytic symbionts to epiphytic microorganisms in general is but a small step. Some epiphytic bacteria, fungi and yeasts are "regulars"; phyllosphere and rhizosphere are their normal habitats. Others are "accidentals". A special class of epiphytes are the sooty moulds (*Capnodium* spp.). They thrive on the excrements of aphids, are non-pathogenic, but can do much damage when they are so numerous that they form a black cover over the leaves which intercepts the

light so completely that no photosynthesis is possible (ZAMBETTAKIS 1964).
Microepiphytes can stimulate or curb the development of pathogens; they play a
role as regulators (see WOLDENDORP in this volume).

The role of microorganisms as reducers must be mentioned for completeness'
sake, but a discussion is beyond the scope of this paper.

The mutual relations between microorganisms are more complex than those
between macroorganisms. There is predation, parasitism, disease (of microorganisms themselves), competition, and antagonism at the microlevel. Some of
these phenomena are so important economically, that they are intensely studied
now (BAKER 1968; PREECE & DICKINSON 1971).

3. HOST DEFENSES

3.1. SPECIALIZATION

In evolutionary time plants have built up defenses against harmful agents.
Simultaneously the parasitic consumers adapted and usually specialized.
Specializing to retain some host plants as food sources they lost others. So
there was and is a coevolution of hosts and parasites. The result is that
parasites are not distributed at random over all possible hosts, but that they
have established specific relations with some hosts. The rust of wheat does
not infect the poplar tree, and the rust of poplars does not infect wheat.

All parasites are specialized somehow, but the degree of specialization
varies. Several parasites accept as hosts a broad spectrum of plant species
from various genera; other parasites are quite particular and accept only one
host species as a host and food source. The former are the polyphagous and
the latter the monophagous parasites.

Looking at the problem of specificity from the other side one might say
that susceptibility of a plant species to a parasite is the exception rather
than the rule. A plant species is specifically susceptible to some identifiable
parasites and resistant to all others. Nevertheless, the catalogue of parasites
of wheat runs up to some 100 items (WIESE, in press).

3.2. ORGANIZATIONAL LEVELS OF DEFENSE

Two questions arise: (1) What are the defense mechanisms that plants have
developed? (2) What determines the recognition of the host by the pathogen,
and *vice versa*? Both questions have been studied extensively in the crop
protection sciences. Entomologists especially have gone a long way in answering
question two. In the following we must limit ourselves to exploring question
one, using the approach of the phytopathologist. But first it must be stated
that defense mechanisms of plants exist at the molecular, cellular, plant, and
population level. The defense mechanisms are often characteristic for the host
species or genus, rarely for the parasite.

3.3. MOLECULAR LEVEL

A typical defense mechanism at the molecular level is that of the phytoalexins.
Phytoalexins are chemical substances, that are formed when the plant is harmed
by a needle, a droplet of salt solution, or a penetrating fungus (KUĆ 1972).
At the site of damage a phytoalexin is formed, a new chemical, aspecific with
respect to the cause of the harm. In peas one phytoalexin is called pisatin.
Pisatin stops the growth of most fungi trying to invade the pea, but a
specific pea pathogen like the fungus *Ascochyta pisi* is tolerant to pisatin so
that it can penetrate and infect the pea. This fungus is not tolerant to the
phaseollin of *Phaseolus* beans, so that it can hardly attack beans.

The phytoalexin defense is one of the active defense mechanisms, one in
which the defense chemical is formed after infection. There are also passive
defenses, where the defense chemical is already present before infection. The
classic example is the red-skinned onion; the red colour of the dead outer
scales is due to certain phenols that inhibit the fungus *Colletotrichum
circinans* causing onion smudge and prevent its penetration. Penetration is not
prevented in white-skinned onions nor in red-skinned onions with scales
artificially perforated by means of a needle. The numerous defense mechanisms
at the molecular level against insects are so well known that "chemical
ecology" came into being.

3.4. CELLULAR LEVEL

At the cellular level other defense mechanisms exist. In some cases a fungus
that penetrates a plant provokes a collapse of the penetrated host cell or
group of host cells. The collapse proceeds so fast that the fungus has no time
to feed on the host cells; so it starves to death. When the resulting lesions are
so small that they cannot be seen by the naked eye, the phenomenon is called
hypersensitivity. Paradoxically, hypersensitivity provides a high level of
resistance; it is commonly used in our major food crops (in potatoes against
Phytophthora infestans causing late blight, in cereals against rusts and mildews).
In other cases the penetrated host cell tries to encapsulate the pathogen. Root
cells of wheat try to encapsulate the penetrating hyphae of the fungus
Gäumannomyces graminis by means of a lignituber, consisting a.o. of lignine-
like substances.

3.5. TISSUE LEVEL

Defenses of plants against fungi at the tissue level are common. There is a
gradual transition from the minute hypersensitivity flecks over large necrotic
flecks to lesions where the fungus reproduces hesitantly. An interesting active
defense mechanism is the formation of new cork cambium by the parenchyma inside
the potato tuber to encapsulate areas invaded by the late blight fungus
Phytophthora infestans.

3.6. PLANT LEVEL

At the plant level the passive defenses are morphological structures. A thick cuticle is a good defense against the mildew fungi (*Erysiphaceae*); in extreme cases phenotypes are selected with relatively small leaves and an early maturing cuticle so that the duration of the period of infectibility is reduced. Dutch rubber breeders have been very successful in selecting a rubber (*Hevea brasiliensis*) clone (Nr. LCB 870) with this type of resistance (YOUNG 1950).

Hairiness of leaves is an excellent defense against many leaf invading or leaf grazing insects. Most cultivated wheat has glabrous leaves for reasons unknown. Therewith they are susceptible to the insect *Oulema melanopus,* a beetle of which the larvae graze on the leaves. The insect does not like hairy wheat and in ovipositing avoids hairy leaves, so that hairiness is a character that can be used in resistance breeding (WEBSTER 1977). Similarly, there seem to be rice cultivars susceptible to the hoja-blanca-virus disease, which nevertheless escape from infection because the vector insect *Sogatodes (Sogata) oryzicola* does not like to land and feed on the extremely hairy leaves of these cultivars.

Besides these passive defenses, plants have active defenses somewhat analogous to immunity. Dutch glasshouse tomato plants are protected against virulent strains of tobacco-mosaic virus (TMV) by inoculating them with a mild strain which does not harm the plant and makes them insensitive to later infections by virulent TMV strains (RAST 1975). This mechanism has been called premunition. The fungus *Colletotrichum lagenarium* causes a serious flecking disease on cucumbers (*Cucumis sativus*). When the first leaf is inoculated with a non-compatible (avirulent) isolate of the fungus, the inoculated cucumber plants are systemically protected against virulent isolates of the fungus during 4 to 5 weeks. Systemic means that the protection was visible also on the non-inoculated leaves, suggesting that the protection was induced by an unknown agent that was transported through the plant to the higher leaves. The protection was expressed as a reduction in the number and size of lesions appearing after inoculation with a virulent isolate of the fungus. A booster inoculation with the non-compatible isolate three weeks after the first inoculation extended the period of protection (KUĆ, in press).

3.7. POPULATION LEVEL

Defenses at the population level are somewhat more difficult to discern. One example comes from peanut growing in West Africa. The groundnut rosette virus is transmitted by aphids, which preferentially land on irregular, patchy fields. A regular, even field is already a means of escape from infection. The more common defenses at the population level will be explained after an introductory discussion of the population genetics of pathosystems (see below).

305

3.8. HERITABILITY

The examples given above were only a sample out of a large set of defense mechanisms. They exist against insects, nematodes, fungi, microbes, and viruses in endless variation. Defenses against grazing vertebrates have not even been mentioned. The development of all these defense mechanisms in various combinations and permutations, and the concomitant development of tricks of the parasites to overcome them, led to the phenomenon of parasitic specialization, at the generic, specific, and intro-specific levels. Defenses of the host and aggressiveness of the parasites have in common, that they are inheritable characters, although their mode of inheritance is rarely known.

4. POPULATION GENETICS

4.1. PATHOSYSTEMS

Whatever the molecular basis of resistance of plants against parasites may be, it is inherited. The same is true for virulence in parasites. Resistance may be complete or partial. In the latter case it is a quantitative phenotypic character which can be expressed in a figure, indicating its position on a scale from fully susceptible (RES = 0) to fully resistant (RES = 1). Similarly, virulence of a parasite is either a qualitative (yes or no) or a quantitative phenotypic character. A fungal isolate may, again, be completely virulent (VIR = 1), completely avirulent (VIR = 0), or anything between.

Natural populations of plants usually show a wide diversity of phenotypes with respect to resistance. Likewise, populations of parasites can have a great diversity of virulence phenotypes. In every host-parasite combination, obviously, the host population and the parasite population interact genetically in one pathogenic system, shortly pathosystem (ROBINSON 1976). Pathosystems are thus sub-systems of ecosystems, consisting of two interacting species, a host and a parasite. Both host and parasite can be partners of more than one pathosystem, because one host species can have many different parasites and one polyphagous parasite species can affect many different host species.

In free nature a parasite does not usually eliminate its host population; to the monophagous parasite this would mean suicide. The host population builds up resistance but usually allows the parasite to survive. So, natural pathosystems are, in the long run, in balance. Agricultural pathosystems, however, are often unbalanced, as is testified by a long record of epidemics and pest outbreaks.

4.2. HOST-PARASITE INTERACTION

There are two major forms of host-parasite interaction in pathosystems. They are called differential interaction and constant ranking (Table 1).

When there is differential interaction the fungus isolates thus differentiated are commonly called physiologic races. These exist in all types of pathogens:

TABLE 1.   *Differential interaction and constant ranking* (ROBINSON 1976).
*For pathodemes read cultivars, for pathotypes read isolates or
physiologic races. The figures 0 to 4 indicate the severity of
disease from healthy to highly diseased*

Differential interaction				Constant ranking			
Vertical pathodemes				Horizontal pathodemes			
	A	B	C		D	E	F
a	4	0	0	d	2	3	4
Vertical pathotypes { b	0	4	0	Horizontal pathotypes { e	1	2	3
c	0	0	4	f	0	1	2

fungi, bacteria (bacteriologists sometimes call them pathovars), insects
(entomologists often speak of biotypes), nematodes (then often called pathotypes), viruses (virologists often indicate them simply as strains), and even
some higher plants (*Striga* spp.). In contrast to the earlier mentioned parasitic
specialization at the genetic or specific level the specialization meant here
is a within-pathosystem specialization at the sub-specific level.

In ROBINSON's (1976) terminology physiologic races become "vertical pathotypes", whereas isolates differentiated according to the constant ranking
interaction become horizontal "pathotypes". In traditional parlance the latter
isolates have no special label.

4.3. MONOGENIC RESISTANCE, THE GENE-FOR-GENE RELATION

When there is differential interaction in the pathosystem, the difference between
complete resistance and full susceptibility often resides in a single host locus.
Similarly, the difference between complete virulence and full avirulence of a
parasite can reside in one parasite locus. If so, there is a one-to-one relation
between the loci for resistance in the host and the loci for virulence in the
pathogen: the gene-for-gene relation (Table 2). The relation can be explained as
that between key and lock. Regard the host locus as a lock. If no R-gene (= gene
for resistance) is present the lock is open and the host can become diseased,
irrespective of the gene in the corresponding pathogen locus. If the R-gene is
present, the lock can be opened only by the corresponding gene for virulence in
the pathogen; if the latter is not available the lock remains closed and the host
stays healthy. When more R-genes are present, they are like as many locks on the
door leading to disease. To open the door, the pathogen must match each lock with
the appropriate key, each R-gene with the corresponding gene for virulence; hence
gene-for-gene relation. Usually, resistance is dominant and virulence is recessive;
there are, of course, many complications and exceptions. The gene-for-gene relation
exists in a considerable number of agricultural pathosystems (SIDHU 1975). As
plant breeders often derive resistance from wild or semi-domesticated ancestors of
crop plants, the gene-for-gene relation apparently also operates in free nature.

TABLE 2. *Gene-for-gene interaction in locus 1 for the flax-rust pathosystem:*
Linum usitatissimum - Melampsora lini;
+ = susceptible reaction; - = resistant reaction

Flax	Rust		
	A_1A_1	A_1a_1	a_1a_1
$R_1R_1^\star$	-	-	+
R_1r_1	-	-	+
r_1r_1	+	+	+

R* = gene conditioning resistance, dominant; r = gene conditioning susceptibility, recessive; A = gene conditioning avirulence, dominant; a = gene conditioning virulence, recessive

4.4. POLYGENIC SYSTEMS

Monogenic inheritance of resistance as in the gene-for-gene relation is certainly not the only possibility. On the contrary, polygenic inheritance of resistance has been encountered frequently. In polygenic systems the gene effects are small and additive. In really polygenic systems the gene effects are so small that individual genes can no longer be recognized. In some oligogenic systems a few genes with small but measurable effects have been identified; their effects are additive. Theoretical considerations and some experimental data suggest that even small quantitative effects operate on a gene-for-gene basis, but convincing evidence is not yet available (PARLEVLIET & ZADOKS 1977). For the time being, the monogenic and the polygenic modes of inheritance of resistance must be regarded as two extremes in a continuum of possibilities.

Polygenic and oligogenic resistance supposedly lead to a pathosystem with constant ranking interaction. It seems that the constant ranking system is more prevalent among plant-insect pathosystems, whereas the differential interaction is more frequently encountered among plant-fungus pathosystems. The monogenic differential resistance is also called vertical resistance, the constant ranking resistance is also indicated as horizontal resistance (VAN DER PLANK 1963). Both types of resistance can concur in a single plant, vertical being epistatic over horizontal resistance.

4.5. EXAMPLES

Two examples must suffice. In the Toluca valley of Mexico the potato *Solanum demissum* is common in the wild. The potato late blight fungus (*Phytophthora infestans*) is ubiquitous but never serious, though the environmental conditions are extremely favourable to disease. Genetic analysis showed the *S. demissum* population to be highly diverse, with many monogenes in various combinations,

polygenic resistance at various levels, and different combinations of monogenic and polygenic resistance. Pathogenic phenotypes there were many but none
was really harmful to *S. demissum,* whereas some of these phenotypes had
disastrous effects on the neighbouring potato crops (*Solanum tuberosum*).
Apparently, the natural *S. demissum - P. infestans* pathosystem had not.

In Israel the oats-crown rust (*Avena* spp. - *Puccinia coronata*) and the
oats-stem rust (*P. graminis*) pathosystems were studied. In both pathosystems
monogenic and polygenic resistance were present, but whereas the oats-crown
rust pathosystem tended towards polygenic resistance, the oats-stem rust
system showed more monogenic resistance. Israel encompassed many different
local climates. Where local climate was conducive to rust disease the frequency
of resistance genes was greater than in areas less favourable to rust.
Evidently, pathosystems interact with environment (BROWNING 1974).

4.6. PLANT BREEDING

Man is part of the environment of pathosystems and, of course, he profoundly
effects them. Agriculture radically changes phenotype distributions and gene
frequencies of the host; the pathogen populations adapt with sometimes
disastrous consequences. Schematically, three periods can be distinguished:
the pre mendelean, the R-gene, and the recent periods.

In the premendelean period there were low nitrogen levels; there was much
mixed cropping, about which we will see later; there were early cultivars
usually indicated as "land races" with a fair degree of intra-specific
diversity; there was at least a subconscious selection which tended towards
polygenic resistance, as in the potato-late blight pathosystem. Polygenic
resistance has been maintained in the maize-rust (*Zea mais - Puccinia sorghi*),
the barley-leaf rust (*Hordeum vulgare - Puccinia hordei*), and the rye-leaf
rust (*Secale sereale - Puccinia recondita*) pathosystems up to this very day
(ZADOKS, in press).

During the R-gene period emphasis was on the selection for monogenic
resistance. This period ranges from roughly 1920 to 1970; it is characterized
by medium nitrogen levels. Application of Mendel's laws to plant breeding, and
the fact that polygenic resistance at the medium nitrogen level seemed to be
less satisfactory than at the low level, induced the change. In some pathosystems plant breeders have been very successful in applying monogenic
resistance, in others the results were variable, as in the wheat-stem rust
(*Triticum vulgare - Puccinia graminis*) pathosystem with successes in North
America and Australia, and failures in Kenya. In other cases there was near-
failure, as in the wheat-stripe rust (*T. vulgare - Puccinia striiformis*)
pathosystem, or complete failure as in the rice-blast (*Oryza sativa -
Pyricularia oryzae*) pathosystem in Japan and Latin America, and in the potato-
late blight pathosystem.

In recent times more research effort is spent on polygenic resistance; successes are few but evident; high nitrogen levels are a draw-back. After a severe epidemic of tropical maize rust (*Puccinia polysora*) on maize in Africa, cultivated maize rapidly developed horizontal resistance, so that the rust situation came under control again, naturally, without intervention of any plant breeder. In the potato-late blight pathosystem plant breeders obtained a high degree of horizontal resistance. In north America horizontal resistance of maize to rust is jealously maintained and adapted to high nitrogen levels; this is also the new trend in Japan with respect to the rice-blast pathosystem.

In addition, experiments with new systems of gene management has been initiated, among which are pyramiding (accumulating R-genes in a single cultivar), mixing (in "multiline varieties") -to be discussed later-, regional deployment, and recycling. Also agronomists now study natural pathosystems in the hope of finding new principles of gene management, *vide* a recently initiated research project on diseases of *Limonium vulgare,* a plant in the tidal zone of the flat coasts of The Netherlands.

5. POPULATION DYNAMICS

5.1. NUMERICAL FLUCTUATIONS

Population genetics cannot be studied independently from population dynamics. The former thinks in terms of frequencies, the latter in terms of absolute numbers; whereas the latter works with phenotypes, the former tries to organize the available information in terms of genotypes.

The fluctuations encountered in the host plant are considerable. A fully developed wheat crop has a leaf area index (LAI) of about 5; at harvest the LAI does not drop to zero, as one might suppose, but to some low value thanks to self-sown volunteer plants. In dry summers, the numerical value can be as low as 5 cm^2 per ha, a LAI of 5.10^{-8} (ZADOKS 1961). Summer ploughing or scaling and deep ploughing in the autumn can each reduce the host population by a hundred-fold. The parasites of wheat have to cope with annual fluctuations of their host in the order of one million- to one billion-fold. In the Australian wheat-stem rust (*Puccinia graminis*) pathosystem the changes are even bigger, and genetic drift in the pathogen population becomes an important factor (MACKINTOSH, personal communication).

Insects survive the unfavourable season in diapause, or in the egg phase, or they bypass it by migratory habits. Among the fungi, many have a resting stage, often with specifically adapted spore forms. Other fungi survive saprobically; they continue their activity at a low pace on dead plant remnants. Some fungi have an alternation of generations, with a sexual phase just before or just after the winter, and specialized spore forms. Only a few of the monophagous, biotrophic fungi have none of this. Yellow rust of wheat (*Puccinia striiformis*) is one of them. It survives in living wheat leaves as long as the wheat survives, but it decreases its activity. Whereas one reproductive cycle takes 10 days at optimum temperature, it may last up to 50 days in winter. The fungus is relatively safe as long as it does not sporulate; the sporulating part of the fungus is killed by frost. During the winter many infected wheat leaves are lost by aging, rotting, frost, or by a rain-splashed cover of dirt; therefore, the rust has to make up for the loss by resuming growth earlier and at lower temperature than the host. The capacity of multiplication is tremendous; it is about 10^9 per season, in some 8 to 10 generations. This means that one single infected leaf in the fall is enough to cause a severe epidemic on a hectare of crop (ZADOKS 1961). Similar figures pertain to the potato-late blight fungus (*Phytophthora infestans*), which can do the same trick, but more than twice as fast. Several of the smaller insects have comparable multiplication rates.

5.2. LIFE STRATEGY AND MIGRATION

The number of individuals formed is prodigal: a well-rusted wheat crop produces 10^{13} spores per ha per day. Rusts are typical *r*-strategists. They invest their biomass mainly in large numbers of small, short-lived diaspores that are blown forth by the wind, and can only multiply on a favourable host given favourable conditions. The monophagous pathogen, that must maintain itself in nature when the compatible host plants are widely dispersed, must be a *r*-strategist. In the fungal world, there are also *K*-strategists, who invest part of their biomass in large and few, durable diaspores, which remain hidden in the soil. They bide their time until activated when growing roots of suitable hosts pass by.

To the *r*-strategists migration is essential. Whereas migration is common in the animal world, migration of fungi, wind- or insect-borne, in free nature is known only at a local scale. It is a random process of the hit-or-miss type. In agro-pathosystems long distance migratory patterns have been discovered among the wind-borne fungi and some viruses enveloped in wind-borne aphids. Usually these migrations are one-way only, from permanent source to temporary target, where great damage can be done in a short time. However, black rust of wheat (*Puccinia graminis*) and crown rust of oats (*P. coronata*) overwinter in Mexico and/or Texas, go north with the growing season carried by the prevailing southern wind well into Canada, and, when in the fall the wind turns, they travel south again (HOGG *et al*. 1969). The result is a closed migratory cycle, which is, however, quite different from migratory cycles of vertebrates. Whereas in vertebrates each individual completes at least one migratory cycle, which is a genetically fixed part of animal behaviour, the rust fungus passes through 10 or more asexual reproduction cycles during one migratory cycle which is undergone passively, without genetic fixation of behaviour.

5.3. PERIODICITY

The annual rhythmicity of parasite populations is obvious; their annual maximum density is variable. Some pathosystems show a long term periodicity with low years and peak years. The pathosystem larch-grey larch bud moth (*Larix decidua - Zeiraphera diniana*) has a seven-years' period (VAN DEN BOS & RABBINGE 1972). In the other pathosystems no apparent periodicity can be detected.

In agropathosystems there can be a man-made pseudo-periodicity. When an outbreak of a new phenotype of the parasite occurs on a hitherto resistant, genetically uniform crop cultivar, the latter will be replaced by another resistant and genetically uniform one. After a few years, the frequency of genes for virulence in the parasite population compatible to the host's new resistance genes has increased. When conditions become favourable, the pathogen population builds up rapidly and destroys the crop. The R-gene itself has not changed, but it has lost its economic value; the cultivar is replaced by the next one, *etcetera,* without end. This is the boom-and-bust cycle, a vicious cycle, governed by the rules of population genetics.

5.4. REGULATION

Boom-and-bust cycles exist in many agricultural pathosystems because there is no natural regulation. Regulation is a major topic in population dynamics, well known to most ecologists; it has been studied extensively in agricultural entomology. As the topic is well known to ecologists, only some results from phytopathology will be discussed after mentioning a few generalities. The regulating mechanism is a.o. lack of host plants, e.g. because of overgrazing, dispersion of host plants, and predation of the parasite. Among the organisms which predate pest insects are real predators, that eat the insects, parasites that live in the insect, and pathogens: fungi, bacteria, and viruses (HUFFAKER 1971). Regulation also occurs in the fungal world. There is again predation by all kinds of smaller animals and by some fungi, probably also by bacteria; there is disease, caused by specific viruses, of which the mushroom viruses have gained economic importance; there is a fierce competition for food; finally, there is antagonism. Some fungi can keep other fungi in check without touching them; antibiotics are probably part of the mechanism.

One example of antagonism must suffice. A disease of cyclamen caused by the fungus *Botrytis cinerea* was well under control by spraying a chemical with the trivial name of benomyl. Suddenly, control was lost because a mutant strain of the fungus appeared, that was tolerant to benomyl. The disease became worse than ever before. Further analysis showed that other fungi than *Botrytis* were also controlled by benomyl, among which an antagonist *Penicillium brevicompactum*. When a benomyl-tolerant strain of the antagonist appeared, disease levels could be reduced to those of the pre-benomyl period (VAN DOMMELEN & BOLLEN 1973). The extra loss of control by killing unsuspected regulating organisms is well known in agricultural entomology; it causes another vicious cycle, that of more and more intense use of pesticides. This phenomenon, which is in part a technical crop protection problem, can also be seen in a wider, social context. In this wider perspective the extra loss of control because of the unexpected killing of an antagonist (or, according to the case, a predator) has been named the pesticide syndrome (HUFFAKER 1971). The cyclamen story contains another message. It suggests that not only the population dynamics of parasite and antagonist must be studied but also their population genetics. Knowledge about the genetics of pesticide tolerance can become a great asset in the prudent and efficaceous application of pesticides.

Agricultural crops are always exposed to the risk of sudden outbreaks of pests and diseases. This risk has been called genetic vulnerability, because it is largely due to the extreme genetic uniformity of crops (COMMITTEE 1972). Another regulating mechanism would be the reduction of uniformity by introducing genetic diversity, intra-specific diversity, maintaining at the same time the minimum uniformity needed for agricultural operations and food processing. Experiments have been done by mixing cultivars or breeding lines to cultivar mixtures and multiline varieties, respectively. The cultivars *cq*. lines are chosen so that each has its own gene for resistance which is different from the others (ZADOKS 1972). The mixing effect is largest in crops where allo-infection is high respective to auto-infection. This means, in simpler words, that plants should infect their neighbours more than themselves. The bushy potato plant does not comply with this requirement, but the slender cereal plant does. In cereals, the effect can be improved by a high seed rate, so that each seed produces one stem only (BROWNING 1974; BROWNING & FREY 1972).

In the mixture, the main effect is a suppression of focal development, a focus being a local concentration of disease started from a single small source (ZADOKS & KAMPMEIJER 1977). Experiments confirm this view. Theoretical considerations show that the impediment of spore diffusion or migration is the underlying mechanism (KAMPMEIJER & ZADOKS 1977). Here is a method to protect crops against that group of parasites that has a typical focal spread. Among these are also a few insects, such as the brown plant hopper (*Nilaparvata lugens*) on rice, against which cultivar mixtures might give protection too (WEERAPAT 1977).

There is a snag in this story of cultivar mixtures and multiline varieties, because population genetics may play the over-enthousiastic plant breeder a dirty trick. By offering the parasite a varied menu its appetite may be increased. In genetic terms this means an accumulation of many genes for virulence in a single phenotype, leading to a super-race, as it is called in the professional jargon. If so, the protective effect of intraspecific diversity is lost as soon as that is created. There is already some experience with the about one million acres of commercially grown multilines of oats. Indeed, accumulation of virulence genes does occur, and the pattern of their phenotype and genotype frequencies changes considerably after the introduction of the multilines (FREY, personal communication). The nett effect is, however, a stabilization of population fluctuations. It is hoped that this situation can be maintained by careful monitoring of frequencies of virulence genes and adjusting the mixtures of lines accordingly. Such management of pathosystems is an example of applied population genetics as a support to practical population dynamics, in the service of disease control.

6. POSTSCRIPT

A synopsis of the biotic environment of plants unveils a multitude of isolated facts, scattered over all integration levels of biology. At the molecular level the entomologists discerned a pattern in the plant-insect interaction; the phytopathologists encountered active defense mechanisms of plants against fungi, but still search for the general pattern. At the population level at least one pattern of pathosystem genetics was clarified, that of the gene-for-gene relation. There is evidence that this pattern occurs among all types of pathogens: plants, animals, microorganisms, and viruses. Another pattern appeared at the population level and was confirmed experimentally: intra-specific diversity leads to stability. This stability refers primarily to the fluctuations in the parasite populations; it must be understood genetically as a dynamic equilibrium. Presently, these patterns, detected and developed in and for agro-pathosystems, are also studied in natural pathosystems. At the same time they are further explored as means of pollution-free biological control in agricultural pathosystems.

7. REFERENCES

ALTA, H. & W.M. DOCTERS VAN LEEUWEN, 1946 - *Gallenboek: Nederlandse zoöcecidiën, door dieren veroorzaakte gallen*. Ned. Nathist. Ver. & Breughel, Amsterdam, 288 p.

BAKER, R., 1968 - Mechanisms of biological control of soil-borne pathogens. *A. Rev. Phytopathol.*, 6, 263-294.

BOS, J. VAN DEN & R. RABBINGE, 1976 - *Simulation of the fluctuations of the grey larch bud moth*. PUDOC, Wageningen, 88 p.

BROWNING, J.A., 1974 - Relevance of knowledge about natural ecosystems to development of pest management programs for agro-ecosystems. *Proc. Am. Phytopathol. Soc.*, 1, 191-199.

BROWNING, J.A. & K.J. FREY, 1969 - Multiline cultivars as a means of disease control. *A. Rev. Phytopathol.*, 7, 355-382.

COMMITTEE ON GENETIC VULNERABILITY OF MAJOR CROPS, 1972 - *Genetic vulnerability of major crops*. Nat. Acd. Sci., Washington, D.C., 307 p.

DAVIS, R.E. & R.F. WHITCOMB, 1971 - Mycoplasmas, *Rickettsiae,* and *Chlamidiae:* possible relation to yellows diseases and other disorders of plants and insects. *A. Rev. Phytopathol.*, 9, 119-154.

DOMMELEN, L. VAN & G.J. BOLLEN, 1973 - Antagonism between benomyl-resistant fungi on cyclamen sprayed with benomyl. *Acta bot. neerl.*, 22, 169-171.

EDMUNDS, J.E. & W.F. MAY, 1966 - Effect of *Trichoderma viride, Fusarium oxysporum,* and fungal enzymes upon the penetration of alfalfa roots by *Pratylenchus penetrans. Phytopathology*, 56, 1132-1135.

HOGG, W.H., HOUNAM, C.E., MALLIK, A.K. & J.C. ZADOKS, 1969 - Meteorological factors affecting the epidemiology of wheat rusts. *Techn. Notes Wld met. Org.* 99, 143 p.

HUFFAKER, C.D. (Editor), 1971 - *Biological control.* Plenum, New York, 511 p.

KAMPMEIJER, P. & J.C. ZADOKS, 1977 - *EPIMUL, a simulator of foci and epidemics in mixtures, multilines, and mosaics of resistant and susceptible plants.* PUDOC, Wageningen.

KUĆ, J., 1972 - Phytoalexins. *A. Rev. Phytopathol.,* 10, 207-232.

KUĆ, J. - Plant protection by the activation of latent mechanisms for resistance. *Neth. J. pl. Path.,* 83 suppl. (in press).

MOOK, J.H., 1971 - Observations on the colonization of the new IJsselmeer-polders by animals. *Misc. Pap. Landb. hogesch., Wageningen,* 8, 13-31.

PARLEVLIET, J. & J.C. ZADOKS, 1977 - The integrated concept of disease resistance; a new view including horizontal and vertical resistance in plants. *Euphytica,* 26, 5-21.

PLANK, J.E. VAN DER, 1963 - *Plant diseases: Epidemics and control.* Academic Press, New York, 349 p.

PREECE, T.F. & C.H. DICKINSON (Editors), 1971 - *Ecology of leaf surface micro-organisms.* Academic Press, London, 640 p.

RAST, A.Th.B., 1975 - Variability of tobacco mosaic virus in relation to control of tomato mosaic in glasshouse tomato crops by resistance breeding and cross protection. *Agr. Res. Rep., Wageningen,* 834, 76 p.

ROBINSON, R.A., 1976 - *Plant pathosystems.* Springer, Berlin, 184 p.

RUINEN, J., 1965 - The phyllosphere. III. Nitrogen fixation in the phyllosphere. *Pl. Soil.,* 22, 375-394.

SCHROTH, M.N., THOMSON, S.V., HILDEBRAND, D.C. & W.J. MOLLER, 1974 - Epidemiology and control of fire blight. *A. Rev. Phytopathol.,* 12, 389-412.

SIDHU, G.S., 1975 - Gene-for-gene relationships in plant parasitic systems. *Sci. Prog., Oxf.,* 62, 467-485.

VERMEULEN, H., 1963 - A wilt of *Coffea liberica* in Surinam and its association with a flagellate, *Phytomonas leptovasorum* Stahel. *J. Protozool.,* 10, 216-222.

WEBSTER, J.A., 1977 - *In:*P.R. DAY (Editor), *The genetic basis of epidemics in agriculture. The cereal leaf beetle in North America: breeding for resistance in small grains. Ann. N.Y. Acad. Sci.,* 287, 230-237.

WEERAPAT, P., 1977 - Mixing rice varieties to combat brown planthopper. *Int. Rice Res. Newsl.,* 2, 3.

WIESE, M.V. - A compendium of wheat diseases. (in press).

WIT, C.T. DE, 1960 - On competition. *Agric. Res. Rep., Wageningen,* 66.8.

YOUNG, H.E., 1950 - Natural resistance to leaf mildew of *Hevea brasiliensis* by clone LCB 870. *Comb. q. Circ. Rubb. Res. Scheme Ceylon,* 1949, 26, 6-12.

ZADOKS, J.C., 1961 - Yellow rust on wheat, studies in epidemiology and physiologic specialization. *Neth. J. pl. Path.,* 67, 69-256.

ZADOKS, J.C., 1966 - On the dangers of artificial infection with yellow rust to the barley crop of The Netherlands; a quantitative approach. *Neth. J. pl. Path.,* 72, 12-19.

ZADOKS, J.C., 1972 - *In:* R.T. BINGHAM *et al.* (Editors), *Biology of rust resistance in forest trees.* Reflexions on disease resistance in annual crops. *U.S. Dept. Agr. Forest Serv., Misc. Pub.,* 1221, 43-63.

ZADOKS, J.C. - Simulation models of epidemics and their possible use in the study of disease resistance. *Proc. Int. Symp. Use of Induced Mutations for Improving Disease Resistance in Crop Plants. FAO - INATOM, Vienna,* (in press).

ZADOKS, J.C. & P. KAMPMEIJER, 1977 - The role of crop populations and their deployment, illustrated by means of a simulator EPIMUL76. *Ann. N.Y. Acad. Sci.,* 287, 164-190.

ZAMBETTAKIS, C.E., 1964 - Une grave perturbation de l'équilibre "faune-flore" dans les oliveraies du bassin méditerranéen due aux insecticides systémiques. *C.R. Acad. Agric. France,* 50, 103-110.

Systematic Index